WIND STRESS OVER THE OCEAN

Parametrization of the wind stress (drag) over the ocean is central to many facets of air–sea interaction, which in turn is vital for models of weather prediction and climate modelling. *Wind Stress over the Ocean* brings together thirty of the world's leading experts in air–sea interaction, under the auspices of the Scientific Committee on Oceanic Research. The book forms a companion volume to *Dynamics and Modelling of Ocean Waves* (Komen et al., Cambridge University Press, 1994).

Wind Stress over the Ocean provides a thorough re-examination of the physical processes that transfer momentum between the atmosphere and the ocean. As well as describing the established fundamentals, the book also explores active areas of research and controversy. The book will form a comprehensive guide and reference for researchers and graduate students in physical oceanography, meteorology, fluid dynamics, offshore and coastal engineering.

Ian S. F. Jones is the Director of the Ocean Technology Group, University of Sydney, Australia and an Adjunct Senior Scientist at Columbia University, NY.

Yoshiaki Toba is a Professor Emeritus at Tohoku University, Japan, and also, an Invited Eminent Scientist at the Earth Observation Research Center of National Space Development Agency of Japan (EORC, NASDA).

WIND STRESS OVER THE OCEAN

IAN S. F. JONES

University of Sydney

YOSHIAKI TOBA

National Space Development Agency of Japan

CAMBRIDGE
UNIVERSITY PRESS

CAMBRIDGE UNIVERSITY PRESS
Cambridge, New York, Melbourne, Madrid, Cape Town, Singapore, São Paulo, Delhi

Cambridge University Press
The Edinburgh Building, Cambridge CB2 8RU, UK

Published in the United States of America by Cambridge University Press, New York

www.cambridge.org
Information on this title: www.cambridge.org/9780521662437

First published 2001
This digitally printed version 2008

A catalogue record for this publication is available from the British Library

Library of Congress Cataloguing in Publication data
Wind stress over the ocean / edited by Ian S. F. Jones, Yoshiaki Toba.
   p. cm.
  Includes bibliographical references and index
  ISBN 0-521-66243-5
  1. Ocean-atmosphere interaction. I. Jones, (Ian S. F.) II. Toba, Y. (Yoshiaki),
1931-

  GC190.2.W57 2001
  551.5′24–dc21   00-040326

ISBN 978-0-521-66243-7 hardback
ISBN 978-0-521-09049-0 paperback

Contents

Preface

Parametrization of the wind stress (drag) over the ocean is an essential issue in numerical analysis of the ocean–atmosphere interactions for climate modelling, satellite observations of air–sea fluxes, and other purposes. While wind stress over the ocean has been the subject of study for 50 years, the parametrization of the ocean surface momentum flux by a drag coefficient is still an uncertain process. There are many uncertainties not only in the way of proper parametrization, but in understanding the physical processes of the generation of stress by the wind system over the complicated nature of the ocean surface.

The drag coefficient has traditionally been treated as a function of the mean wind speed at a certain level, say at 10 m. Alternatively the coefficient can be represented by an aerodynamic roughness parameter. However, the spread of the observed values indicates that the question is not so simple. In 1955 Henry Charnock proposed a disarmingly simple expression for aerodynamic roughness, which was expressed in terms of the air friction velocity and the acceleration of gravity, independent of the state of ocean waves. While this expression has been widely used, there are important deviations and these were studied as a function of the wave age, a parameter representing the state of growth of wind waves relative to the local wind speed. Alarmingly, the trend of the observational values of aerodynamic roughness could be interpreted as opposite to the theoretical prediction according to the experiments considered.

It was with this background that a workshop was organized during the International Union of Geodesy and Geophysics Assembly in Vienna in August 1991. About twenty people interested in drag over the sea attended and contributed to the discussion. During the workshop, six questions were posed: Are there different dynamic regimes in the wind-wave interactions and generation processes? Is there a laboratory-field continuum? What wave scale supports momentum flux? What is the role of wind unsteadiness? What is the influence of swell? What new experiments are needed?

Another difficulty highlighted by the workshop was the observation that numerical modellers assume most of the drag is supported by the longer waves, while most of the literature suggests that the short waves carry the flux of momentum. This emphasizes our lack of a fundamental understanding of the processes involved. The Scientific Committee on Oceanic Research recognized the importance of this topic and appointed a working group, WG101 in 1993, to survey the state of knowledge and suggest where research should be directed. The title of the working group is "Influence of the sea state on the atmospheric drag coefficient". The members were Y. Toba, Ian S. F. Jones, M. A. Donelan, N. E. Huang, S. E. Larsen and Y. Volkov. The Working Group convened a Workshop in Avignon in 1994, and working group meetings in Marseilles in 1994, Tokyo in 1995, and Moscow and St. Petersberg in 1996.

This volume is the final report of the working group and was authored by the working group members with contributions from a worldwide selection of experts. We have attempted to produce in the first chapter an overview of the subject that might be accessible to a wide range of readers and that contains a summary at its end. The rest of the volume consists of two parts. Part 1 is concerned with basic issues of the dynamics related to the wind stress over the ocean; Part 2 deals with the uncertainties in parametrizing drag over the ocean, and describes issues of a more specific nature which have been recognized by the working group as crucial in understanding the physics to be clarified.

The order of contributors for each chapter is that the lead author, who is frequently one of the WG101 members, comes first, and co-authors follow in inverse alphabetical order. Readers may find some variations of opinion in different chapters written by different authors. However, it is natural that subjects like the present contain material unable to be unified at this time. In these points of variation there should be clues for further progress of the science. The editors hope that this monograph will inspire many new research projects which lead to a deeper understanding of the subject.

This book forms a companion volume to the earlier SCOR monograph *Dynamics and Modelling of Ocean Waves* by Komen, Cavaleri, Donelan, Hasselmann, Hasselmann and Janssen (1994).

Support for the Avignon workshop in 1994 was provided by the University of Sydney and was organized by Helen Young. Support for the Tokyo workshop, which was convened in May 1995 at Earth Observation Research Centre (EORC) of National Space Development Agency of Japan (NASDA), was provided by EORC/NASDA and was organized by Sally Hill. Support for the Moscow and St. Petersberg workshop in 1996 was organized by Yuri Volkov. We take this opportunity to express our appreciation to all the enthusiastic contributors to this monograph, SCOR, NASDA, Helen Young and Sally Hill.

Ian S. F. Jones, Sydney
Yoshiaki Toba, Tokyo

Acknowledgements

Carl Friehe wishes to acknowledge the support of the Office of Naval Research and the National Science Foundation. The calculations of the Ekman model runs were done by B. W. Berger at UC Irvine; the data from the MBL *FLIP* experiment were processed by T. Hristov at UC Irvine; S. Miller of UC Irvine and J. Edson of Woods Hole Oceanographic Institution were co-investigators on *FLIP*.

H. Murakami EORC/NASDA kindly enhanced Figs. 13.2 and 13.3.

The activities of Working Group 101 were supported by the Scientific Committee on Oceanic Research (SCOR) and by a grant to SCOR from the US National Science Foundation, Division of Ocean Sciences.

Contributors

Professor Michael L. Banner
Department of Applied Mathematics
University of New South Wales
PO Box 1 Kensington
NSW 2033 AUSTRALIA

Dr Eric J. Bock
Institut fur Umweltphysik
Im Neuenheimer Feld 229
69120 Heidelberg
GERMANY

Professor Robert A. Brown
Department of Atmospheric Sciences
University of Washington AK-40
WA 98195
USA

Dr John A. T. Bye
School of Earth Sciences
The University of Melbourne
Victoria, 3010
AUSTRALIA

Dr Gabe T. Csanady
408 Jefferson Dr. W.
Palmyra
VA 22963
USA

Dr Fred W. Dobson
Bedford Institute of Oceanography
Department of Fisheries & Oceans
PO Box 1006 Dartmouth NS
CANADA B2Y 4A2

Professor Mark A. Donelan
RSMAS/Applied Marine Physics
University of Miami
Miami Florida FL 33149
USA

Professor Naoto Ebuchi
Graduate School of Science
Tohoku University
Sendai 980-8758
JAPAN

Professor Carl A. Friehe
Department of Mechanical Engineering
University of California
Irvine CA 92697-3975
USA

Dr Gerald L. Geernaert
Ministry of Environment
National Environmental Research
Institute
PO Box 358
DK-4000 Roskilde
DENMARK

Dr Jean-Paul Giovanangeli
IRPHE
Laboratoire 10A
Centre de Luminy-Case 903
163 Avenue de Luminy
Marseille
FRANCE

Dr Lutz Hasse
Institut fur Meereskunde an der
Universitat Kiel
Dusternbrooker Weg 20
2300 Kiel
GERMANY

Dr Norden E. Huang
Code 971
NASA Goddard Space Flight Centre
Greenbelt, MD 20771
USA

Dr Bernd Jähne
Physical Oceanography Resources
Division
Scripps Institution of Oceanography
University of California, San Diego
La Jolla CA 92093-0320
USA

Dr Alistair Jenkins
Nansen Environmental & Remote
Sensing Centre
Edvard Griegs vei 3a, N-5037
Solheimsvinken/Bergen
NORWAY

Assoc. Professor Ian S. F. Jones
The Ocean Technology Group J05
University of Sydney NSW 2006
AUSTRALIA

Dr Kristina B. Katsaros
Director, Atlantic Oceanographic &
Meteorological Laboratory
4301 Rickenbacker Causeway
Miami FL 33149
USA

Dr J Klinke
Physical Oceanography Research
Division
Scripps Institution of Oceanography
University of California, San Diego
La Jolla CA 92093-0230
USA

Dr Soren E. Larsen
Riso National Laboratory
Meteorology and Wind Energy
Department
PO Box 49 DK-4000
Roskilde
DENMARK

Dr Vasylii N. Lykossov
Institute of Numerical Mathematics
Russian Academy of Sciences
Gubkina Str. 8
Moscow, GSP-1, 117930
RUSSIA

Dr Vladimir K. Makin
Ministerie van Verkeer en Waterstaat
Koninklijk Nederlands Meteorologisch
Instituut
Postbus 201
3730 AE De Bilt
Wilhelminalaan 10
NETHERLANDS

Professor Hisashi Mitsuyasu
Professor Emeritus
Kyushu University
Kasuga
Fukuoka 816-8580
JAPAN

Dr Karl F. Rieder
Marine Physical Laboratory
Scripps Institution of Oceanography
University of California, San Diego
La Jolla, CA 92093-0213
USA

Dr Jerry A. Smith
Marine Physical Laboratory
Scripps Institution of Oceanography
University of California, San Diego
La Jolla, CA 92093-0213
USA

Dr Stuart D. Smith
Department of Fisheries and Oceans
Bedford Institute of Oceanography
PO Box 1066
Dartmouth N.S.
CANADA B2Y 4A2

Dr Peter Taylor
James Rennell Division (254/27)
Southampton Oceanographic Centre
Southampton SO14 3ZH
UK

Professor Yoshiaki Toba
Invited Eminent Scientist
EORC/National Space Development
Agency of Japan

1-9-9 Roppongi Minato-ku
Tokyo 106-0032
JAPAN

Dr Yuri A. Volkov
Chief, Laboratory of Atmosphere
Ocean Interaction
Pyshevsky 3 Moscow
RUSSIA

Dr. Margaret Yelland
James Rennell Division (254/27)
Southampton Oceanographic Centre
Southampton SO14 3ZH
UK

Dr Zheng Shen
Department of Earth & Planetary
Sciences
Olin Hall
The Johns Hopkins University
Baltimore MD 21218
USA

Frequently Used Symbols

C_D	drag coefficient
c	phase speed
e	turbulent kinetic energy
d	water depth
E	elastic modulus, Ekman number
f	Coriolis parameter
$F(\omega)$	frequency spectra
g	acceleration of gravity
H	significant wave height
k	wavenumber
k_B	Boltzmann's constant
K_m	eddy viscosity
L	Monin–Obukov length
p	pressure (subscript, peak)
q	humidity
$S(k)$	slope spectra
t	time
u_*	friction velocity
u	x-component of velocity
v	y-component of velocity
w	z-component of velocity
U	x-component of mean velocity
V	y-component of mean velocity
W	z-component of mean velocity
x	horizontal coordinate
y	horizontal coordinate
z	vertical coordinate
α	spectral constant
β	spectral constant

β_*	Charnock's constant
Γ	wave growth rate
Γ_i^G	Gibbs surface excess
γ	surface tension
δ	nondimensional depth
ε^2	density ratio
ζ	stability
η	wave profile
θ	angle, potential temperature
κ	von Karman constant
λ	wavelength
μ	viscosity
ν	kinematic viscosity
π	pi
ρ	density
σ	frequency
τ	stress
ϕ	stability function
ϕ_M	stratification
χ	similarity parameter
ψ	stratification function
Ω	vorticity, angular velocity, roughness function
ω	frequency (rad/sec)

1 Overview

Ian S. F. Jones, Y. Volkov, Y. Toba, S. Larsen and N. E. Huang

1.1 Exchanges at the Sea Surface

While the boundary between the ocean and the atmosphere has been extensively studied it is still not well understood. Heat, mass and momentum cross this boundary at a rate determined by many features of not only the sea surface motion but also the properties of the atmosphere and the ocean boundary layers on each side of the interface. There are some simplifications that can be made because the sharp variations are predominantly in the vertical, but there is a hierarchy of scales and processes at play which cannot be ignored in many applications. Central to understanding the processes at the boundary is gaining knowledge about the flux of momentum between the water and the air. The flux, which is the rate of transport of momentum across unit area, can be in either direction. In a frame of reference fixed to the earth, flux is mostly from the wind to the ocean currents, but less frequently, the flux is from the *wind-waves* or currents to the atmosphere.

The winds and currents have gradients of horizontal momentum. If we treat the sea surface as a sharp boundary between two fluids of different properties, we can model the flux of momentum from one of the fluids to the other as a drag force per unit area at the sea surface. This is the surface shear stress. As well, the ocean waves that propagate on this sharp boundary can transport momentum horizontally. If there is variation in the wave properties, we can expect a horizontal gradient of momentum flux. For propagating waves, such a gradient can also be modelled as a force on the boundary. This force per unit area is known as a radiation stress.

Direct measurement of the force on the boundary is too difficult to consider. Therefore parametrization between the force per unit area and an easily measurable quantity like the wind speed some distance above the sea surface is required. The parametrization using a drag coefficient is sometimes called the bulk aerodynamic method of estimating surface stress.

1.2   Background

Since the *wind-waves* and currents at the sea surface are of practical importance to
mariners, information has been built up over the centuries. Civil engineering works
required wind-wave information and by the middle of the nineteenth century
Stevenson (1852) had produced a relationship between height and fetch based
on observations. The determination of drag coefficient came later when the understanding of turbulent boundary layers was enough to allow Jeffreys (1925) to
calculate a drag due to sheltering (flow separation) and for Montgomery (1936)
to make one of the first measurements of drag coefficient. Sixty years have now
passed.

Early in the twentieth century aerodynamicists had successfully related the drag
over a solid surface to its physical roughness, and meteorologists had, with less
success, used these concepts to calculate atmospheric drag over land. When interest
shifted to the sea surface, much lower drags were observed than expected from the
physical roughness. Why did the sea surface look so rough but support so little
drag?

While this question still remains to be fully answered, it soon became obvious
that the drag coefficient of the sea was not a constant but varied with wind speed.
Such a dependency was established by the time Deacon and Webb (1962) wrote
their review of air–sea exchanges. The nature of the surface roughness was shown
to be important by Van Dorn (1953) who reduced the drag on a pond by suppres-
sing the short waves with a surfactant.

1.3   Fluid Mechanics

Let us begin our investigation by assuming we are on a rotating earth with a
horizontally homogeneous fluid (water) overlaid by air. Let the boundaries of
this fluid be at infinity and the fluid at rest, relative to the earth. The air starts to
move. It cannot slip at the sea surface so a velocity gradient must develop in the air.
Since the air is viscous, momentum is transported towards the surface which itself
starts to move. Let us consider that the wind well above the surface, suddenly takes
on a velocity of magnitude U_0. Soon the earth's rotation has influenced the flow, as
it subjects the air to an acceleration. In the frame of reference at the surface of the
earth this can be modelled as a Coriolis force, which is written $2\boldsymbol{\Omega} \times \mathbf{u}$, where $\boldsymbol{\Omega}$ is
the angular velocity of the earth and $\mathbf{u} = (u, v, w)$.

In order to maintain this flow we need a pressure gradient $\partial p/\partial x$ or other
"external force". The simplified momentum equation allows us to obtain an
expression for the flow velocity as a function of height above the sea surface.
The coordinates and symbols are shown in Fig. 1.1 while the density of the fluid
is ρ.

$$\rho\frac{\mathrm{d}\mathbf{u}}{\mathrm{d}t} + 2\rho\boldsymbol{\Omega} \times \mathbf{u} + \nabla p - \rho\mathbf{g} = \text{frictional force per unit volume}$$

The discussion of such equations is deferred to Chapter 3.

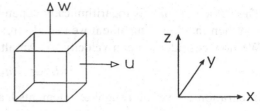

Figure 1.1.   The coordinate system used.

We look for a steady solution, that is one that applies after a long time has elapsed, so that

$$\frac{\partial u}{\partial t} = \frac{\partial v}{\partial t} = 0$$

Let us initially neglect viscosity and assume no turbulence. The sea is "flat" and lies in the $z = 0$ plane so there is no upward flow for a horizontally uniform situation, i.e. $w = 0$. We can find a solution where $f = 2\Omega \sin\phi$ where ϕ is latitude.

$$U_g = \frac{1}{\rho f}\frac{\partial p}{\partial y} \tag{1.1}$$

The flow which is at right angles to the pressure gradient is known as geostrophic flow.

However, we know there is viscosity and if we are prepared to parametrize its influence we can solve the simplified equations and find solutions that apply after a long time.

When the Reynolds number is large, turbulent instabilities set in and the viscosity must be replaced by an eddy viscosity K_m. Ellison (1956) has presented a solution to the simplified momentum equation where K_m varies with distance above the sea, z, and with some velocity scale u_*. He writes

$$K_m = \kappa u_* z$$

where $\kappa \approx 0.4$ is known as von Karman's constant. In the direction of the surface stress, Ellison (1956) found

$$U = U_g - \mathrm{Real}\left[\frac{i\pi u_*}{\kappa}\mathrm{Ho}^{(1)}\left\{\sqrt{\frac{4fzi}{\kappa u_*}}\right\}\right]$$

where $\mathrm{Ho}^{(1)}$ is the Hankel function. Near the sea surface where z is small

$$\mathrm{Ho}^{(1)} \doteq \tfrac{1}{2} + \frac{2i}{\pi}\ln\sqrt{\frac{4fz}{\kappa u_*}}$$

so near the surface

$$U = U_g + \frac{2u_*}{\kappa}\ln\sqrt{\frac{4fz}{\kappa u_*}}$$

$$U = U_g + \frac{u_*}{\kappa}\ln\frac{4fz}{\kappa u_*}$$

The velocity profile is logarithmically dependent on height.

When the flow is turbulent we need to treat the velocity as a statistical quantity. We now consider a mean velocity, U. An alternative is to assume that

$$dU/dz = u_*/\kappa z$$

integration above the roughness elements leads to

$$U/u_* = (1/\kappa)\ln(z/z_0)$$

where z_0, the integration constant, is a measure of aerodynamic roughness. This log law, as Fig. 1.2 shows, is well supported by observations. Von Karman's constant, κ, was used to equate $u_* = \sqrt{\tau/\rho}$, where τ is surface stress with the velocity gradient.

When the mean and fluctuating velocity are substituted in the momentum equation and the mean taken, quantities $u'w'$ and $v'w'$ appear in the mean momentum equation. The steps needed to expose this Reynolds stress can be seen in Lumley and Panofsky (1964). The average of quantities such as $u'w'$ is not zero in general. If the eddies which cause the velocity fluctuation have some structure, brought about say by the gradient of the mean velocity, then they can produce non zero average quantities $\langle\ \rangle$.

We know that turbulence, being a nonlinear process, transfers energy from the larger to the smaller eddies. This transfer process causes a loss of structure and it is well established that there is a tendency for the smaller eddies in the flow to be isotropic, that is, to have no characteristic direction. This implies that for small eddies the average of any velocity product, e.g. the Reynolds stress should be zero.

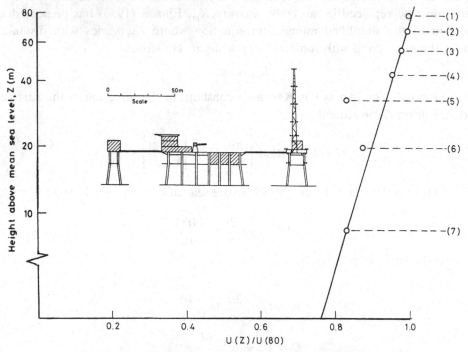

Figure 1.2.   Velocity profile measured over the sea.

This can be shown by recognizing that the average of velocity products should not change for different alignments of the axis.

It can be shown for horizontally homogeneous flow in the y direction, that $\langle v'w' \rangle = 0$. Such conditions might be expected near the surface over a boundary without directional texture where the x axis is aligned with the wind. On a rotating earth the Coriolis force encourages the shear stress to rotate with height. Even at distances such as 10 m above the sea surface, the Ekman solution predicts lateral Reynolds stress as small (order 4% at mid latitudes, 10 m height and wind speed 10 m/s) and on this basis the lateral stress has often been neglected. This is an important issue that we will return to later in this chapter and again in Chapter 12. Remember that the symbol $\langle \; \rangle$ implies a mean over a period long compared with the eddies which contribute to the momentum flux.

This flux of momentum is drawn from the mean velocity gradient. Thus we expect a momentum deficiency to develop near the sea surface. With a suitable external pressure gradient to balance the surface stress, there is no need for the momentum deficit to grow and we can have a horizontally homogeneous boundary layer.

While we know that our interface soon becomes ruffled with waves, let us go ahead and assume it is smooth. Laboratory experiments show this is possible for wind speeds below a few metres per second.

There is a surface stress τ_a on the water interface and a corresponding water stress on the air τ_w. We can define nondimensional drag coefficients

$$\tau_a = C_D \rho_a U_a^2 \qquad \tau_w = C_D^w \rho_w U_w^2 \tag{1.2}$$

where the reference air velocity and the water velocity need to be specified closely since the velocity varies with the distance from the surface.

The velocity is not expected to be parallel to the stress on a rotating earth, so we now define the x axis in the direction of the stress vector. Then we have the opportunity to use either U at height z above the surface or $\sqrt{U_z^2 + V_z^2}$ as the reference velocity. When interfacial waves are present there is no need for τ_a and τ_w to be equal as momentum can be radiated away by the waves.

We need to remember that the interface is moving with a mean velocity U_s, which we hope is in the same direction as the wind stress. [At this stage we are not admitting any water currents driven by a pressure or density gradient, and the flow is steady.] Let us define our reference velocities in the air and the water as

$$U_a^2 = (U - U_s)_z^2 + V_z^2 \qquad U_w^2 = (U_s - U)_{-z}^2 + V_{-z}^2$$

a choice that makes sense via reference to Fig. 1.3.

We do not need to make the height of the reference velocity, z, equal above and below the surface, but there is no reason to do otherwise. Many authors neglect the rotation of the wind and the current and use $\tau_a = C_D \rho U_z^2$. The angle between τ_a and U_z is addressed in Chapter 12 and found to be significant in unsteady conditions and situations with the waves not aligned with the wind.

An alternative way to parametrize the ocean surface is to introduce the aerodynamic roughness length, z_0. This has the advantage of not needing to introduce

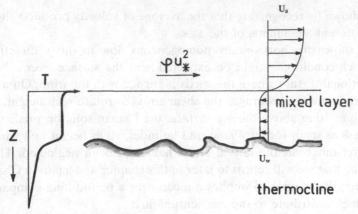

Figure 1.3. Schematic of the flows near the sea surface. Velocities are shown relative to the interface.

the height above the surface to define a measure of drag. For a mean flow that can be represented by a log law, it can be shown that C_D and z_0 are uniquely related, by

$$\sqrt{C_D} = \frac{u_*}{U_z} = \frac{\kappa}{\ln(z/z_0)} \tag{1.3}$$

If the world was as simple as the above and C_D were a constant, we would not need a monograph on drag over the ocean. The results plotted in Fig. 1.4 persuade us there is much more to the problem of describing the drag at the sea surface. It is the differences from the above idealization of the air–sea interface that will fill the remaining pages of this monograph.

1.4 Irrotational Waves and Forced Waves

Possibly the first restriction to remove is that of horizontal homogeneity near the sea surface. Wind flow above a few metres per second induces surface waves that

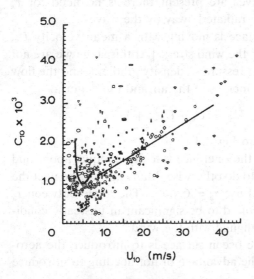

Figure 1.4. Drag versus wind speed after Huang et al. (1986). The solid line is the drag coefficient for a smooth plate at low wind speed.

can be idealized as an interplay between potential energy (gravity) and kinetic energy, or between the stretching of the surface in the presence of a surface tension force and displacing it in a gravitational field.

First we will assume these interfacial waves are inviscid and irrotational. Solutions to the equations, with appropriate boundary conditions, allow the properties of such surface waves to be determined. For details readers are referred to Phillips (1977). Infinitesimal irrotational waves of wave number k in deep water propagate at a speed

$$c = \omega/k \tag{1.4}$$

where the frequency is given by

$$\omega^2 = gk(1 + \gamma k^2/\rho_w g) \tag{1.5}$$

and where γ, the surface tension, has a typical value of $\sim 7 \times 10^{-2}$ N/m for sea water and air. Equations (1.4) and (1.5) can be used to show that there is a phase speed minimum of 0.23 m/s. For long waves, that is for waves where the wave number k is small, the first term dominates and the disturbances are called gravity waves. In this case the longer the wave, the lower its frequency and the faster its phase speed.

The phase speed given by Eq. (1.4) assumes the underlying water is stationary. However, when a spectrum of irrotational waves are present, the shorter waves are swept backwards and forwards by the orbital motions of the longer waves. Now the phase speed is modulated by the motion of the longer waves and the dispersion relation of Eq. (1.5) cannot be used.

When irrotational waves become steeper, the nonlinear nature of the surface boundary condition induces noticeable bound harmonics. These are waves that travel with the same speed as their fundamental but have wavenumbers that are integer multiples of the fundamental. In wind wave tanks, the harmonics of the characteristic wave are often clearly seen in the frequency spectra.

When the waves are forced by the wind, they can no longer be accurately modelled as irrotational. We will use the term *wind-waves* to describe this class of wave motion.

In an infinite ocean the fetch is infinity and for long duration winds, when we have a steady state in the air and water away from the interface, there is a tacit acceptance that there might be fully developed *wind-waves*. That is to say that the (oscillating) *wind-wave* energy density, E, might be constant. It can be made non-dimensional by using the surface stress and gravity, g.

$$\frac{g\rho_a E}{\tau^2} = \frac{gE}{\rho_a C_D^2 U_a^4}$$

People can find evidence that this nondimensional number is a constant and so for a fully developed sea, significant wave energy would rise with wind speed. Remember the duration of wind at constant speed may need to be days. The weather patterns in the atmosphere do not allow us to test the above belief.

If the *wind-waves* were irrotational they would play no direct part in the transfer of momentum. Pressure unlike shear stress acts normal to a surface. If the surface wave is thought of as a sine wave (or the sum of a number of sine waves) then the displacement away from the mean surface can be written

$$\eta = a\cos(kx - \omega t)$$

for a wave propagating in the x direction with a phase speed ω/k. When there is a flow of velocity U over the surface, the pressure induced by streamline fluctuations is the rate of working of surface forces per unit wavelength $\dot{E}$

$$p = \rho_a(U - c)^2 ak^2 e^{-kz}\cos(kx - \omega t)$$

and the net *momentum flux* in the x direction, averaged over the wave ($z \approx 0$), becomes

$$\int_0^{2\pi/k} p\frac{\partial \eta}{\partial x}\mathrm{d}x = \int_0^{2\pi/k} \rho_a(U - c)^2 a^2 k\cos(kx - \omega t)\sin(kx - \omega t)\mathrm{d}t = 0 \qquad (1.6)$$

There is no force on the wave unless the pressure and the displacement are out of phase, as could be the case for flow separation. Then the drag produced by a wave of slope ak is proportional to $\rho a U^2(1 - c/U)^2$, the slope and the phase angle between the pressure and the displacement.

Wind-waves however are not irrotational waves but rather part of a coupled turbulent boundary layer system. The *wind-waves* are not propagating freely but are forced by the air and water flows. Thus while some properties of irrotational waves are relevant for a windsea, most are not.

The surface has another interesting characteristic. The disturbances travel relative to the mean motion of the surface (see Fig. 1.3). Let us assume that there is no local distortion of the surface due to the air flow. Then it looks like a noncompliant surface to the air. The orbital motions move the surface fluid forward and backward and a process called Stokes drift may induce some mean flow. The wind also induces a mean surface flow, as noted earlier. Typically the surface velocity is a few per cent of the geostrophic wind speed. We know empirically this is much less than the phase speed of roughnesses that rush ahead of the surface particles. These roughnesses act like rollers moving below a sheet of rubber. In fact this experiment has been performed by Kendall (1970) and is most revealing. He shows that the aerodynamic roughness of his sheet of rubber depends on the speed at which the roller moves, relative to the wind.

It is movement of the roughness that many authors have thought is important in producing the low values of atmospheric drag coefficient. While the sea can be very rough, with rms displacements of metres, C_D is observed to be only a little above that of a smooth plate.

1.5 *Wind-wave* Sheltering and Wave Age

When air flows over a steeply deformed water surface it is not able to follow the surface but instead separates. Jeffreys (1924) suggested that such separation of the airflow might occur at each wave crest and produce a region of low velocity and low pressure downwind of the crest. Thus there would be a difference in pressure between the upwind and downwind face of the wave able to transfer energy from the wind to the *wind-waves*. When Jeffreys' sheltering hypothesis was popular the concept of wave age was developed to express the relative speed of the wind and "characteristic" *wind-wave*. The observed rate of growth of *wind-waves* is lower than Jeffreys' calculation and Belcher and Hunt et al. (1993) introduced the concept of nonseparated sheltering where the turbulence stresses in the atmospheric boundary layer de-accelerate the air flow near the crest of the wave and produce asymmetric displacement of the streamline without separation. This concept leads to drag coefficients more in line with observations. Early in the development of a wave field, the characteristic *wind-wave* is of short length and travels more slowly than the wind (at some height above the surface). The phase speed of the most energetic or characteristic *wind-wave* will be designated c_p. As the storm progresses, the characteristic frequency decreases and the speed of propagation increases until it becomes equal to the wind. Under the sheltering hypothesis there is no forcing mechanism to drive faster travelling waves so c_p/U equal to one would be a mature wave. Life is more complicated because of nonlinear transfer between spectral components, but it is evident that in a general sense the growth of a wave field after $c_p/U = 1$ is small.

The term "wave age" of the sea was introduced to designate the velocity c_p normalized by a measure of the wind speed. In the past people used the wind velocity at a height (often 10 m) above the sea surface U_{10}. We will use u_* rather than the wind velocity U since, amongst other things, it avoids taking into account the influence of stratification on the velocity profile and the issue of the appropriate height above the sea at which to measure the wind speed. Note that the direction of propagation of the characteristic *wind-wave* does not have to be in the wind direction. As we are treating only an idealized situation at present we will return to this issue later. Thus our wave age is defined as c_p/u_*.

With this velocity scale, a mature *wind-wave* has an age of about 30. Smaller values of c_p/u_* imply the wind relative to the wave profile flowing in the direction of wave propagation. Larger values of c_p/u_* imply the wave, relative to the wind, moves in the opposite direction to the wind and one would expect a transfer momentum from the wave to the wind. Wave age is a general measure which expresses the state of *wind-wave* development and is now used to indicate conditions with respect to energy input, dissipation and non-linear transfer in the wave field. See Fig. 1.5.

The characteristic *wind-wave* becomes less steep with wave age in the manner shown in Chapter 4 (Fig. 4.19), and possibly many other effects are a function of wave age. One question we will return to both in this chapter and in Chapter 10 is the dependence of the sea surface drag coefficient on wave age.

1.6 Drag Generating Mechanisms

A turbulent boundary layer over a smooth solid surface has a viscous sublayer adjacent to the surface. If the small roughnesses on such a surface do not protrude through the sublayer, the surface is called aerodynamically smooth. The

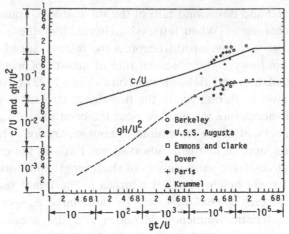

Figure 1.5. Wave age as a function of nondimensional time (duration), after Sverdrup and Munk (1947).

surface stress is provided by skin friction, defined by the velocity gradient at the interface, that is:

$$\tau_s = \mu \left(\frac{\partial U}{\partial z} \right)_0$$

The log law for velocity over a smooth surface has $z_0 = 0.11 \nu / u_*$, where ν is the kinematic viscosity of the air (about 1.5×10^{-5} m^2/s for air and an order of magnitude less for water) and 0.11 is an empirical constant determined from wind tunnel experiments. The equivalent atmospheric drag coefficient is shown in Fig. 1.4 where one sees that such a C_D decreases with wind speed. The aerodynamic roughness beneath the sea surface when it is smooth is about one third the value on the air side due to the different values of u_* and kinematic viscosity. There has been some discussion of super smooth water surfaces and we consider this again at the end of the chapter.

On rough surfaces when there are substantial elements protruding through the viscous sublayer, separation can occur round these elements. If this happens some of the momentum is transferred to the surface by pressure forces with a horizontal component. When these elements move, relative to the surface, in the direction of the wind, as they do for *wind-waves*, the element is not exposed to the full force of the wind.

Let us consider the drag over a fixed train of two dimensional sinusoidal waves. The surface of the waves may have roughness elements but if we go some distance above the crest of the waves, the log law approximation to the wind profile will produce an apparent aerodynamic roughness incorporating the form drag on the waves. When the wave steepness is significant this form drag can dominate over the surface or skin friction roughness. Gong et al. (1996) measured values of z_0/a of order 0.1 and we have plotted this in Fig. 1.6.

When c_p/u_* is positive there is a region of flow near the wave surface that is in the opposite direction to the wave speed that is expected to reduce the form drag.

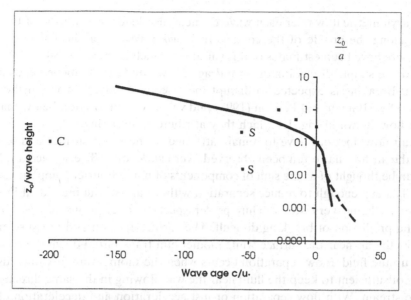

Figure 1.6. Some z_0/a results above and below the sea surface. Below the surface c/u_* is negative as the flow, relative to the surface, is against the wave propagation. Only aerodynamically rough cases, that is $z_0 u_*/v > 2.3$ have been used. S = Shemdin (1972); C = Csanady (1984); broken line = Donelan et al. (1993).

As the value of c_p/u_* increases further, the whole of the flow is against the wave propagation direction. When c_p/u_* is negative as it is in the water boundary below the waves, the flow relative to the backward moving wave crests is greater than over the stationary wave, $c_p = 0$. Now we would expect the form drag of this faster flow to be larger, increasing the drag coefficient and the value of C_d over that of the stationary wave.

While much too simple to be realistic description of flow over *wind-waves*, some understanding of the role of a moving roughness can be gained by examining the form drag on an idealized sinusoidal surface on the assumption that the flow is asymmetric and the resultant phase angle between the pressure and the wave is constant. In Eq. (1.6) the form drag was shown to depend on $\rho_a U^2 (1 - c)^2$. With the use of the log law expression (1.3) we can relate z_0 on the unsubstantiated assumption (see below) that all the drag is form drag with this c/u_* dependence. Then

$$(1 - c/U) \propto \frac{\kappa}{\ln(z/z_0)}$$

This expression is shown in Fig. 1.6 as a solid line for values of c both in the direction of the wind and for negative values when the waves are moving against the wind. By reference to Fig. 1.3 one can see that for the boundary layer below the surface c/u_* is negative for steady state situations. The constant was determined by matching the fixed surface values of Gong et al. (1996). All the data points for negative c_p/u_* are from the marine boundary layer. This view of the world does not support the assumption sometimes made that the aerodynamic roughness is equal above and below *wind-waves*.

The asymmetric flow over each wave element also leads to variation of the skin friction along the profile of the characteristic *wind-wave*. For short young *wind-waves* there have been estimates of this variation, which is shown in Fig. 1.7. Notice that now the simple discussion of form drag that we undertook above needs modification. Breaking is expected to disrupt the viscous sublayer on which the skin friction relies. Banner and Peirson (1998) find values of skin friction below that for smooth flow shown in Fig. 1.4 which they attribute to breaking.

The air flow does not have to remain attached to the water surface and separation of the air flow has often been observed over *wind-waves*. The surface displacement can be thought of as the sum of components of many Fourier components and be locally steep enough to induce separation without any of the individual Fourier components being very steep. Thus power spectral descriptions of *wind-waves* make the prediction of breaking difficult. The surface friction and pressure gradients slow the air near the surface while momentum transported down the gradient speeds up the fluid. Flow separation occurs when the momentum flux towards the wall is not sufficient to keep the fluid near the wall flowing in the same direction as the free stream. With flow separation or just acceleration and deceleration of the flow over the roughness elements (waves), the horizontal component of the normal pressure on the surface does not have to average to zero. This is the process that leads to form drag, as discussed above.

In an open ocean situation there are many different wavelengths present. Which ones carry most of the form drag? In Chapter 4 on ocean waves, the short *wind-waves* (order 1 cm wavelength) are shown to increase in amplitude with increasing wind speed. They are believed to quickly become fetch limited. Do these shortest *wind-waves* carry most of the stress? When there are surface slicks on the ocean the mean square slope of the surface, a measure of roughness, is reduced and Van

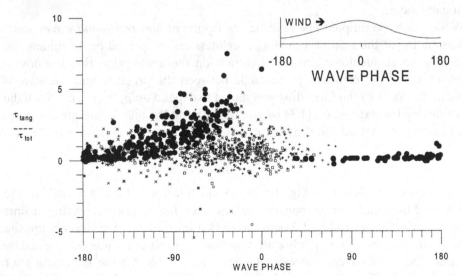

Figure 1.7. The distribution of skin friction along a wind-wave after Banner and Peirson (1998) showing low values down wind of the wave crest. The solid dots are from Okuda et al. (1977). Wave ages are small for these data. The average of τ_f over the wave is of order $0.5\tau_t$.

Dorn (1953) showed that the drag coefficient was reduced. This is evidence that the short *wind-waves* carry some, but not most, of the surface stress.

The fraction of drag supported by these two processes appears to vary with the state of development of the sea. One way to estimate the partition between form drag and skin friction is to measure the velocity gradient in the water very near the interface and so calculate the skin friction. Such attempts in laboratory situations by Okuda et al. (1977) illustrated in Fig. 1.7 first reported that skin friction dominated in short laboratory waves but Banner and Peirson (1996) found that skin friction was more like half of the drag and that form drag was more important at higher wind speed.

Another approach is to calculate the momentum flux to the *wind-wave* and assume the remaining stress is due to skin friction. Such a calculation is discussed in the next section.

While it is clear that the aerodynamic roughness over a complex surface such as the sea is not simply related to the physical roughness, there have been attempts to model the moving roughness by decomposing the surface into Fourier components and attributing a phase speed to each component. Under a hypothesis that the velocity of the roughness relative to the wind speed is important, one can construct models for the aerodynamic roughness. This is an extension of the discussion that was used to construct the solid line in Fig. 1.6.

The largest Fourier components travel close to the wind speed for many open ocean situations, so their contribution to z_0 would be low. Shorter *wind-waves* with lower phase speeds, while not of such large amplitude, move relative to the wind and so have some contribution to z_0. This can be expressed mathematically by following Kitaigorodskii (1973) as:

$$z_0^2 = \int f\left(\frac{c}{u_*}\right)\Phi(k,\omega)dk d\omega \tag{1.7}$$

where Φ is the power spectral density of surface displacements.

It has already been noted, Eq. (1.3), that $u_* = \kappa U_z / \ln(z/z_0)$ so Eq. (1.7) provides an expression for

$$\tau = f_1(U_z, \Phi) = \rho C_D U_z^2$$

There is no obvious way to recognize in this formulation that the surface stress and the wind may not be aligned.

Nordeng (1991) used this concept and produced an expression for z_0 which will be cited later. For the present it is sufficient to note that such models predict that the most effective *wind-waves* in transferring momentum are about half the wavelength of the characteristic *wind-wave*.

1.7   Aerodynamically Rough Flow

When the surface roughness elements protrude through the viscous sublayer on a flat rough plate, the nature of the drag coefficient changes. Reynolds number becomes unimportant as momentum is transferred to the surface irrespective of the viscosity. As the *wind-waves* become larger, the value of z_0 increases and the surface becomes more aerodynamically rough. Since the kinematic viscosity and friction velocities are different on the air and water side of the interface, the thicknesses of the sublayers are different with the water boundary layer having a thinner sublayer. With increasing wind speed, the water boundary layer is expected to become aerodynamically rough before the air boundary layer. Since the sea surface tends to have sharper crests than troughs there is uncertainty about this issue.

On an aerodynamically rough ocean, it is the surface roughness (the short *wind-waves*) which is the presumed principal recipient of the momentum. If we continue to assume steady state, then the drag could be calculated by summing the momentum flux to each Fourier component of the surface roughness. To estimate the momentum flux we turn to the experiments on the growth of *wind-waves*. For short *wind-waves* such results have been obtained in wind wave tanks because in the ocean the complexity of making the measurements allows estimates of growth rate to be made only for longer *wind-waves*.

The experiments consist of rapidly turning on the wind and studying the increase in energy in a particular frequency component (sometimes wavenumber component). The growth starts off slowly, then goes through an exponential phase before steadying out (after overshooting its steady-state value). The momentum flux to the growing component can be estimated from its rate of growth, Γ, after making an allowance for the viscous dissipation of energy. This value can be used as an estimate of the momentum flux to the steady-state value of this frequency component in the presence of a whole spectrum of wavelengths. Whether the flux of momentum can be treated as a Fourier decomposition problem is a matter of speculation. It is well known that air flow separation depends on the local slope of the surface and this cannot be deduced from the wave height power spectra. The Fourier coefficients, not the power spectra are required to reconstruct the surface. The second difficulty is that there is no guarantee that the rate of transfer of momentum to the dominant wave in the spectrum, used to determine growth rate, is the same as to the component that is in near steady state and immersed in a field of much larger *wind-waves* of lower frequency. Not to be deterred by these doubts, we will press ahead to examine the consequences of assuming that the momentum to the short *wind-waves*, in an aerodynamically rough flow, can be modelled by this approach.

Wave drag becomes

$$\tau^w = \rho \int \frac{\Gamma}{c} \Phi(\mathbf{k}, \omega) d\mathbf{k} d\omega \qquad (1.8)$$

where the growth rate expression of Komen et al. (1984) can be used.

It is generally believed that the sea surface is not aerodynamically rough, in the sense used above and that only a fraction of the wind stress is supported by form drag (i.e. τ_f) which increases with wave steepness (Mitsuyasu 1985). These calculations for the fraction of *wind-wave* drag, however, depend upon the *wind-wave* growth expression used and on the nature of the wavenumber spectrum $\phi(k)$ discussed below. As the *wind-wave* stress τ^w should not exceed the wind stress, these calculations place some bounds on the wave spectrum and growth expressions.

It is important to note that the term "*wind-wave* drag" does not imply that the momentum goes to make the *wind-waves* themselves grow. The *wind-wave* momentum is the flux of momentum to the sea surface that is associated with the presence of *wind-waves*. Breaking and other turbulent processes transfer much of this *wind-wave* momentum straight into the currents below the surface. Also, by discussing the drag in terms of the momentum fluxes in the air, we are implying that the same discussion could be undertaken by considering the momentum fluxes in the water. The fluxes each side of the surface are of the same order of magnitude.

Even if we were confident of the value of Γ to use in Eq. (1.8) the wave age dependence of the short *wind-waves* portion of the spectrum is not at all clear. It is difficult to determine the dependence of wave drag coefficient on wave age. The experimental results such as HEXOS cast some serious doubts on "*wind-wave* growth" parametrization. HEXOS (Smith et al. 1992) shows the total drag coefficient decreasing with wave age (over the narrow range, 10 to 27). If the wave drag component of the total drag were to provide a significant fraction of the drag, then it would need to decrease with wave age. The lower wavenumbers of the surface displacement spectrum are known to increase with wave age and so the integral Eq. (1.8) could be expected to increase. (See the discussion around Fig. 1.9.) There is no evidence that the high wavenumber components decrease with wave age to reverse the impact of the low frequencies. Note that the discussion revolves around wavenumber spectra. In frequency space the impact of the dispersion relationship can produce different spectral dependencies, as illustrated in Banner et al. (1989). Unless this is some effect of water depth, it is hard to reconcile the HEXOS drag observation with the form drag approach.

The determination of drag is far from resolved. If Eq. (1.8) provided the dominant term, the problem of describing drag over the ocean would be mostly reduced to that describing the two dimensional wavenumber spectra of the sea surface. The nature of momentum transport is addressed in more depth in Chapter 6.

1.8   Ocean *Wind-wave* Spectra and *Wind-wave* Equilibrium

The *wind-waves* on the ocean surface are neither unforced nor irrotational as stressed in previous discussions in Section 1.4. The wind forces them, they are embedded in a turbulent field that extracts their energy (Olmez and Milgram 1992), they break and their statistical properties are unsteady due to the constantly changing wind.

If we consider a subset of environmental situations we may be able to make progress in the face of some complexity. Dimensional analysis can be used when the relevant variables can be identified. Let us consider situations where the *wind-waves* are in *local equilibrium with the wind*. This is a situation where some measure of the wind velocity, say u_*, is all that is necessary to characterize the forcing of the *wind-waves*. In situations with rapidly changing wind stress, which are discussed in Chapter 9, this is unlikely to apply as the history of u_* and the direction of the stress would appear to be important variables. Also the influence of the sea floor must be negligible. In such situations where there is *local equilibrium with the wind* the spectra might be a function of g, u_* and ω. Such an expression attributed to Toba (1973) which has good empirical support is:

$$\phi(\omega) = \alpha_s g u_* \omega^{-4} \qquad \omega > \omega_p$$

It seems that the sea quickly comes to a local equilibrium with the wind and takes this spectral form – in a time much less than the "duration time". Notice this expression implies there is a wind speed dependence for the frequency spectrum. The experimental support is presented in Chapter 4.

The concept of a *wind-wave* equilibrium recognizes that the water boundary layer below the waves, the air boundary above the waves and the waves themselves are in a complex nonlinear relationship. This does not mean that the waves are statistically steady. The interaction is such that the *wind-wave* frequency spectrum takes on a level for $\omega > \omega_p$ that depends only on u_*.

For steady situations, with increasing fetch, the *wind-wave* spectral peak moves to lower values as shown in Fig. 1.8 and the phase speed of the peak components travel more quickly. This effect has been called "downshifting". The empirical rules show that the characteristic *wind-wave* becomes less steep with increasing fetch.

It is not the frequency spectrum that is actually needed in the form drag calculation of Eq. (1.8) but rather the wavenumber spectrum. Let us define the spectrum $F(k)$ as that dependent on the wavenumber magnitude and averaged over all wave directions. This quantity is discussed in some detail in Chapter 4. Near the peak wavelength (that is the wavelength that corresponds to the spectral peak) and near the minimum phase speed, the spectrum $F(k)$ depends upon the wind speed or friction velocity. In between there is a region of wind speed independence. The directional dependence of the *wind-waves* on the angle to the wind also seems to be high at long and short waves and weak at intermediate wavelengths. The wave spectral values decrease rapidly for *wind-waves* shorter than 6 mm (see Chapter 4). Viscosity and temperature do not seem important at any wavenumber range although this is surprising.

For wavenumbers near the spectral peak there is fetch dependence and the spectrum depends on the wave age c_p/u_*. However as one moves to shorter *wind-waves* the evidence suggests that this dependence diminishes. Figure 1.9 shows the above ideas schematically.

The surface wave spectrum $\Phi(\omega, k)$ has been traditionally described by a differential equation:

$$\frac{\partial \Phi}{\partial t} + \mathbf{c} . \nabla \Phi = S_{\text{in}} + S_{\text{ds}} + S_{\text{nl}}$$

where S_{in}, S_{ds} and S_{nl} are the wind inputs, dissipation and nonlinear transfer respectively. The terms on the right hand side are independent and the reader is

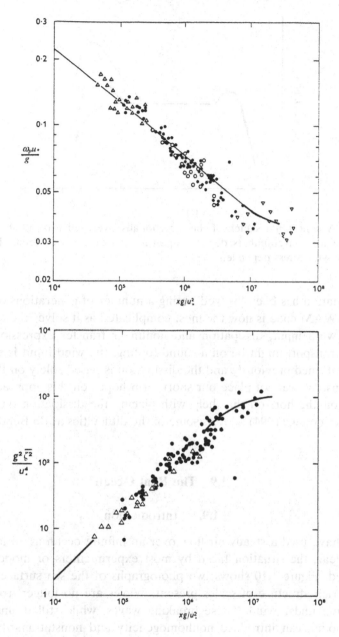

Figure 1.8. Fetch and duration rules after Phillips (1977) using the Kitaigorodskii (1973) scaling.

referred to Chapter 9 for more discussion. If we could solve this equation and had a set of initial conditions we could know the physical roughness of the sea surface and use Eq. (1.8) to estimate the flux of momentum to the *wind-waves*. Remember that correct expression for Γ is still uncertain. All we would need to do would be to add the drag due to tangential stress, τ_s, and the problem of drag determination would be resolved.

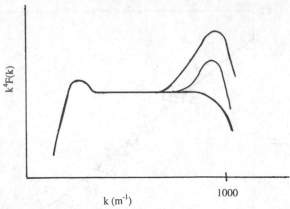

Figure 1.9. A conceptual sketch of the directionally averaged wavenumber spectra. The shape at the low wavenumbers depends upon u_* and duration or fetch. The high wave numbers are wind stress dependent.

This equation has been "solved" using a number of generations of approximations. The WAM code is now the most complicated as it solves the equation for a presumed wind input, dissipation and nonlinear transfer expression. While the nonlinear transport might be on a sound footing, the wind input is subject to the doubts mentioned previously and the dissipation is based solely on the need to fit observations. We cannot place our short term hopes on this approach. There are few ideas on the horizon that help with placing the dissipation on a firm basis. Banner and Young (1994) address some of the difficulties as do Bender and Leslie (1994).

1.9   The Real Ocean

1.9.1   Introduction

While we have used a steady air flow over an infinite ocean as an idealization to develop ideas, the situation faced by most experimenters or modellers is more complicated. Figure 1.10 shows two photographs of the sea surface that demonstrate the rich structure of scales present. Waves are no longer irrotational, but are breaking under wind. These breaking waves, while still ill-understood and hard to model, can introduce nonhomogeneity and nonstationarity, cause flow separation and enhance momentum flux by the injection of mass from the breaking wave tips into the water column (see for example Huang 1986). The extra turbulence below waves is one of the reasons for believing the water boundary layer sees the surface as having a larger z_0 than the air boundary layer. The ocean is not infinite and in particular the atmospheric forcing is of finite extent and never truly steady.

Fetch is the distance either from the shoreline or the edge of the storm to the point of discussion since we are still concerned with steady state conditions. The wave field, being the physical roughness of the sea surface, depends on this quantity

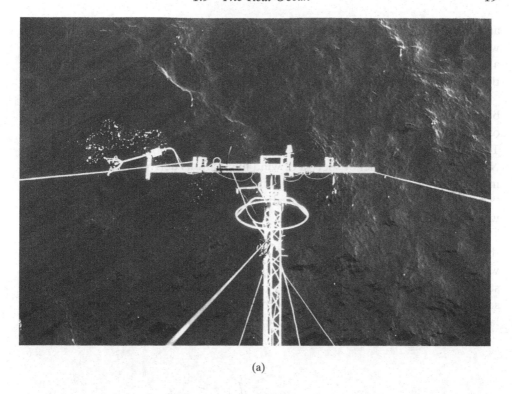

(a)

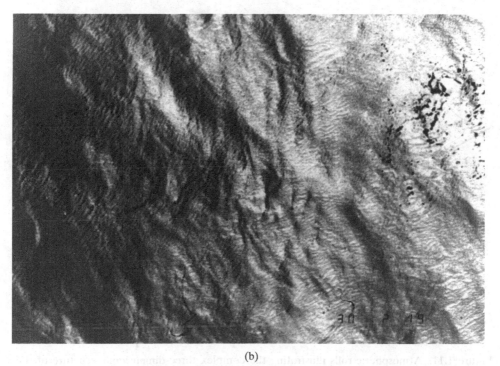

(b)

Figure 1.10.   Photographs of the sea surface. $U = 13.3$ m/s. SWH = 2.5 m.

and so we need an understanding of the development of the *wind-wave* spectrum
with fetch. While the directional power spectra is difficult enough to describe, it
alone may not be a sufficient description of the sea surface to provide the information needed for modelling momentum transfer at the surface.

Also, the atmosphere is seldom horizontally homogeneous. Large planetary
boundary layer vortices or atmospheric rolls are often seen in satellite pictures
of clouds and their footprints can be seen in radar reflectance of the sea surface.
They take the form of Fig. 1.11 and can be seen in the horizontal wind spectrum.
Nor is it steady. Frequency spectra of wind fluctuations derived from long series of
measurements over water are continuous and pink. There is however often a
"spectral valley" for disturbances of about 20 minutes that allow the fluctuations
within 20 minute records to be treated as steady.

Unsteadiness in wind forcing means that except for short fetches, most of
the ocean is covered with *wind-waves* which are still evolving due to changing
wind. There are two classes of problems for *wind-waves* previously in steady
state. The first is where the wind has increased. This leads to a young sea and
seems better understood than the second where the wind has abated. In the
latter case, when there is a slight decrease in the wind, the sea becomes a little

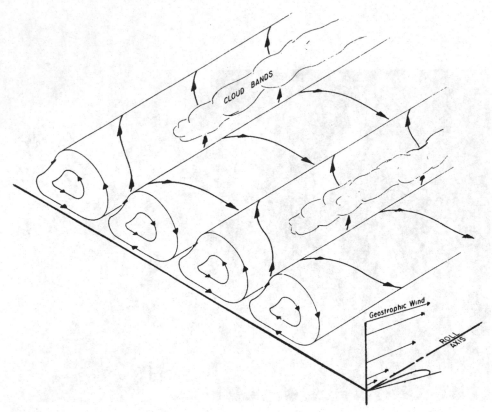

Figure 1.11. Atmospheric rolls illustrating the complex three-dimensional structure of the
atmospheric boundary layer over the ocean.

older, but when there is a large change in wind speed (or direction) the sea becomes very old. These *wind-waves* are "over saturated", and the adjustment process leads to the propagation of the lowest frequencies as swell and to the higher frequencies being subjected to dissipation to bring them into equilibrium with the wind. Often the fluctuation in the mean wind is rapid enough that the previous change is still being assimilated by the wave field. Many of the fluctuations in wind stress are due to shifts in wind direction. This issue is discussed in Section 1.9.3.

Not all *wind-waves* are a function of the local wind. Once we admit nonuniform wind fields, for example we treat storms of finite extent, the wave field under observation can be distorted by swell that has propagated from an earlier disturbance some distance away. How much influence does swell have on the drag of the sea? Nor need the two boundary layers be uniform in density. Temperature gradients exist above and below the surface. There are a number of additional complications but we will defer these till the above four difficulties have been addressed.

1.9.2 Atmospheric Rolls

The large organized structures in the atmosphere and the water can both be expected to influence the drag coefficient. In the water the presence of lines of foam in the direction of the wind, signal the presence of Langmuir circulations in the water. The presence of foam implies that the Langmuir cells cause a secondary flow at right angles to the wind. The foam lines appear where the secondary circulation is downward and the buoyant foam, which has been concentrated by the converging surface flow, cannot follow the water. The spatial scale of Langmuir cells is of the same order as the height of measurement of the stress and many cells drift through the observation area during the typical sampling period of an air–sea interaction experiment. Their impact on the drag coefficient averaged over the period of minutes is not expected to be large.

In the case of atmospheric rolls their time and space scales make their influence more obvious. The rolls are inclined to the wind at a small angle and so drift past a fixed observation point with a time scale much longer than the atmospheric boundary layer thickness time scale. These secondary flows upset local homogeneity and can be expected to induce variations in the drag coefficient.

Such a variation in drag has been found by Jones and Negus (1996) in an experiment where the Reynolds stress, defined as the fluctuations about a 15 minute mean, was measured from an oil production platform in Bass Strait, Australia. They found the drag coefficient, 53 m above the sea determined from the Reynolds stress, was high when there was an updraft and lower during a period of downdraft. Atmospheric rolls, slowly drifting past the observation site were proposed as a possible mechanism for the updrafts. Such rolls would also transport momentum (towards the surface in downdrafts, away in updrafts) that is not included in the Reynolds stress calculations.

1.9.3   Unsteady Turbulent Boundary Layer

If the roughness of the sea surface changes or if the velocity at the outer edge of the log law changes, either in speed or direction, it takes some time for the flow to adjust to a new steady state. When the flow is from land to the sea, an "inner layer" develops from the point of zero fetch as shown in Fig. 1.12. Above the inner layer the log law will be uninfluenced by the change in surface roughness (although the streamlines will be displaced). Much the same can be expected when there is a shift in wind. One consequence of this is that using a drag coefficient based on a reference height above the inner layer leads to confusion. Aerodynamic roughness, z_0, is spared this problem but it must be obtained from that part of the log law that is within the inner layer.

When there is a sudden change in the wind (or the current) the sea state does not immediately adjust to be in local equilibrium. The surface roughness change lags the wind shift. However one can think of cases where the *wind-waves*, encountering a change in current, for example near a river mouth, would change before the wind. There is a time scale for the diffusion of momentum to (or from) the sea surface and so the adjustment process is not immediate.

When the wind is steady, the windsea takes a universal form of spectral distribution that depends on friction velocity. If the spectral levels, as a result of unsteadiness, are larger than the steady state value the windsea is described as "oversaturated". When it is less than the steady state value for the local wind speed it is termed "undersaturated". There appears to be a rapid response to over or undersaturation. Initially the *wind-waves* change quickly to come into *local equilibrium with the wind* and then more slowly as an evolution of the wave field occurs as in other duration-limited situations. It is during the relatively short period of rapid re-adjustment to the wind shift that one can expect the drag coefficient to be strongly different from the steady state value. There is further discussion of unsteadiness in Chapter 9.

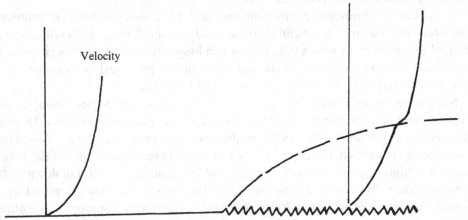

Figure 1.12. Inner layer. The inner layer is found to grow after a change in surface roughness.

Wind-waves have a directional structure which in steady state is believed to have a maximum in the wind direction. When the wind changes direction, the sea will initially retain a direction of propagation of the characteristic *wind-waves* which is inclined to the new wind direction. This provides an opportunity for the surface stress to take on a significantly different direction to the wind some 10 m above the sea. Being a coupled system, as described in Chapter 4, the turbulent boundary layers each side of the interface and the *wind-waves* will adjust to restore the system to a new local equilibrium.

1.9.4   Drag Dependence on Stability

In the earlier parts of this chapter, we have ignored the issue of density stratification in the atmosphere and the ocean. In the atmosphere, the density depends mostly on the temperature and humidity while in the water it depends on the salinity and temperature. The transport of momentum away from the surface is controlled by the turbulent velocity fluctuations and these are in general increased in density unstable conditions and decreased in stable conditions. Suppression of turbulence reduces the momentum flux (in general) and so the drag coefficient is less in stably stratified situations. Models of the changes in drag coefficient with atmospheric stability are described in Chapter 7.

The influence of solar heating and fresh water input on the upper ocean will produce strong stratification which can be expected to influence the drag coefficient defined in terms of the currents. This is a problem yet to be studied in detail.

1.9.5   Role of Swell

The low frequency waves generated remotely are termed "swell". Such waves have the possibility of influencing both the magnitude of the surface stress and the angle between the stress and the wind at 10 m.

When the wind speed is low, the influence of the swell is likely to be more pronounced. Considering first the magnitude of the drag coefficient, when the wind and the swell are in opposite directions, the drag coefficient is enhanced. Here the swell, travelling in the opposite direction to the wind, provides an additional form drag that is expected to increase with the swell steepness. The momentum transferred to the surface is expected to attenuate the wave train. When the wind is at right angles, the impact of the swell is small; while when the swell and the wind run together, there is a reduction in the drag coefficient. The steeper the swell and the closer its frequency to the characteristic windsea frequency, the more influence it can be expected to have. These issues are discussed in Chapter 8.

1.9.6   Drag Dependence on Wave Age

Within the context of a steady, finite extent, unstratified ocean and atmosphere, we can look for simple expressions for the drag coefficient that consider the state of the sea. One approach, first proposed by Charnock (1955), is to use dimensional analysis to choose a simple parameter to characterize the aerodynamic roughness, z_0. The friction velocity together with gravity, g, provides the nondimensional number,

$$\frac{z_0 g}{u_*^2} = \text{constant}$$

This expression proposes that the aerodynamic roughness is proportional to u_*^2/g. The sea surface physical roughness depends in part on u_* and so (g being a constant on the surface of the earth) some consideration of the sea state has been taken into account. The constant is now known as the Charnock constant and is of order 0.015.

If we generalize this expression to

$$\frac{z_0 g}{u_*^2} = f(c_p/u_*)$$

we now have an expression that depends both on the physical roughness and the speed of the *wind-wave* crests relative to the wind stress. Kitaigorodskii and Volkov (1965) were the first to explore such concepts. We could look for support in experimental results to prescribe the form of the function of wave age. For a steady wind, the wave height (roughness) increases with fetch as does wave age and $z_0 g/u_*^2$ is found experimentally to initially increase with fetch. At longer fetches the wave age grows larger and the speed of the characteristic wave crests approaches that of the wind. The input of wind momentum to the characteristic waves drops off and the value of $z_0 g/u_*^2$ is found to decrease again. Equation 1.8 reflects such behaviour. The possible reasons for this behaviour are discussed in Chapter 10.

For unsteady increasing air flow, at a fixed fetch, we also suspect that $z_0 g/u_*^2$ changes with wave age. Initially the *wind-wave* spectrum is undersaturated and demands more momentum. The phase speed of the peak frequency slowly increases and the windsea ages. For unsteady falling winds, oversaturation follows and there seems to be some empirical evidence, discussed in Chapter 9, that $z_0 g/u_*^2$ takes on a lower value than in the steady case. One would think this question could be resolved from an examination of the measured drag values.

1.10   Drag Observations

Figure 1.13 shows a selection of drag measurements, with steady and unsteady winds, young and old *wind-waves*. Confusion! Have these results been averaged long enough to be statistically reliable? For young *wind-wave* situations brought about by unsteadiness, the wave age may have changed but the aerodynamic roughness may still be evolving. In fetch-limited cases, where wave age could remain constant, the wind may have changed before a stable statistic can be obtained.

Certainly there is some new evidence that some measurements may be too short to sample the atmospheric rolls adequately. Jones and Negus (1996) show large variations in the aerodynamic roughness as updrafts and downdrafts pass over the observation site. Some measurements may not be accurate because of the difficulties of correcting for atmospheric stability or distortion of the supporting structures. Some measurements, taken in wind wave tanks, may be influenced by the tank walls.

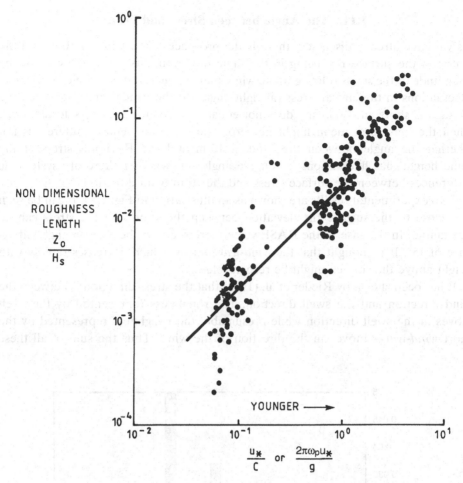

Figure 1.13. The nondimensional aerodynamic roughness as a function of wave age, after Jones and Toba (1995). H_s is the significant wave height, defined as 4 times the rms of the surface displacement.

A number of the difficulties we are facing have been illustrated by a series of aircraft flights over the Southern Ocean undertaken by Banner et al. (1996). They show for cross wind flights that the wind velocity at 50 m or so above the wave field varies on scales of 10 to 20 km. This is a larger scale than the atmospheric rolls discussed in Chapter 11. The mean square slope of the sea surface follows the wind speed variations rather well, but measures of the larger *wind-waves*, say the significant wave height, do not. The latter is to be expected because the significant waves are the result of wind input over hundreds of kilometres. The drag coefficient does not appear to follow the mean square slope. A model of the stress, dependent on a *wind-wave* growth expression and wave spectra would predict an increase in drag for high mean square slope where none is seen. Are there subtle changes in directional spectra? Or more likely, is the *wind-wave* growth too complex to parametrize as above?

1.11   The Angle between Stress and Wind

As we have already discussed, there is the prospect of the wind at 10 m and the stress on the surface not being in the same direction. Just as in the case of the magnitude of the stress relative to the wind, there is the question of averaging time. Fluctuations in the lateral stress (at right angles to the wind) can occur when the average is short and so $\langle v'w' \rangle$ does not equal zero for the sample selected, even when the long term mean might be zero. The question of more interest is not whether the angle between the wind at 10 m and the Reynolds stress at the same height, has fluctuations in relative angle but whether there are systematic differences between the surface stress and the 10 m wind direction.

Two experimental results are shown as an illustration in Fig. 1.14. It can be seen that close to the surface the deviation between the wind and the stress can be substantial. In the case of the RASEX site there seems to be a mean angle difference of 15°. It is thought that the *wind-wave* stress, which decreases rapidly with height above the surface might be responsible.

It has been shown by Rieder et al. (1994) that the stress direction is between the wind direction and the swell direction. The roughness represented by the swell moves in the swell direction while presumably the roughness represented by the short *wind-waves* moves in the direction of the wind. Thus the sum of all these

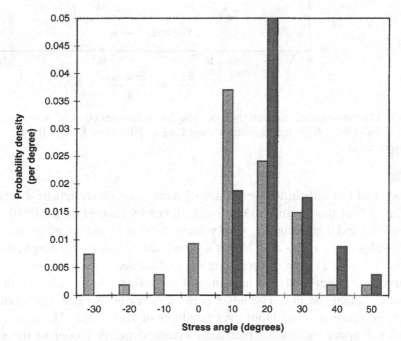

Figure 1.14. The angle between the Reynolds stress and the wind at a height z above the sea surface. The RASEX results are from Johnson et al. (1998) while the Bass Strait results are from Jones and Negus (1996). Fetches are typically 10–20 km for RASEX, 300 km for Bass Strait. The measurement height is 3 m for RASEX, shown as the darker bars, and 59 m for Bass Strait. RASEX is in the northern hemisphere, Bass Strait is at 38°S.

components has a direction between the two components. There is more discussion of this issue in Chapter 12.

1.12   A Representative Expression for the Drag Coefficient

The Charnock expression for aerodynamic roughness length is able to produce a drag coefficient of the right order of magnitude that increases with wind speed. It can be written

$$\frac{gz_0}{u_*^2} = \beta_* = \text{constant}$$

The expression proposed in Chapter 10 relates nondimensional roughness to wave age and recognizes two observed trends. In laboratory scale experiments nondimensional roughness increases with wave age while at large wave ages, more typical of the ocean, many observations show the roughness decreases with wave age. Figure 1.15 represents such trends for situations *in local equilibrium with the wind*. It is assumed that for moderate to high winds there is a maximum wave age that can be observed in nature since the *wind-wave* spectral peak that determines the phase speed cannot grow to travel much faster than the wind. Sverdrup and Munk (1947), for unlimited fetch and duration, thought the wave speed of the spectral peak was limited to $1.3U$. This value seems to be near $c/u_* = 35$. Steady state waves can be the result of light winds over swell older than $c/u_* = 35$ and is

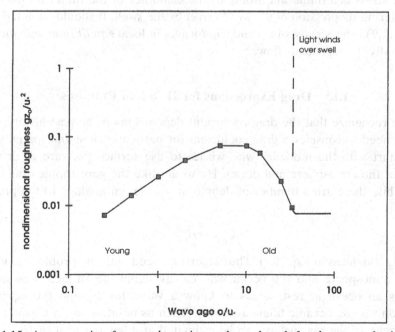

Figure 1.15. An expression for aerodynamic roughness length for the atmospheric surface layer for situations in *local equilibrium with the wind*. Comparison with observations are shown in Chapter 10 (Fig. 10.4).

shown with a nondimensional roughness near 10^{-2}. For modest winds, a *wind-wave* field in local equilibrium with the wind develops and it has higher frequency peak than swell. The *wind-wave* field then is typically of wave age much less than 35 and rides on the back of the swell. The impact of swell on drag coefficient is discussed in Chapter 8.

For light winds over swell the surface can be smooth and so the smooth flow drag coefficient can be used. The Charnock expression has been used for light winds that are not strong enough to produce a recognizable low wave age. Notice it is not the best fit Charnock coefficient for *all* wind speeds.

Thus as discussed in Chapter 10, a representative expression for aerodynamic roughness where the wind speed is known in detail and atmosphere rolls are not very strong is

$$
\begin{aligned}
\frac{gz_0}{u_*^2} &= 0.03\left(\frac{c}{u_*}\right)\exp\left\{-0.14\frac{c}{u_*}\right\} \qquad &\sim 0.35 < \frac{c}{u_*} < 35 \\
&= 0.008 \qquad &\frac{c}{u_*} \geq 35
\end{aligned}
\tag{1.9}
$$

For the youngest waves the dependence in Eq. (1.9) on the wave age is nearly linear. This was a form proposed by Toba and Koga (1986).

Examination of the fit of this data to observation at wave ages 20–30 shows a large number of observations below the expression of Eq. (1.9). The reasons for this are many. Experimental error is substantial as pointed out in Chapter 7. The aerodynamic roughness is the exponential of the drag coefficient and so magnifies errors in stress determination. Many of the examples in the literature may have been suffering suppression of the *wind-waves* by the swell. It should be noted again that Eq. (1.9) is for steady winds and *wind-waves* in local *equilibrium with the wind* in a laterally homogeneous flow.

1.13   Drag Expressions for Different Purposes

Since we recognize that the drag coefficient depends upon the past history of the wind we need to consider a drag coefficient for particular time and spatial scales. Let us start with the modeller who wishes to use surface pressure to drive his models of the atmosphere and ocean. He would like the geostrophic drag coefficient. While there are a number of definitions in use, an analogy to the previous sections is

$$
C_g = \tau/U_g^2
$$

where U_g is defined in Eq. (1.1). Thus a person, faced with the problem of modelling the atmosphere and the ocean who has available the surface pressure and considers an ocean at rest, needs to know a value for C_g and the equivalent expression for the oceanic boundary layer. Then as pointed out in Chapter 3 the drift velocity of the sea surface can be estimated at a few per cent of U_g.

To relate C_g to previous estimates of the aerodynamic roughness, a model for the planetary boundary layer velocity profile is needed. In particular the depth of the boundary layer, H_b, is needed to relate U_g to values such as U_{10}. Most expressions non-dimensionalize the problem in terms of the Rosby number, U_g/fz_0. More details are provided in Chapter 10.

Next there is the global climate modeller who may want a drag coefficient appropriate for a one degree square averaged over a month. During this period the wind will have fluctuated. Let us use $\overline{U}$ to represent the hourly mean wind averaged over one degree square of the ocean. Then

$$\overline{\tau} = \overline{C_D} \rho \overline{U}^2$$

where we use the over bar to emphasize the average over one hour and one degree square of the ocean. Let us write the monthly mean as $\langle \ \rangle$

$$\langle \overline{\tau} \rangle = \rho \langle \overline{C_D} \ \overline{U}^2 \rangle$$

where we assume fluctuations in density are unimportant. Firstly let us assume wind direction and $\overline{C_D}$ are a constant over the month. It is clear that

$$\langle \overline{\tau} \rangle = \rho \langle \overline{C_D} \rangle \langle \overline{U}^2 \rangle \neq \rho \langle \overline{C_D} \rangle \langle \overline{U} \rangle^2$$

If we choose to express the hourly mean wind in terms of the monthly mean and U' fluctuation in the hourly mean as $\overline{U} = \langle U \rangle + U'$ we can see by substitution

$$\langle \tau \rangle = \rho \langle C_D \rangle \langle U \rangle^2 \left(1 + \frac{\langle U'^2 \rangle}{\langle U \rangle^2} \right)$$

Thus the monthly average drag is larger by $\langle U'^2 \rangle / \langle U \rangle^2$ than would be calculated by using the monthly average velocity. The fluctuation term, $\langle U'^2 \rangle$, can be large, especially if the flow reverses. There often needs to be more careful analysis to take into account that $\langle \tau \rangle$ is the average of a vector. For further discussion the reader is referred for example to Hanawa and Toba (1987).

To summarize, the recommended drag coefficient for a monthly mean, expressed in terms of the monthly mean velocity, is

$$C_D = 1.3 \times 10^{-3} \left\{ 1 + \frac{\langle U'^2 \rangle}{\langle U \rangle^2} \right\}$$

where U' is the fluctuation in hourly mean wind $\overline{U}$ around the monthly average, $\langle U \rangle$. This coefficient, suitable for climatological analyses, will be inappropriate for weather models or for process studies or engineering models.

For people interested in processes on the time scale of hours and spatial scales of kilometres, who have wave age information, Eq. (1.9) is appropriate. At constant wind speed the drag coefficient is a function of age of the sea, as Fig. 1.16 shows. If the atmosphere is stratified, a correction increasing the drag coefficient in unstable situations and decreasing it for stable cases is discussed in Chapter 7.

For people studying unsteady flow problems, the drag coefficient needs to be adjusted for rapid changes in wind speed or direction. When there is strong swell and light winds, the direction of the stress can be significantly different from that of the wind. For people concerned with horizontal scales of less than one kilometre,

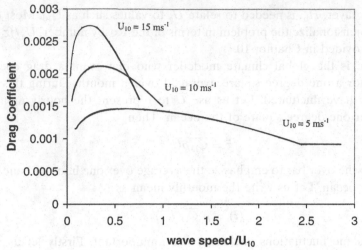

Figure 1.16. The neutral stability drag coefficient based on the 10 m wind using the expression in Eq. (1.9).

the relative position of organized turbulent motion in the atmosphere needs to be taken into account. It seems the drag coefficient increases under updrafts and decreases under downdrafts. If there is evidence of surface contaminants on the sea the drag coefficient should also be reduced.

1.14   Further Complications

As well as unsteadiness, atmospheric stability and changing fetch, there is a long list of other variables that occur over the ocean that may be important for determining the drag coefficient. These include rain, snow and hail. Rain has been considered by Manton (1973) and is discussed in Section 3.4. It carries horizontal momentum which is an additional term to the form drag and skin friction discussed above. As well, rain is known to suppress *wind-waves* in light winds and so change the form drag. Intense spray at speeds above 15 m/s carries momentum from the surface to the air and will play a part in the drag.

The turbulence in the marine boundary layer is not all generated locally but can result from processes higher in the atmosphere. This is discussed in Section 3.2 and implies that the local wind speed may not be able to parametrize the atmospheric stress.

There is the suggestion by Csanady (1974) that for low winds the interface may act as a flexible surface that can damp turbulence and reduce the drag coefficient. Drag coefficients less than that of a smooth flat plate have been observed. Incipient ice formation in cold conditions may be another complication.

The surface tension at the air–water interface is influenced by biological processes. Waves of length order 2 cm travel at the slowest speed and shorter waves travel faster and are increasingly dominated by surface tension. In this manner surface tension has a role to play in the form drag. The whitecaps, manifestation of rising bubbles entrained in breaking events, are also believed to be influenced by surfactants which change the viscoelastic nature of the water surface as well as lower the surface tension.

The next fourteen chapters have been provided for those who wish to probe deeper into the complexities of momentum transfer at the sea surface. The rules for adjusting drag coefficients for various environmental conditions are not at all well established. Many issues are not clear and the following chapters give only preliminary suggestions of how to model the wind stress over the ocean.

1.15 Summary

The atmosphere and the ocean exchange momentum over 70% of the surface of the earth by physical processes that involve viscosity, surface tension, gravity, turbulence and *wind-waves*. The change in density across the air–sea interface ensures that the motion is greater in the horizontal than in the vertical.

The large length scales involved allow the flows to be turbulent while the disparities of vertical and horizontal scale allow the boundary layer approach to be used to advantage. In fixed boundaries the physical roughness is an important parameter and the magnitude of the surface stress can, in the absence of density gradients, be shown to depend on the geometry of the roughness. A mobile boundary between the air and water makes the problem of describing the physical processes transferring momentum much more complex than over a solid surface, even though the main features are common to both situations.

Relative to the air–water interface, the *wind-waves* which are present when the air–water velocity difference is greater than a few metres per second, travel principally in the same direction as the wind. Some portions of the complex shape of a windsea are steep enough to induce separation of the flow and the resultant pressure forces have a horizontal component known as form drag. Even in the absence of separation, the pressure fluctuations above the undulating surface due to the asymmetric streamlines can transport momentum to the surface wave field. Finally, very close to the surface, the turbulence is suppressed and a viscous sublayer forms, to be destroyed by breaking. Turbulent Reynolds stresses, pressure transport and viscous transfers all play a part in producing a drag force at the sea surface.

Much uncertainty remains but a picture is emerging where it is recognized that roughness elements travelling at an angle to the wind induce a surface stress also at an angle to the wind. The phase speed of the most prominent roughness, compared with a measure of the wind speed (say friction velocity, u_*) provides the wave age on which many qualities such as the wave steepness and drag coefficient depend. The wave age is an important reflection of the relationship between the wind and the local state of the *wind-waves*. A representative expression for aerodynamic roughness (and drag coefficient) has been provided. It has a maximum value at c_p/u_* around 10 and for c_p/u_* greater than 35 (light wind over swell) the sea surface becomes smooth. (Such situations are only possible in steady state for very low wind speeds.) Although we have presented a representative expression for aerodynamic roughness Eq. (1.9), we recognize that there are many inconsistent observations over the real ocean.

After 50 years of measuring the drag coefficient over the ocean, the data spans a large range with only the weakest hints to the causes of the variations. The measurements are difficult and rely on assumptions of stationarity which may not be realistic. Inertial dissipation techniques make assumptions about terms that are not measured and in common with eddy correlation measurements, extrapolate the observation to the sea surface. Errors of 10 or 20% in vertical gradients of friction velocity can explain most of the observational variation in drag.

In unsteady situations there is tentative evidence that the shortest waves steepen or relax to change the amount of microbreaking. With changing atmospheric forcing *wind-waves* deviate from *local equilibrium with the wind*, with the high frequency part of the wind-wave spectra becoming over- or under-saturated, raising the prospect of variations in the state of breaking. To correlate wave breaking with variations in momentum transfer remains a challenge for the future.

The causes of variations in the statistical measures of the marine boundary layer are not well understood. Ordered motions, discussed in Chapter 3 and "inactive motions" from afar are complicating factors. Variations occur even under ostensibly similar conditions. One manifestation of this is the lateral drift of atmospheric rolls. They migrate laterally past a fixed observation point and aircraft measurements may be the only practical way to address this inhomogeneity.

The fluctuations in wind direction mean that the *wind-wave* field is always attempting to come to equilibrium with the forcing from a new direction. A rapid adjustment period for direction may be more common than we assume. Stress measurements at many levels may be necessary to average the short term (15 minute) fluctuations in stress direction; and complementary detailed wave directional measurements to examine the temporal changes in *wind-wave* patterns may be required.

While progress has been substantial over the last few decades much remains to be understood. The role of breaking in the direct transfer of momentum and in the enhancement of turbulence below the sea surface needs further clarification. How do we calculate the wave induced stress over a complex wave surface? Can we sum the momentum flux over the individual Fourier components? What role does the variation in drag coefficient with sea state play in weather forecasting and climate studies? Not till we resolve these and other unstated difficulties can oceanographers and meteorologists realistically couple their two domains.

In summary this monograph represents an effort to discriminate between the various elementary processes that make up the very complicated processes occurring in the real ocean. Chapters 2 to 7 review the present state of knowledge of these component processes while Chapters 8 to 15 look at regions of uncertainty in terms of the influence of swells, unsteadiness and so forth. Here we hope are some clues for further studies by the younger generation. Investigations of the stress tensor in relation to directional wavenumber spectra, the variation of stress with height in veering winds and the link between breaking waves and drag coefficient could be important issues which are not adequately addressed in this book.

Also to put investigations of the global scale geophysical problems on a solid foundation, it is emphasized that, on one hand, we should advance our understanding of the elementary processes described above, while on the other, we should study the linkages between microscale physics or local dynamics and the bulk parametrization used today in global atmosphere–ocean models.

PART ONE

2 Historical Drag Expressions

Y. Toba, S. D. Smith and N. Ebuchi

2.1 Introduction

The drag coefficient of the sea surface is

$$C_D = |\tau|/\rho U_{10}^2 = (u_*/U_{10})^2 \tag{2.1}$$

where ρ is the density of air, U_{10} is the wind speed at a 10 m reference height above mean sea level, $|\tau| \equiv \rho u_*^2$ is the wind stress on the surface, and u_* is called the friction velocity. The difference between the magnitude of the stress vector and its component in the direction of the wind was often ignored in early work. The drag coefficient varies with the state of the surface waves, which in turn are generated by the wind. Until recently the drag coefficient has been parametrized mainly as a function of the wind speed. However not only are waves generated by the local wind, but also waves generated in the past arrive as swell, and the wave generation process itself depends not only on the wind but also, in a complex way, on its interaction with pre-existing waves and swell. A more complete representation of the drag coefficient must consider not only local wind speed but also sea state.

In Section 2.2 we will give a brief history of sea surface wind stress studies, particularly from the point of view of field experiments. This part has been contributed principally by S. D. Smith. If the results from various experiments are combined, with C_D expressed as a function of U_{10} alone, there remains a large scatter. Some of this is due to experimental error (see Chapter 7), and some is statistical scatter associated with estimating the wind stress on an area from measurements at a point and from a finite sample of a continuous process. Some is due to treating a continuously changing geophysical flow as steady. In this review we will attempt to avoid methods used in the past to estimate wind stress which were subsequently shown to be flawed.

After allowing for scatter that is inherent in the experimental techniques, the simple representation of C_D as a function of wind speed alone, ignoring variations of sea and swell, leaves considerable variability unexplained. The full two-dimensional wave spectrum is too complex to use directly, and efforts to investigate the dependence on waves, based on finding a simple parameter that can represent the development of the waves with respect to the local wind, are reviewed in Section 2.3 principally prepared by N. Ebuchi and Y. Toba. One approach is to represent a nondimensional roughness parameter gz_0/u_*^2 (see Eqs (2.4) and (2.7a) to follow) as a function of the inverse wave age $u_*\omega_p/g$, or u_*/c_p, where σ_p and c_p are the radian frequency and the phase velocity of waves at the peak of the wave spectrum. This approach has the potential of explaining the scatter of measured drag coefficients due to variations in sea state that is otherwise hidden in C_D versus U_{10} plots. In Section 2.3 drag formulas which have been proposed in the past are reviewed, with emphasis on gz_0/u_*^2 versus $u_*\omega_p/g$ forms. The discrepancies among various proposed formulas are surprising and this confusing situation, as presented in 1992 at the SCOR WG 101 Workshop in Avignon, France, became the starting point of the present efforts. In this chapter, we review drag expressions from an historical point of view to serve as an introduction to the more detailed discussions in the following chapters. Discussion of the wave age dependence itself of the drag coefficients will be found in Chapter 10.

2.2 A Brief History of Sea-Surface Wind Stress Field Experiments

2.2.1 Early Work

Based on theories and hypotheses proposed in the pioneering works of Reynolds (1883), Taylor (1915), Prandtl (1924), Richardson (1920), von Karman (1930), Monin and Obukhov (1954) and others, experimentalists were able to construct field projects to measure and parametrize the drag coefficient of the sea surface. From data of the Deutsche Seewarte, Rossby and Montgomery (1935) determined an average drag coefficient, $C_D = 0.0013$, that remains valid for moderate winds; but the higher value of 0.0026 found by Sverdrup et al. (1942) was widely used for years. Francis (1951) and van Dorn (1953) found from lake data that C_D seemed to exhibit a dependence on wind speed.

2.2.2 The 1950s

During the postwar years interest in sea surface wind stress grew with the development of boundary-layer theory and of capabilities to record and analyse turbulent time series. Among the earliest open-ocean measurements were those of Sheppard and Omar (1952), for low wind speeds. There was a growing need for data in higher wind speed conditions to determine the dependence of C_D on the wind speed over the ocean. More experiments were conducted (Hay 1955; Fleagle et al. 1958; Brocks 1959), but there were still no data at wind speeds above 13 m/s. By the end of the 1950s, there was some agreement on the value of C_D but there was still no consensus on its wind speed dependence. It began to appear that open-ocean coefficients were slightly lower than those from semi-enclosed waters. In a review of early determinations of transfer coefficients from profiles and a small quantity of eddy correlation data, Deacon and Webb (1962) found for the neutral drag coefficient (see Chapter 1)

$$10^3 C_{10N} = 1.00 + 0.07 U_{10N} (1 < U_{10N} < 13 \text{ m/s}) \qquad (2.2a)$$

Although this is in fact quite similar to modern results, the existence of a dependence on wind speed was not well established from the data then available.

2.2.3   The 1960s

In the 1960s an era of large international field projects began, with an emphasis on tropical seas. In 1964 the International Indian Ocean Expedition (IIOE) yielded the first large data sets of fluxes based on profiles of wind, temperature, and humidity (Badgley et al. 1972). Smaller-scale experiments continued, e.g. in the Mediterranean Sea (Volkov 1970) and in Bass Strait (Hicks and Dyer 1970; Hicks et al. 1974; Dyer 1967). Deacon (1962), with data at wind speeds up to 14 m/s, was able to demonstrate an increase of C_D with wind speed. By the late 1960s there were suggestions that the drag coefficient should depend inversely on wave age (Stewart 1974; Kitaigorodskii and Volkov 1965; Volkov 1970), and the first use of hot-wire and sonic anemometers to measure wind stress by the direct eddy-correlation and the indirect dissipation methods (Weiler and Burling 1967). Other indirect methods for estimating fluxes were also employed, notably the ageostrophic method (Charnock et al. 1956) by which Hawkins and Rubsam (1968) reported drag coefficients rising from 0.001 for light winds to 0.004 for a wind speed of 45 m/s; there is still no eddy flux data at such high speeds. Modern measurement techniques are fully described in Chapter 7.

The Barbados Oceanographic and Meteorological Experiment (BOMEX) in 1969 was the first intensive oceanographic and meteorological experiment to include studies of air–sea fluxes from an array of ships (Davidson 1974; Pond et al. 1971).

Atlantic Trade Wind Experiment (ATEX). ATEX was a study of the development of the boundary layer in the trade winds on their way towards the intertropical convergence zone (ITCZ). In 1969 a triangle of ships drifted with the NE Trades. Air–sea fluxes were measured by the profile and the eddy correlation methods on two buoys: a surface-following buoy for profiles and a servo-stabilized buoy for eddy fluxes (Dunckel et al. 1974). The average neutral drag coefficient from the eddy correlation data was $C_{DN} = 1.39 \times 10^{-3}$, and the evaporation coefficient, see (Eq. 3.36), from profiles, $C_{EN} = 1.28 \times 10^{-3}$. These flux measurements in the open ocean demonstrated the need to consider the influences on profile measurements of waves (Hasse et al. 1978b) and of the flow distortion effects of even a slender mast (Wucknitz 1977). Comparison of fluxes from eddy correlation and dissipation techniques led Wucknitz (1979) to discover a systematic variation of the ratio of C_D from eddy correlation and dissipation methods.

Kansas Experiment. The similarity theory of Monin and Obukhov (1954), see Chapter 3, Section 3.4, was tested and an empirical determination of the dependence of the drag coefficient on stability was made in the Kansas Experiment in 1968. The land surface provided a less difficult means to obtain flux profiles, and a larger range of atmospheric stratifications was encountered. Empirical flux-gradient formulas derived from this experiment (Businger et al. 1971) have been widely used in air–sea interaction studies to calculate the influence of stratification on wind, temperature and humidity profiles in the surface layer. Not until 1994 was a comparable experiment, MBLP, conducted over water.

2.2.4   The 1970s

In 1970, there was excitement over the prospects of data collected during ATEX, BOMEX, and other projects. Planning was under way for experiments making use of fast-response wind and temperature sensors, and of the ability to compute spectra and eddy fluxes from time series data.

GARP Atlantic Tropical Experiment (GATE). GATE, a study of the energetics and dynamics of cloud clusters that drift from the African continent over the Atlantic Ocean where they modulate the Inter-Tropical Convergence Zone convection, was the first large-scale international field experiment of GARP (Global Atmospheric Research Programme) and the largest air–sea interaction field experiment to date. BOMEX, ATEX and the boundary layer programme of the IIOE can be seen as precursors of GATE that developed the tools for cooperative air–sea interaction work at sea.

In the boundary-layer subprogramme (Volkov et al. 1982) both radiative and turbulent fluxes were measured. Turbulent fluxes from boom-mounted sensors on ships and from a stable buoy were intercompared (Hasse et al. 1975). One group used the dissipation method and the other groups the eddy correlation method, while both eddy correlation and profile measurements were conducted at a buoy. Average neutral drag and evaporation coefficients for a 10 m reference level were $C_{DN} = 0.00125$, $C_{HN} = 0.00134$ and $C_{EN} = 0.00115$ (Businger and Seguin 1977; Hasse et al. 1978b). During outbreaks of cold air the surface layer profile up to 10 m height was seen to establish a new equilibrium within a few minutes. No wind speed dependence of the coefficients was observed at the moderate wind speeds encountered. GATE saw the first measurements of air–sea fluxes from aircraft (Nicholls and Readings 1979). It also included extensive radiosonde measurements, and observations of the ocean mixed layer. The combination of surface and aircraft data provided unique insight into the interaction between surface fluxes and convective elements in the boundary layer (Hasse et al. 1978a, b; Khalsa and Businger 1977; Galushko et al. 1975, 1977; Volkov et al. 1974, 1976; Müller-Glewe and Hinzpeter 1975).

Air Mass Transformation Experiment (AMTEX). On the other side of the globe near Japan in 1974–75, fluxes were measured within a larger-scale experiment,

AMTEX. Progress was made in estimating fluxes from a ship. It was found that the fluxes needed to be understood in terms of co-spectra and the eddy processes, and a coupling became apparent among surface waves, surface fluxes and turbulent eddies at the scale of the the boundary layer (Mitsuta 1977–79). Among the AMTEX contributions the parametrization scheme of Kondo (1975) gives drag, evaporation and sensible heat flux coefficients as a function of wind speed and sea–air temperature difference, which remain widely used.

Joint Air–Sea Interaction (JASIN) Experiment. The JASIN experiment was conducted in 1978 north and west of Scotland, from ships, aircraft and buoys (Pollard et al. 1983; Guymer et al. 1983). Both air and ocean boundary layers were studied with the intention of closing a momentum and heat budget. Surface fluxes measured from aircraft showed that roll vortices were able to systematically modulate fluxes, even in the neutral marine boundary layer (Nicholls 1985; Shaw and Businger 1985), a topic taken up in Chapter 11. Inertial-dissipation flux measurements from Meteor are reported in the often-quoted paper of Large and Pond (1982). The brief flight of SEASAT in 1977 demonstrated the potential power of remote sensing of the oceans, but unfortunately it failed before a planned validation experiment could take place. Consequently remote sensing studies in JASIN became an important source for SEASAT validation (Guymer and Taylor 1983; Katsaros et al. 1981; Liu and Large 1981; Taylor et al. 1981, 1983a).

Maritime Remote Sensing (MARSEN) Experiment. Late in the 1970s, a series of international experiments began over the mid-latitudes of the northeast Atlantic and North Sea, taking advantage of research platforms off Germany and the Netherlands. MARSEN, an integrated remote sensing and surface flux experiment, was conducted in the North Sea in 1979 to extend physical understanding of the relation of remote sensing signatures to surface waves and wind stress. The remote sensing emphasized synthetic aperture radar (SAR) and scatterometry using radars on aircraft and on towers. From wind stress measurements at an offshore mast, Geernaert et al. (1986) found that C_D was larger for steeper waves, and extrapolated this result to suggest that, for the same wind speed, shallow-water waves should produce a larger stress than deep-water waves. Chapter 15 looks further at this issue.

Regional Flux Experiments. A series of experiments led by the Bedford Institute of Oceanography at a small island in the North Atlantic and at stabilized offshore towers yielded the largest compilations of eddy drag coefficients to date (Smith and Banke 1975; Smith 1980; Large and Pond 1981). For long fetch (onshore winds) the increase of C_D with wind speed,

$$10^3 C_{10N} = 0.61 + 0.063 U_{10N} \quad (5 < U_{10N} < 22 \text{ m/s}) \tag{2.2b}$$

(Smith 1980) was slightly greater than predicted by the dimensional analysis of Charnock (1955), but less than that found by compilations of data from coastal and shallow sites (Garratt 1977; Wu 1980). Donelan (1982) found a strong wave dependence of the surface roughness of Lake Ontario.

2.2.5   The 1980s

Following MARSEN a series of four experiments, designed to study SAR imaging of waves, included surface flux measurements. Two TOWARD (Tower Ocean Wave and Radar Dependence) experiments took place off San Diego in 1984 and 1985, with an emphasis on Ku-band imaging. Two SAXON (SAR X-band Ocean Nonlinearities) experiments were conducted: off Virginia at the Chesapeake Light Tower in 1990, and at the North Sea Platform in 1992.

Byrne (1982) reported an analysis of aircraft flux data collected over the Pacific. A dependence of fluxes on waves generally was revealed when swell was not present. Since then it has been found that the presence, size, and direction of swell can significantly affect both the high-frequency part of the wave spectrum and the drag coefficient (Geernaert et al. 1993; Dobson et al. 1994). More detail is provided in Chapter 8. In general, short fetch and/or shallow water areas have larger drag coefficients than the open ocean.

Canadian Atlantic Storms Program (CASP). In 1985–6 CASP, coordinated with the USA Genesis of Atlantic Lows Experiment (GALE), studied the growth of winter storms and their interaction with the upper ocean (Smith and Stewart 1989). This was followed in 1992–93 by CASP II, a study of the mature stages of explosive cyclogenesis in winter storms and their influence on ocean circulation and sea ice on the Newfoundland continental shelf and the Grand Banks (Smith et al. 1994).

Humidity Exchange over the Sea (HEXOS) Programme. In 1986 an international team participated in the Main Experiment of the HEXOS Programme at and in the vicinity of the Dutch Noordwijk Platform. This coordinated series of field, laboratory and modelling studies provided the largest range of wind speeds (up to 19 m/s) for simultaneous measurements of momentum, sensible and latent heat fluxes, and provided the largest data set to date for the water vapour exchange coefficient, C_E (DeCosmo et al. 1996). From a subset of HEXOS data selected with single-peaked wave spectra (to minimize the influence of swell) the aerodynamic roughness of the sea surface, z_0, was found to have a reciprocal dependence on wave age,

$$z_0 = 0.48(u_*^2/g)(c_p/u_*)^{-1} \tag{2.3}$$

(Smith et al. 1992). Wind stress and heat flux measurements collected in 1985 from the German North Sea Platform in support of a scatterometry remote sensing project (Geernaert et al. 1987) generally corroborated those in HEXOS. Donelan et al. (1993) showed that a larger data set from Lake Ontario supports the HEXOS result, but because the variability of u_* is larger than the variability of c_p in these (and in all other) field data sets, the derivation of the wave age dependence is statistically weak.

Specifying the surface roughness is equivalent to specifying the neutral drag coefficient, which is uniquely but nonlinearly related to the aerodynamic roughness, as explained in Chapter 1:

$$C_{10N} = [\kappa / \ln(10/z_0)]^2 \qquad (2.4)$$

Here the reference height is 10 m, z_0 is in meters and $\kappa = 0.4$ is the von Karman constant. The water depth at Noordwijk platform is 18 m, and at this depth the wave phase velocity at the spectral peak, c_p, is limited to 13.3 m/s.

Coastal Ocean Dynamics Experiment (CODE). CODE, a regional coastal oceanography experiment off California in 1985, emphasized ocean temperature response to wind stress patterns and needed detailed information on wind forcing. The wind stress direction was found to deviate from the wind direction, and this was attributed to cross-shelf pressure gradients (Zemba and Friehe 1987; Enriquez and Friehe 1997).

Frontal Air Sea Interaction Experiment (FASINEX). FASINEX was designed to evaluate the air–sea interaction processes around an oceanic surface temperature front in the subtropical convergence zone of the western North Atlantic. Fluxes were measured both from aircraft and from a ship near the Gulf Stream southwest of Bermuda in 1986 (Li et al. 1989). The fluxes in the unstable atmospheric boundary layer on the warm water side of surface temperature fronts were expected to be larger than those on the cold water side, but the difference was larger than predicted using the coefficients of Businger et al. (1971). This suggested that additional processes may be acting on the marine surface layer. Differences in radar backscatter cross-section across a front were more consistent with the variation in wind stress than with the smaller variation in wind speed, supporting an hypothesis that radar remote sensing may respond more to wind stress than to wind speed over the oceans.

2.2.6   The 1990s

TOGA Coupled Ocean–Atmosphere Experiment (COARE). The pivotal role of tropical sea-surface temperature and the El Niño Southern Oscillation in interannual climate variability led to establishment of the Tropical Ocean Global Atmosphere (TOGA) program; the subsequent identification of the dominant influence on atmospheric convection of a warm pool of surface water in the western tropical Pacific resulted in the COARE (Webster and Lukas 1992). Interfacial fluxes are one of four elements of the COARE, which was focussed on the role of large atmospheric convective systems (westerly bursts) in modifying the equatorial air–sea coupling. COARE has produced an enormous quantity of direct air–sea flux data over the open ocean in a variety of regimes, from week-long highly suppressed periods to two full-blown westerly bursts associated with the Madden–Julian oscillation.

A large quantity of dissipation wind stress data at relatively low wind speeds of 1–10 m/s supports use of the COARE 2.0 bulk flux algorithm with a combined smooth-surface and Charnock formula as proposed by Smith (1988),

$$z_0 = 0.11\nu/u_* + 0.011 \; u_*^2/g \qquad (2.5)$$

(Fairall et al. 1996), where ν is the kinematic viscosity of air and g is gravity. The first term in Eq. (2.5), based on wind tunnel studies, determines the friction velocity mainly at low wind speeds (<5 m/s) and had not been previously verified in a marine application.

Marine Boundary Layer Project (MBLP). To address intermittency and smaller-scale boundary layer processes whose existence was suggested from FASINEX, and to reduce uncertainty in flux models, the MBLP was launched in 1994. The first MBLP experiment was the Risø Air–Sea Experiment (RASEX) described in Johnson et al. (1998) at an offshore mast north of Lolland, Denmark, with supporting information on spatial variability from a University of Kiel ship. Atmospheric and oceanographic instrumentation was deployed to measure bulk quantities and fluxes of momentum, heat, moisture, CO_2, ammonia, nitric acid, and aerosols. Aerosol composition and supporting marine biological and surface wave data were also compiled. All terms of the turbulent kinetic energy and flux budgets were measured, and intermittency of the fluxes was assessed. The second series of MBLP experiments took place west of Monterey, California in April and May 1995 using the RASEX instrumentation, and in addition the floating instrument spar buoy, (FLIP), two ships, an airplane, and LIDAR and radar remote sensing.

2.2.7 Wind Stress and Sea State

It is now well established that the drag coefficient (or equivalently the aerodynamic roughness z_0) increases with increasing wind speed, and that this increase is greater at shallower sites (e.g. Geernaert et al. 1986, 1987; Smith et al. 1992). Clearly the sea state has an important influence on the wind stress and on all air–sea fluxes (Donelan 1990; Geernaert 1990; Fairall et al. 1990).

Some consider "young" waves that grow by extracting energy and momentum from the wind should require greater momentum flux from the wind (i.e. wind stress) than "mature" waves that are approaching equilibrium with the wind. All spectral components of the waves participate in wind–wave interactions, but for the simplest case of a "pure" locally-generated sea with a single spectral peak it is possible to represent the wave field by the phase velocity c_p at the peak of the wind-driven wave spectrum. The sea state can then be represented by a dimensionless "wave age", c_p/u_*. After many years, there is still lively debate over the relation between wind stress and sea state. The exact nature of this influence is the subject of this entire book.

We have at present a debate between two approaches: Smith et al. (1992) and Donelan et al. (1993), using field data from the HEXOS Main Experiment in the

North Sea and from a platform in Lake Ontario, found the normalized aerodynamic roughness gz_0/u_* to be inversely proportional to wave age (Eq. 2.3). Donelan et al. (1995) derived a relationship in which the drag coefficient of young waves ($c_p/U_{10} = 0.2$) is about 50% higher than that of mature waves (Fig. 2.1). To describe the normalized roughness, Toba et al. (1990) fitted an empirical power law between a cluster of field data from several experiments (including some situations possibly affected by the sea floor) and a cluster of data from several sets of much younger laboratory waves. They obtained a +0.5 power law (their Eq. 30) with the consequence that for any given wind speed their young (growing) waves have less wind stress than their mature waves (Fig. 2.2). This follows from using a single power law in a z_{0*} versus σ_{p*} diagram (see Section 2.3.2 to follow) to describe both field and laboratory waves. It now seems from Chapter 10 that a more complex relation than either of the above may be needed to describe the roughness over the entire range of field and laboratory wave ages.

Even though it describes the expected general pattern of variation of effective roughness over the range of wave ages observed in the field, there are arguments against the general application of the HEXOS formula (Eq. 2.3). First, only a limited range of wave ages is observed at sea. The initial stages of growth of extremely young waves take only a few seconds to a few minutes, while in aver-

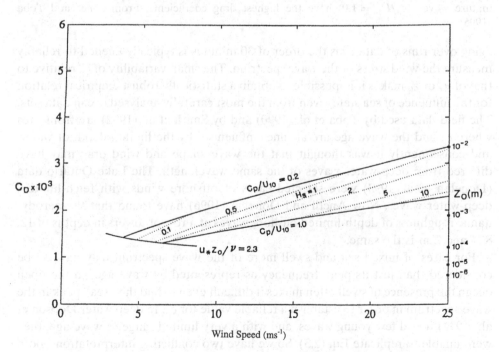

Figure 2.1. The drag coefficient for "equilibrium" waves from Eqs (9) and (10) of Donelan et al. (1993). The solid lines are for very young waves and for mature waves ($c_p/U_{10} = 0.2$ and 1.0 respectively). The dotted lines are lines of constant significant wave height in metres. Roughness lengths corresponding to the drag coefficient values are shown on the right-hand axis. The line $u_*z_0/\nu = 2.3$ is the limit of fully rough flow. Here the young waves have the highest drag coefficients. From Donelan et al. (1995).

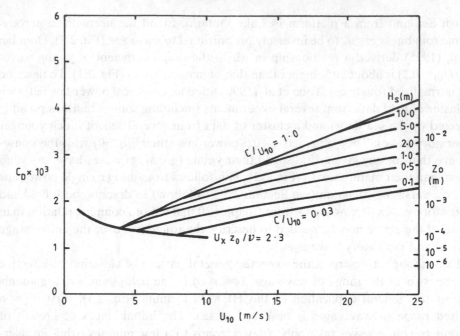

Figure 2.2. As in Fig. 2.1, but from Eq. (30) of Toba et al. (1990) or Eq. (2.14). Here the mature waves $(c_p/U_{10} = 1.0)$ have the highest drag coefficients. From Jones and Toba (1995).

aging over runs of data runs the order of 30 minutes is typically needed to reliably measure the wind stress or the wave spectrum. The small variability of c_p relative to that of u_* or z_0 makes it impossible to obtain a statistically robust empirical relation for the influence of sea state, even from the most carefully analysed ocean data sets. The field data used by Toba et al. (1990) and by Smith et al. (1992) are from sites where c_p and the wave age are at times influenced by the limited depth of water, and consequently it was thought that the wave shape and wind drag may have differed from deep-water waves of the same wavelength. The Lake Ontario data (Donelan et al. 1993) were selected cases of offshore winds with fetch-limited, deep-water waves. Now Anctil and Donelan (1996) have found that the aerodynamic roughness of depth-limited onshore waves at a row of towers in depths of 12, 8, 4 and 2 m is the same.

For cases of mixed sea and swell more of the wave spectrum may have to be considered than just its peak frequency as represented by wave age. In the open ocean the presence of swell often makes it difficult even to find the "sea" peak in the wave spectrum in order to establish a reliable value for c_p. In deep water Dobson et al. (1994) found few young waves, and with a very limited range of wave age, they were unable to replicate Eq. (2.3). So we have two conflicting interpretations, both of which are known to be subject to criticism. The purpose of this monograph is to resolve the differences where possible, and to recommend ways to find definitive answers, as will be discussed in more detail in Chapters 8 and 15.

Yelland and Taylor (1996) offer hope for simplification in some applications. For a set of "inertial dissipation" data taken in deep water at all latitudes, the neutral drag coefficient was found to be well represented as a function of wind speed alone. With corrections for flow distortion (Yelland et al. 1998), a linear regression ($\pm$ standard error) is

$$10^3 C_{10N} = (0.50 \pm 0.02) + (0.071 \pm 0.002)U_{10N} \quad (6 < U_{10N} < 26 \text{ m/s}) \quad (2.6)$$

with a correlation coefficient of 0.80 for 1111 data points. This agrees closely (within 0.1×10^{-3}) with Eq. (2b) from Smith's (1980) eddy correlation data at a stabilized platform in the North Atlantic. For this large data set the standard errors of the intercept and the slope are much smaller than in previous studies, although there is still considerable scatter among individual values of the drag coefficient results (e.g. Yelland & Taylor (1996) shown at Fig. 7.5). So for large-scale or climatological applications involving averages over many hourly values it may be possible to ignore the enhancement of wind stress in young waves. For local studies, for studies of transient events and for modelling wave spectra (where initial growth cannot be ignored) a more detailed understanding is still needed.

At low wind speeds (<5 m/s) Yelland and Taylor (1996), and also Dupuis et al. (1995) in results of SOFIA-ASTEX (Surface of the Ocean, Fluxes and Interactions with the Atmosphere/Atlantic Stratocumulus Transition Experiment) and SEMAPHORE (Structure of the Exchanges in a Marine Atmosphere, Properties of Heterogeneities in the Ocean, Research and Experiment) report a rise in the drag coefficient with decreasing wind speed. Both of these results should be viewed with caution because the assumptions required in the inertial-dissipation method to be found in Chapter 7 (i.e. existence of an inertial subrange, local isotropy, homogeneity) are less well satisfied as the wind speed decreases.

2.2.8 Visions of Future Research: Smaller and Larger Scales

On the scale of the local boundary layer, a myriad of eddy scales influences fluxes, and intermittency is hidden and ignored by many conventional analysis techniques. For example, Toba et al. (1996) have pointed out that natural gustiness of the wind speed over time scales of several minutes results in fluctuations in wind stress associated with the adjustment process of the waves, including the high-frequency tail of the wave spectrum (see Chapter 9). These fluctuations are concealed by the averaging time of measurements of wind stress and of wave spectra. Weissman et al. (1996) find that the microwave cross-section of gravity-capillary waves of 0.6 and 2.1 cm wavelength respond to wind variations with periods of 100 s and longer, while longer waves (9.2 and 14.1 cm) do not. The shortest waves, responsible for the obvious roughness of the sea surface, are thus seen to adapt to changes in wind much more quickly than the rest of the wave spectrum.

On the larger scale, boundary layer processes such as coherent structures influenced by the mesoscale affect the fluxes. FASINEX, for example, was an intensive study of the variability of fluxes around an oceanic front. Once we achieve a better understanding of wind stress in terms of wave age (see Chapter 10), it will become necessary to consider the more complex problem of the influence of sea that is interacting also with swell (see Chapter 8). There is a requirement for global coupled atmospheric and oceanic models for investigation of future climate change and for long-term forecasting. A precise knowledge of wind stress on the sea surface will be needed for these models.

2.3 A Review of the Form of Nondimensional Roughness Length

2.3.1 Nondimensional Roughness Length versus Wave Age

We restrict ourselves in this section to the case of neutral stratification. Also, we exclude very low wind speeds, i.e. cases with very small roughness Reynolds number, $u_* z_0 / \nu$ (ν is the kinematic viscosity of air) where the air–sea interface is aerodynamically smooth and we have a different regime in which the roughness and C_D are no longer controlled by wind waves. In other words, we restrict ourselves to cases where the surface can be considered aerodynamically rough, as discussed in Chapter 1.

Neglecting for the moment momentum transfer to the waves via pressure fluctuations, the profiles of mean wind and current in the coupled turbulent boundary layers in air ($z > 0$) and in water ($z < 0$; subscript w) follow classical logarithmic profile laws,

$$U(z) = (u_*/\kappa) \ln(z/z_0) \qquad z > 0 \qquad (2.7a)$$

$$U(-z) = (u_{*w}/\kappa) \ln(-z/z_{0w}) \qquad z < 0 \qquad (2.7b)$$

Here the wind speed and current are taken relative to the average velocity of the surface. In equilibrium conditions the portion of the wind momentum flux that supports the growth of wave momentum is small and the stress is nearly continuous across the interface, $\rho u_*^2 \approx \rho_w u_{w*}^2$, and likewise z_0 is related to z_{0w}.

For *wind-waves* in idealized *local equilibrium with the wind*, a 3/2 power law relates the nondimensional significant wave height $H_* = g H_s / u_*^2$ and the nondimensional wave period $T^* = g T_s / u_*$ (Toba 1972).

$$H_* = B T^{*^{3/2}} \qquad B = 0.062 \qquad (2.8)$$

Consequently in these conditions the system can be described by four variables: u_*, z_0, ω_p and g, the acceleration of gravity. The expression *local equilibrium with the wind* is used in the sense explained in Chapter 4.

Since C_D has a one-to-one correspondence with z_0 for neutral stratification (Eq. 2.4), z_0 has often been used instead of C_D. Stewart (1974) proposed on the basis of "similarity" (i.e. by arguing that all purely wind-driven wave spectra should have similar shape) that the wave roughness of the sea surface should be mainly a function of the phase speed c_p of wind waves at the spectral peak frequency. It follows by a dimensional argument that a general form of the wave-dependent nondimensional roughness length, z_0^*, can be expressed as a function of the wave age (nondimensional phase speed) c_p / u_*,

$$z_0^* \equiv g z_0 / u_*^2 = f(c_p / u_*) \qquad (2.9)$$

If we assume for the characteristic wave the linear dispersion relationship of deep water waves, the inverse wave age equals the nondimensional angular peak frequency of wind waves, $\omega_p^* \equiv \omega_p u_*/g$, with ω_p the angular peak frequency. By using this relation, Eq. (2.9) can be expressed by a relation between the two nondimensional variables z_0^* and ω_p^* as,

$$z_0^* = f(\omega_p^*) \tag{2.10}$$

Equation (2.10) can also be obtained by applying the Buckingham Pi-Theorem to the system of four variables pertinent to the air–sea boundary processes: u_*, z_0, ω_p and g. Because z_0 has a highly nonlinear (exponential) dependence on C_D, this expression has the effect of emphasizing the scatter of the data as compared to the conventional C_D versus U_{10} diagram.

2.3.2   A Synthesis of Historical Formulas and Data on a z_0^* versus ω_p^* Diagram

Among formulas which have been proposed so far for the wind dependence of C_D, only a few include wind-wave parameters explicitly. Table 2.1 (on pp. 52–53) is a compilation of those formulas expressed in the form of Eq. (2.10). The formulas are also shown by curves on a z_0^* versus ω_p^* diagram in Fig. 2.3b.

Masuda and Kusaba (1987) assumed, as a simple form of Eq. (2.10):

$$z_0^* = n\,\omega_p^{*m} \tag{2.11}$$

with n and m as constants. Then the classical formula of Charnock (1955):

$$z_0^* = \beta_* \tag{2.12}$$

corresponds to the case of $m = 0$ and $n = \beta_*$. In this case the wind-wave measure does not appear explicitly. Various candidates for the constants have been proposed by several authors (e.g., Kitaigorodskii and Volkov 1965; Garratt 1977; Wu 1980; Geernaert et al. 1987). A formula proposed by Toba and Koga (1986):

$$z_0\omega_p/u_* = \Omega \tag{2.13}$$

corresponds to $m = -1$ and $n = \Omega$. Equation (2.13) corresponds to the case where g has been dropped from the above-mentioned system of four variables, e.g. as if the processes concerned are of purely turbulent transfer, regardless of the existence of gravity waves. That is, the system of three variables, z_0, ω_p and u_*, has only one nondimensional variable, $z_0\omega_p/u_*$ which is set to a constant Ω in Eq. (2.13). This case may set an upper limit on the value of z_0^* in Fig. 2.3.

A composite data set from references in which values of z_0, u_* and ω_p were available, and in which ω_p was estimated for wind waves (not for swells), is shown in Fig. 2.3a in the form of a z_0^* versus ω_p^* diagram. The data points are distributed in a triangular area between Eq. (2.13) with, say, $\Omega = 0.03$ as the upper envelope, and Eq. (2.12) with, say, $\beta_* = 0.005$ as the lower limit. z_0^* does not seem to be a unique function of ω_p^* as expected from Eq. (2.10). The scatter seems to be due to devia-

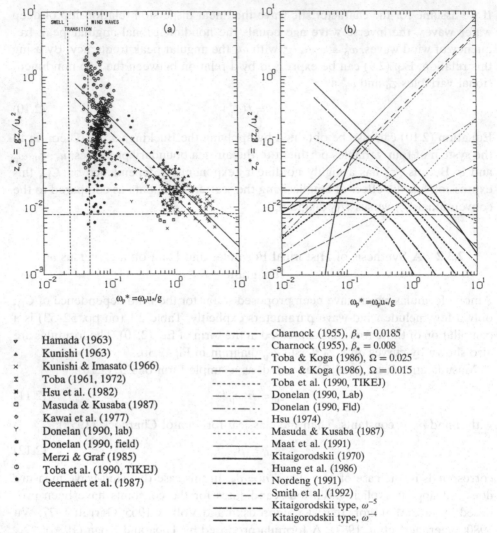

Figure 2.3. (a): A composite data set plotted on the z_0^* versus ω_p^* diagram, from Toba and Ebuchi (1991). (b): A composite set of historical formulas on the z_0^* versus ω_ρ^* diagram.

tions from the idealized local equilibrium of wind and windsea, as described in the first part of this section. The many causes for this are discussed in later chapters.

Toba et al. (1990) proposed a formula representing an overall average of the scatter of the data within the triangular region as:

$$z_0^* = 0.020 \ \omega_p^{*-0.5} \tag{2.14}$$

Donelan (1990) and Smith et al. (1992) proposed formulas which gave slopes of $m \approx 1$

$$z_0^* = 0.42 \ \omega_p^{*1.03} \tag{2.15}$$

and

$$z_0^* = 0.48 \, \omega_p^*$$ (2.16)

respectively. Since their field data have a slope $m = 1$ and their laboratory data also have positive-m slope, but with a discontinuity between them, Donelan (1990) and Donelan et al. (1993) argued that laboratory experiments could not represent field conditions for the same wave ages, and that one should not draw a straight line (power law) which passes through the two clusters of the field and the laboratory data.

However, another interpretation is possible. As for the positive slope for field data, Toba and Ebuchi (1991) presented data showing that wind fluctuations cause large fluctuations of z_0, related to the adjustment processes of the high-frequency, equilibrium range of the wind-wave spectra to changing winds. Toba et al. (1996) discussed a relation between data points on the z_0^* versus ω_p^* diagram and the degree of under- and over-saturation of the energy level of wave spectra under changing winds. From this result, it is expected that a single series of field data would result in a formula with a slope m close to unity. However, for different series of data associated with different mean values of wave age, the locations of the lines will differ from one to another, as seen in Fig. 2.3b.

Since the present data of Fig. 2.3a are not sufficient, we should await further accumulation of data before a final conclusion is given. Especially, we need data for intermediate wave ages where data points are scanty in Fig. 2.3a. The effects of changing winds and gustiness will be discussed further in Chapter 9.

Kitaigorodskii (1968) proposed a form of z_0 as an integration of roughness components which were expressed by a weighting function of the wave age, of the form

$$z_0^2 = A^2 \int_0^\infty F(k) \exp(-2\kappa c/u_*) \mathrm{d}k$$ (2.17)

where $F(k)$ is the wave-number spectrum, c is the phase speed of waves whose wave number is k, A is a constant, and κ is the von Karman constant (= 0.4). The integration with k gives a form of z_{0*} which approaches asymptotically to a constant value with the increasing wave age as shown in Fig. 2.3b, where self-similar wave spectral forms are assumed for the high frequency range.

Kitaigorodskii (1973) replaced the integrated form with

$$z_0^* = 0.068(u_*/c_p)^{-3/2} \exp(-\kappa c_p/u_*)$$ (2.18)

which is not necessarily consistent with the integration of Eq. (2.17). Equation (2.18) has a maximum value at middle wave ages, as seen in Fig. 2.3b.

2.3.3   The z_0/H versus ω_p^* Diagram and its Conversion with the z_0^* versus ω_p^* Diagram

Another approach was proposed by Huang et al. (1986), who normalized z_0 by the significant wave height H_s. The same formulas and data as in Figs 2.3a and 2.3b are plotted as a z_0/H_s versus ω_p^* diagram in Figs 2.4a and 2.4b. This is similar to the

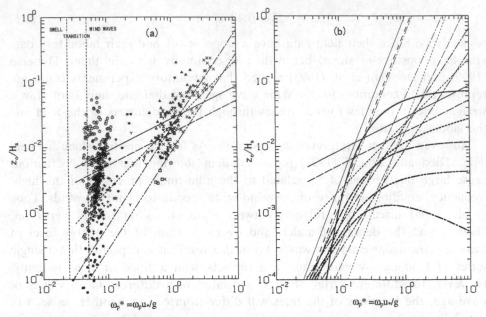

Figure 2.4. (a), (b): The same data and formulas as in Fig. 2.3 plotted as z_0/H versus ω_ρ^*.

presentation of Donelan (1990), who used the root-mean-squared surface displacement ($0.25\ H_s$) instead of H_s.

The z_0^* versus ω_p^* diagram (Figs 2.3a and 2.3b) suffers the danger of "spurious self-correlation" (Kenney, 1982) since u_* is used in normalizing both the axes. The z_0/H_s versus ω_p^* diagram in Figs 2.4a and 2.4b does not suffer from this effect provided the normalizing parameters for the two axes are independent, i.e. provided that H_s is not correlated with u_*. Unfortunately H_s is in general strongly dependent on u_*. A disadvantage of the latter representation (Fig. 2.4a) is that the data points are spread out from the bottom left to the top right, so that details of the distribution of the data points are not easily discriminated.

There is a way to transform between the z_0^* versus ω_p^* and the z_0/H_s versus ω_p^* diagrams. For *wind-waves* in local equilibrium with the wind, a 3/2-power law (Eq. (2.8)) relates the nondimensional significant wave height and period. In the dimensional form Eq. (2.8) is

$$H_s = B(gu_*)^{1/2} T_s^{3/2} \tag{2.19}$$

Dividing z_0 by Eq. (2.19), we obtain

$$z_0/H_s = (1/B)(1.05/2\pi)^{3/2} \omega_p^{*3/2} z_0^* \tag{2.20}$$

where $(2\pi T_s)/\omega_p = 1.05$, as used by Toba et al. (1990). By using Eq. (2.20), Eq. (2.11) is expressed as

$$z_0/H_s = (n/B)(\alpha/2\pi)^{3/2} \omega_p^{*(m+1.5)} \tag{2.21}$$

The slope of m on the z_0^* versus ω_p^* diagram (Figs 2.3a and 2.3b) is converted to a slope of $m + 1.5$ on the z_0/H_s versus ω_p^* diagram (Figs 2.4a and 2.4b). Charnock's formula (2.12) becomes

$$z_0/H_s = b' \omega_p^{*3/2} \tag{2.22}$$

with $b' = 0.0204$, corresponding to $B_* = 0.0185$ as proposed by Wu (1980). Toba and Koga's (1986) formula (2.13) becomes

$$z_0/H_s = d' \omega_p^{*1/2} \tag{2.23}$$

with $d' = 0.0275$ or 0.0165, corresponding to $\Omega = 0.025$ or 0.015, and Eq. (6) of Toba et al. (1990) is

$$z_0/H_s = 0.022 \; \omega_p^* \tag{2.24}$$

Comparing the distributions of the data points in Fig. 2.3a and Fig. 2.4a, it seems that effect of spurious self-correlation may not be large for this particular case, since the variation of the z_0^* is mostly determined by the overwhelming variation of z_0 itself.

In addition to the above-mentioned diagonal spreading of the data points, there is a second and more serious drawback of the form of Fig. 2.4a. When the waves are purely driven by the local wind, or when there is no significant swell, H_s represents wind waves. However, where there are some significant components of swell, then by definition H_s includes the effect of swell that is believed to have little or no influence on the surface roughness, see Chapter 8. In the form of Fig. 2.3a we can, by exercising skill and care, usually find ω_p for the wind-wave part of ocean wave spectra, but in the form of Fig. 2.4a, in the open ocean H_s is more often than not dominated by swell.

A formula proposed by Nordeng (1991), somewhat similar to that of Huang et al. (1986), is shown in Figs 2.3b and 2.4b. This is based on the integration of wave-induced drag components. Nordeng assumed a σ^{-5} spectral form with Phillips' constant β varying with the wave age, resulting in very low energy levels for very high frequency range as the wave age becomes large. This may not be consistent with the view of Chapter 4 that the energy level of the very high frequency range is proportional to around a 2.5 power of u_* (e.g., Mitsuyasu and Honda 1974; Jähne and Riemer 1990).

The above has given an introduction to the present state of observation and its interpretation. The reader is invited to continue on to the following chapters where more detailed insights will be presented.

Table 2.1. A synthesis of historical nondimensional roughness parameter as expressed in the form of Eq. 2.10 and in the right hand column after conversion using Eq. 2.21 (these relationships are drawn in Fig. 2.3b and Fig. 2.4b)

Author(s)	$z_0^* \equiv \dfrac{g z_0}{u_*^2}$	$\dfrac{z_0}{H_S}$
Charnock (1955)	$z_0^* = \beta_*$ $\beta_* = 0.0185$ (Wu 1980) 0.035 (Kitaigorodskii and Volkov 1965) 0.0144 (Garratt 1977) 0.0192 (Geernaert et al. 1986)	$\dfrac{z_0}{H_S} = 1.10 \beta_* \left(\dfrac{u_*}{c_p}\right)^{3/2}$
Kitaigorodskii (1968)	$z_0^2 = A^2 \displaystyle\int_0^\infty F(k) \exp\left(-\dfrac{2\kappa c}{u_*}\right) dk$ $c = c(k)$	
Kitaigorodskii with $F(\omega) = \beta g^2 \omega^{-5}$ $\beta = 0.012$	$z_0^* = 0.012 \Phi(x_0)$ $\Phi(x_0) \equiv \left[1 - e^{-x_0}\left(1 + x_0 + \dfrac{x_0^2}{2} + \dfrac{x_0^3}{6}\right)\right]^{1/2}$ $x_0 \equiv 2\kappa c_p u_*$	$\dfrac{z_0}{H_S} = 0.013 \left(\dfrac{u_*}{c_p}\right)^{3/2} \Phi(x_0)$
Kitaigorodskii with $F(\omega) = \alpha_s g u_* \omega^{-4}$ $\alpha_s = 0.062$	$z_0^* = 0.014 \Phi(x_0)$ $\Phi(x_0) \equiv \left[1 - e^{-x_0}\left(1 + x_0 + \dfrac{x_0^2}{2} + \dfrac{x_0^3}{6}\right)\right]^{1/2}$ $x_0 \equiv 2\kappa c_p u_*$	$\dfrac{z_0}{H_S} = 0.015 \left(\dfrac{u_*}{c_p}\right)^{3/2} \Phi(x_0)$
Kitaigorodskii (1970)	$z_0^* = 0.068 \left(\dfrac{u_*}{c_p}\right)^{-3/2} \exp\left(-\kappa \dfrac{c_p}{u_*}\right)$	$\dfrac{z_0}{H_S} = 0.075 \exp\left(-\kappa \dfrac{c_p}{u_*}\right)$
Hsu (1974)	$z_0^* = 0.144 \left(\dfrac{u_*}{c_p}\right)^{1/2}$	$\dfrac{z_0}{H_S} = 0.159 \left(\dfrac{u_*}{c_p}\right)^2 = \dfrac{1}{2\pi} \left(\dfrac{u_*}{c_p}\right)^2$
Toba and Koga (1986)	$z_0^* = \Omega \left(\dfrac{u_*}{c_p}\right)^{-1}$ $\Omega = 0.025$ (Toba and Koga 1986) 0.015 (Toba et al. 1990)	$\dfrac{z_0}{H_S} = 1.10 \Omega \left(\dfrac{u_*}{c_p}\right)^{1/2}$
Huang et al. (1986)	$z_0^* = 0.085 \left(\dfrac{u_*}{c_p}\right)^{1/2} \Phi(x_0)$ $\Phi(x_0) \equiv \left[1 - e^{-x_0}\left(1 + x_0 + \dfrac{x_0^2}{2} + \dfrac{x_0^3}{6}\right)\right]^{1/2}$ $x_0 \equiv 2\kappa c_p/u_*.$	$\dfrac{z_0}{H_S} = 0.06 x_0^{-2} \Phi(x_0)$
Geernaert, Larsen and Hansen (1987)	$z_0^* \equiv \dfrac{10g}{u_*^2} \exp\left(-3.65 \left(\dfrac{u_*}{c_p}\right)^{1/3}\right)$ $C_D = 0.012 \left(\dfrac{u_*}{c_p}\right)^{2/3}$	

Author(s)	$z_0^* \equiv \dfrac{g z_0}{u_*^2}$	$\dfrac{z_0}{H_S}$
Masuda and Kusaba (1987)	$z_0^* = 0.0129\left(\dfrac{u_*}{c_p}\right)^{1.10}$	$\dfrac{z_0}{H_S} = 0.0142\left(\dfrac{u_*}{c_p}\right)^{2.60}$
Donelan (1990) Field	$z_0^* = 0.42\left(\dfrac{u_*}{c_p}\right)^{1.03}$	$\dfrac{z_0}{H_S} = 0.46\left(\dfrac{u_*}{c_p}\right)^{2.53}$
Donelan (1990) Lab	$z_0^* = 0.047\left(\dfrac{u_*}{c_p}\right)^{0.68}$	$\dfrac{z_0}{H_S} = 0.051\left(\dfrac{u_*}{c_p}\right)^{2.18}$
Toba et al. (1990) [TIKEJ]	$z_0^* = 0.020\left(\dfrac{u_*}{c_p}\right)^{1/2}$	$\dfrac{z_0}{H_S} = 0.022\left(\dfrac{u_*}{c_p}\right)$
Mast, Kraan and Oost (1991)	$z_0^* = 0.8\left(\dfrac{u_*}{c_p}\right)$	$\dfrac{z_0}{H_S} = 0.88\left(\dfrac{u_*}{c_p}\right)^{5/2}$
Nordeng (1991)	$z_0^* = 0.11\left(\dfrac{u_*}{c_p}\right)^{3/4}\Phi(x_0)$	$\dfrac{z_0}{H_S} = 0.073 x_0^{-9/4}\Phi(x_0)$
	$\Phi(x_0) \equiv \left[1 - e^{-x_0}\left(1 + x_0 + \dfrac{x_0^2}{2} + \dfrac{x_0^3}{6}\right)\right]^{1/2}$	
	$x_0 \equiv 2\kappa c_p/u_*$	
Smith et al. (1992)	$z_0^* = 0.48\left(\dfrac{u_*}{c_p}\right)$	$\dfrac{z_0}{H_S} = 0.53\left(\dfrac{u_*}{c_p}\right)^{5/2}$

3 Atmospheric and Oceanic Boundary Layer Physics

V. N. Lykossov

3.1 Introduction

The globe of the earth is surrounded by a gaseous atmosphere which is always in motion. When in contact with the land or the water surface of the earth the flow is reduced to zero, relative to the underlying surface, and it is this boundary flow that interests us here. As well as the planetary boundary layer in the air, also known as the Ekman layer, there is an oceanic boundary layer which interacts with the air above. The thermal structure through these two regimes is shown schematically in Fig. 3.1. An adequate description of physical processes and mechanisms that determine the structure of the interacting atmospheric and oceanic boundary layers as well as a theoretical background is needed for developing parametrization schemes. The more general features of this problem are treated in the monograph by Kraus and Businger (1994).

One of the most important problems is the parametrization of the turbulent fluxes of momentum, latent and sensible heat at the sea surface. The oceans are the major source of atmospheric water and a major contributor to the heat content of the atmosphere. Most of the solar energy is absorbed by the oceans, and this energy becomes available to maintain the atmospheric circulation only through turbulent fluxes of latent and sensible heat. Radiative, sensible and latent fluxes determine the ocean surface energy flux and, consequently, the vertical structure of the upper ocean. On average, surface buoyancy fluxes are stabilizing over vast oceanic areas (Gargett 1989). Momentum fluxes act as a drag on the atmospheric motions and induce the so-called "wind-driven" component of the ocean currents which produce considerable horizontal transport of energy and momentum. When surface buoyancy fluxes are destabilizing, i.e. a heavier fluid overlies a lighter one, convection in the upper part of the ocean can occur. This process can lead to the sinking of near-surface water to greater depths.

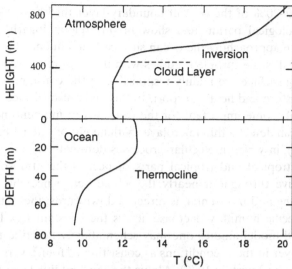

Figure 3.1. Typical thermal profile through the atmosphere and the upper ocean.

The turbulent exchange processes at the air–sea interface are strongly influenced by the state of the sea surface which is varying in time. However, the sea surface temperature changes little over a diurnal cycle because water has a large heat capacity. The roughness of the sea surface depends on the atmospheric surface layer parameters and consequently, on the processes in the whole atmospheric boundary layer. It is believed that the sea surface roughness influences the variability of the surface drag coefficient (Kitaigorodskii 1970). It is very important that the atmospheric motions generate surface waves which also contribute to a turbulent mixing of the atmosphere and ocean.

Thus, one can recognize that the wind stress over the ocean, which is the main interest of this monograph, is not an isolated characteristic of the sea surface state but rather an indicator of the coupled atmospheric and oceanic boundary layer dynamics. It is outside the scope of this chapter to consider the dynamic interaction of the atmospheric boundary layer with the troposphere, and oceanic boundary layer with deep ocean, and so it is assumed that all the necessary characteristics are known at the outer boundaries of the corresponding boundary layers.

3.2 Marine Atmospheric Boundary Layer

3.2.1 Vertical Structure

In general, the atmospheric boundary layer may be more or less arbitrarily split into two regions: the region immediately adjacent to the air–sea interface which is called the constant-flux layer (Monin and Yaglom 1971), and a free-atmosphere-topped interfacial (or "transition") layer over it. Formally, the former is defined as the layer where vertical variations of the turbulent fluxes do not exceed, say, 10% of their surface values. Typically, this layer is about 10–100 m thick which makes approximately 10–20% of the overall boundary-layer thickness. Vertical distributions of meteorological parameters show in this region logarithmic asymptotic dependence when approaching the ocean surface and influenced by the air density stratification. Small scale turbulent eddies with the sizes restricted by the distance to the underlying surface are mainly responsible in the constant flux layer for the momentum, moisture and heat transport. In this layer also surface waves influence the air flow and become important for the transport of momentum. Very close to the surface one can detect a thin microlayer with height of order 1 cm (the so-called viscous sublayer), in which molecular processes dominate.

Over the subtropical and tropical parts of oceans the atmospheric boundary layer is convective throughout nearly the whole year since the surface density flux due to heating and moistening is directed downwards; the potential temperature and the specific humidity decrease across the constant flux layer. Based on experimental and model studies one may schematically describe the structure of the interfacial layer in these conditions as consisting of four layers, each governed by different physics (Augstein 1976). Above the constant flux layer there is a mixed layer with thickness of order 1 km. The change of potential temperature with height is small here, and mixing is dominated by convectively-driven organized motions (large eddies). Atop the mixed layer one can detect the so-called entrainment zone which is 100–500 m thick. In this layer turbulence is intermittent, the air stratification is stable with regard to potential temperature, and internal waves and sometimes small clouds are observed. A cloud-topped boundary layer includes an additional layer of broken or uniform clouds. This cloud layer connects to the free atmosphere via an inversion layer, i.e. a layer of increasing potential temperature with height.

The most variable part of this idealized structure is the cloud layer. If no clouds are formed, then the atmospheric boundary layer terminates at the entrainment zone. When very deep clouds are developed and extended through the whole troposphere (for example, in cyclonic conditions of surface convergence and large-scale upward motion), the top of the boundary layer is not defined. However, in many cases the free atmosphere is turbulently decoupled from the boundary layer, typically at the ocean on the rear side of depressions and in the Inter-Tropical Convergence Zone.

In middle and high latitudes, the balance between the pressure gradient, Coriolis and turbulent stress divergence mainly determines the structure of the boundary layer. A barotropic atmosphere is one where constant density surfaces coincide with constant pressure surfaces. The special but sometimes observed case of a steady state, horizontally homogeneous, neutrally stratified, barotropic atmosphere where the stress can be represented by a height-independent eddy diffusivity, results in the well-known Ekman spiral wind profile (Brown 1974). The characteristic feature of the Ekman profile is that, due to friction, winds in the boundary layer cross the isobars from high towards low pressure. In the case of low or high pressure systems, the cross-isobaric component of flow induces upward or downward vertical motions, respectively. Such a process known as Ekman pumping (Stull 1988) is very important for linking the boundary layer with the free atmosphere.

The one-dimensional representation of the boundary layer structure was found to be useful in many cases, but it may become incorrect when horizontal advection is dominant, in particular, in the vicinity of oceanic fronts. Observations have shown (Guymer et al. 1983; Khalsa and Greenhut 1989; Rogers 1989) that the spatial variability of the sea surface temperature on a scale of 200 km or less causes horizontal variability on similar scales in the atmosphere. It was also found that the turbulent structure of the marine atmospheric boundary layer has different scales on opposite sides of a sea surface temperature discontinuity. Effects of oceanic fronts on the wind stress are discussed by Gulev and Tonkacheev (1995).

3.2.2 Turbulence

Since Reynolds numbers for atmospheric motions are very large (of order 10^7), turbulence in the boundary layer is fully developed and three-dimensional. The vertical turbulent transport of momentum, heat and moisture is the main process which links the large scale motions in free atmosphere with the surface. Small-scale turbulence consists of a set of disturbances with scales which do not exceed the distance to the surface. Besides the classical descriptions of turbulent flows which are connected with the names Reynolds and Taylor, a new approach was introduced by the discovery of ordered motions in many turbulent shear flows. It has been understood that traditional parametrization schemes do not describe some essential features of the atmospheric boundary layer, in particular, non-local nature of turbulence caused by the presence of coherent structures (large eddies). Following Stull (1991), one can define large eddies as turbulent structures with size of the same order as that of planetary boundary layer, or of the same order as that of mean flow.

Spectral decomposition of turbulent fields (see, for example, Pennel and LeMone 1974) indicated that large eddies contain most of the turbulent kinetic energy. Examples of these coherent structures (convective thermals of the same 1 to 2 km diameter as the mixed-layer depth, well-ordered roll vortices, mechanical eddies of the same size as the 100 m thick shear region of the surface layer and convective plumes) are given in reviews by Mikhailova and Ordanovich (1991), Etling and Brown (1993) and Stull (1993). Byzova et al. (1989) have also discussed turbulent cell convection with quasi-ordered convective structures of the 3 to 10 km size and small-scale cell convection with coherent motions of a few hundred metres. All these structures coexist with small-scale turbulence.

In particular, it was discovered (Wyngaard and Brost 1984) that the vertical diffusion of a dynamically passive conservative scalar through the convective boundary layer is the superposition of two processes. These processes are driven by the surface flux, "bottom-up" diffusion, and by the entrainment flux, "top-down" diffusion. It was also found that the vertical asymmetry in the buoyant production of the turbulent kinetic energy caused the top-down and bottom-up eddy diffusivities to differ.

At the very small scale, other coherent motions play an important role in the surface layer near the wall. Many laboratory experiments (e.g. Kline et al. 1967;

Corino and Brodkey 1969; Narahari Rao et al. 1971; Brown and Roshko 1974) have demonstrated the significance of so-called bursting processes for turbulence production. Following Narasimha (1988), bursting processes may be considered as coherent, quasi-periodic cycles of events. These events can be described as a retardation of the near-surface fluid in forms of streaks, a build up of a shear layer leading to a violent ejection of fluid from the surface and a sweep of faster fluid from the outer layer towards the surface. It was found by Corino and Brodkey (1969) that nearly 70% of the shear stress could be due to the ejection process. Sweep events play a major role in the bedload transport in rivers and oceans (Heathershaw 1974; Drake et al. 1988). It was shown (Wallace et al. 1972) that an interaction between ejections and sweeps accounts for a substantial part of the momentum flux to balance its excess produced by motions of these two categories. Intermittent coherent motions have been detected in the turbulence measurements as periodic, large-amplitude excursions of turbulent quantities from their means. It was discovered that this process strongly influences the turbulent transport through cycles of ejections and sweeps also within the atmospheric surface layer (Narasimha 1988; Mahrt and Gibson 1992; Collineau and Brunet 1993) and, in particular, in the marine boundary layer (Antonia and Chambers 1978).

However, it seems that turbulence production close to the surface is a process which is nearly independent of large-scale processes in the outer layer (Kline and Robinson 1989). Most of the turbulence and most of the momentum flux is produced by the near-wall process, and this phenomenon is usually called active turbulence (Townsend 1961). He has also suggested the concept of inactive turbulence, according to which the scale of turbulence is of the boundary layer size and does not produce the momentum flux at the surface. Such frictional decoupling has often been observed at the sea surface (e.g. Volkov 1970; Makova 1975; Chambers and Antonia 1981; Smedman et al. 1994). It was derived from these investigations that momentum transfer from the decaying surface waves to the atmosphere ("old wave age" conditions, where the wave speed is greater than the wind speed) can be suggested as the possible mechanism causing the frictional decoupling. If by that time the surface buoyancy flux is relatively small, and relatively large wind shear is maintained in the upper part of the boundary layer (for instance, due to the development of a low-level jet), then turbulent energy produced in this region can be brought down, including the surface layer, by pressure transport (Smedman et al. 1994). This imported-from-above turbulence is an example of turbulence of the inactive kind. It was found that the turbulence statistics of the boundary layer resemble in these conditions those of a convective boundary layer but with different scaling, since the buoyancy production is small. Even if these conditions are not very typical for the ocean, the frictional decoupling mechanism must be taken into account when the wind stress parametrizations are developed for use, for example, in global climate models.

Additionally, drag reduction in flows with suspended particles is another well known phenomenon (Toms 1948). It is recognized now that a major dynamical effect of suspended fine particles is stabilization of the secondary inflexional instability, suppression of the intense small-scale turbulence and decrease of turbulence production (Lumley and Kubo 1984; Aubry et al. 1988). It was also found that particles influence the flow turbulence in two contrary ways – at the expense of the energy of turbulent fluctuations to suspend particles, and by destabilization of the flow, when a significant slip between the two phases exists. In the case of coarse particles such a flow destabilization is the main factor, leading to the additional turbulence production, since each particle sheds a wake disturbance to the flow like a turbulence-generating grid (Hinze 1972; Tsuji and Morikawa 1982).

Over oceans, at wind speeds above, say, 15 m/s intensive spray is detected near the air–sea interface. Spray droplets, detached from the sea surface, carry their instantaneous momentum with them. The biggest of these droplets fall down into the sea and return their momentum back to the sea surface. On their way through the air, they may interact with the air and exchange momentum with the air. This process seems to be not very important, since spray is generated mainly from the larger waves which have an orbital velocity near to the wind velocity. However, the small bubble-derived water droplets and salt particles can be suspended in the air flow, and carried by turbulence higher up to cloud heights (de Leeuw 1986). If their concentration is large enough, the density stratification can be remarkably altered and hence this can influence the momentum transfer. At the present time, it is not clear how important this mechanism could be for the observed wind stress over the ocean. Due to additional evaporation from the spray droplets, a modification of the temperature and humidity gradients might also be expected.

There are studies (e.g. Lumley 1967; Shaw and Businger 1985; Narasimha and Kailas 1987; Narasimha 1988; Mahrt and Gibson 1992; Collineau and Brunet 1993), in which the intermittent nature of the near-surface turbulence is discussed with respect to the atmospheric boundary layer. Its possible connection with drag reduction phenomena due to the presence of suspended particles in the air (in particular, sea spray) seems also to be important. An expansion of the Monin–Obukhov similarity theory for the case of a two-phase flow (e.g. Barenblatt and Golitsyn 1974; Wamser and Lykossov 1995) allows one to describe generally the drag reduction due to more stable density stratification, but without regard to details caused by the intermittent nature of turbulence and by the non-regular loading of particles into air.

3.2.3 Turbulence Closure

Let a be any meteorological variable: horizontal components of the wind velocity (u and v for the alongwind and crosswind directions, respectively), potential temperature (θ), specific humidity (q), etc. The Reynolds equation for a conservative statistically averaged quantity reads

$$\frac{\partial a}{\partial t} = -\frac{\partial \overline{a'w'}}{\partial z} + (\cdots) \tag{3.1}$$

where t is time, z is the vertical coordinate, and w is the vertical component of the wind velocity. As usual, the potential temperature, θ, is defined as

$$\theta = T\left(\frac{p_0}{p}\right)^{0.286} \tag{3.2}$$

where T is the absolute air temperature, p is air pressure and p_0 is a reference pressure, which is usually set to 1000 hPa. The overbar denotes the average values, primes stand for the turbulent fluctuations, and the terms responsible for the contribution of the other (non-turbulent) processes into dynamics of the boundary layer are marked by dots. To make the turbulence problem more tractable, it is often assumed that the ergodic condition is satisfied, i.e. the ensemble, space and time averages are equal (Stull 1988). This is true for turbulence that is both homogeneous and statistically stationary. In practice, temporal averages are frequently used. For simplicity in notation, the bar signifying mean values is omitted everywhere except for the notation of the turbulent covariances.

Under the assumption of horizontal homogeneity, the budget equation for the turbulent kinetic energy (e) may be written as follows (Monin and Yaglom 1971):

$$\frac{\partial e}{\partial t} = -\left(\overline{u'w'}\frac{\partial u}{\partial z} + \overline{v'w'}\frac{\partial v}{\partial z}\right) - \frac{g}{\rho}\overline{\rho'w'} - \frac{\partial \overline{w'e}}{\partial z} - \frac{1}{\rho}\frac{\partial \overline{p'w'}}{\partial z} - D \tag{3.3}$$

where g is the acceleration due to gravity, ρ is the air density, p is the pressure, D is the dissipation rate, and

$$e = (\overline{u'^2 + v'^2 + w'^2})/2$$

The terms on the right-hand side of Eq. (3.3) describe, in order, the shear production, the buoyancy production/destruction, the vertical turbulent transport, the pressure transport, and the viscous dissipation (the conversion into heat) of the turbulent kinetic energy. Mainly, there are two sources of turbulence in the atmospheric boundary layer: the wind shear and the buoyancy flux near the surface, and the wind shear at the top of the boundary layer. The frequent presence of clouds in the marine boundary layer leads to a strong change of the radiation budget at the surface. In this case there is also an additional production of turbulence due to long-wave cooling at the top of the cloud layer, short-wave heating of the inner part of clouds and phase changes of the water.

Most of parametrization schemes are based on the K-theory stability-dependent eddy diffusivity closure, which assumes that fluxes are associated with small-size eddies only in a manner similar to molecular diffusion, and that the static stability is estimated on the basis of the local lapse rate. The turbulent fluxes in the boundary layer are calculated in this case as suggested by the Boussinesq (1877) hypothesis [1]

$$\overline{a'w'} = -K_a\frac{\partial a}{\partial z} \tag{3.4}$$

where the eddy diffusivity K_a, having a positive value by its physical sense, is evaluated by means of the mixing length theory (Monin and Yaglom 1971):

---

[1] Strictly speaking, Boussinesq had applied this approach to the turbulent transport of momentum. However, it was found that such a closure can also be used for the description of the turbulent transport of any passive scalar.

$$K_a = \alpha_a l^2 \left| \frac{\partial \mathbf{V}}{\partial z} \right| F_a(Ri) \tag{3.5}$$

Here α_a is a universal constant, l is the integral turbulence scale, and F_a is a universal non-dimensional function, depending on the gradient Richardson number,

$$Ri = g/\theta_v \frac{\partial \theta_v / \partial z}{|\partial \mathbf{V}/\partial z|^2}$$

where $\theta_v = (1 + 0.61q)\theta$ is the virtual potential temperature, $\mathbf{V}$ is the horizontal wind velocity vector with the components u and v. To calculate l, the following formula, suggested by Blackadar (1962), can be used

$$l = \frac{\kappa z}{1 + \kappa z / l_\infty} \tag{3.6}$$

where $\kappa = 0.4$ is von Kármán's constant, and l_∞ is some function of the external parameters (see Holt and SethuRaman 1988, for the review). The eddy diffusivity coefficients K_a are also often related to the (mean) turbulent kinetic energy and dissipation rate (Monin and Yaglom 1971)

$$K = le^{1/2} = Ce^2/D \qquad K_a = \alpha_a K \tag{3.7}$$

where C is a universal constant.

At the same time much experimental data (e.g. Budyko and Yudin 1946; Priestley and Swinbank 1947; Deardorff 1966) showed that sometimes the atmospheric boundary layer is neutrally or weakly stably stratified (for example, the convective mixed layer), but the heat flux is directed upward. This corresponds to negative heat diffusivity which is not consistent with the basic molecular diffusion analogy. Moreover, the vertical transport of the turbulent kinetic energy is directed upward throughout the whole depth of the convective layer (see, for example, Andre 1976; Kurbatskii 1988).

The K-theory also has essential shortcomings in application to jet-like flows. For example, in the case of a jet which was observed in the fair-weather trade wind boundary layer (Pennel and LeMone 1974) a countergradient transport of momentum was found. Other examples of such phenomena for channel flows are given by Narasimha (1984), Yoshizawa (1984) and Kurbatskii (1988). In such flows the point of maximum velocity and that of zero momentum flux do not coincide. Thus, there is a region where positive momentum flux is accompanied by positive velocity gradient. This means that momentum is transported up from this region, but not down to the surface, as is expected for the situation without jet. Persistent countergradient fluxes of momentum and heat (density) have also been observed in homogeneous turbulence forced by shear and stratification: at large scales when stratification is strong, and at small scales, independently of stratification (Gerz and Schumann 1996).

A number of reviews of the state-of-the-art in the turbulence closure problem for the boundary layer with coherent structures have been published (e.g. Stull

1993, 1994; Lykossov 1995). An hierarchical description is usually used to classify the numerous approaches presented in the literature. From a most general point of view, these approaches can be subdivided into two broad classes which describe the local and non-local closures, respectively. The former is based on the assumption that the turbulent fluxes depend only on the mean quantities. The latter requires that the turbulent fluxes are described as more or less arbitrary functionals of the mean flow parameters. There is no strong separation between these two classes, since some characteristics of non-locality can be also found in the local closures.

In order to account for the countergradient heat transport in the boundary layer, the Boussinesq hypothesis may be generalized as follows (e.g. Budyko and Yudin 1946; Priestley and Swinbank 1947; Deardorff 1966):

$$\overline{\theta' w'} = -K_\theta \left(\frac{\partial \theta}{\partial z} - \gamma_\theta \right) \tag{3.8}$$

where K_θ is, as before, the eddy diffusivity coefficient, and the term γ_θ is a counter-gradient correction term. The expressions for this term can be derived from the equations for higher order moments and from the large eddy simulation data.

For example, the Deardorff (1972) formula reads

$$\gamma_\theta = (g/\theta_v)(\overline{\theta'^2}/\overline{w'^2}) \tag{3.9}$$

Assuming, in particular, that $\overline{\theta'^2}$ and $\overline{w'^2}$ are constant throughout the convective boundary layer, Deardorff (1973) suggested

$$\gamma_\theta = (g/\theta_v)(\theta_*^2/w_*^2) = \overline{w'\theta'}_0/w_* h \tag{3.10}$$

where $\overline{w'\theta'}_0$ is the surface kinematic heat flux, $w_* = (g/\theta_v \overline{w'\theta'}_0 h)^{1/3}$ is the convective velocity scale, h is the height of the convective boundary layer, and $\theta_* = \overline{w'\theta'}_0/w_*$ stands for the convective temperature scale. Another modification of the formula ((Eq. (3.9)) can be found in the overview by Lykossov (1995).

Since turbulence in the convective boundary layer of the atmosphere is usually characterized by the narrow, intense, rising plumes and by broad, low-intensity subsiding motions between the plumes, the vertical velocity field is strongly skewed. Wyngaard (1987) suggested that this skewness is responsible for the difference between the top-down and bottom-up scalar (e.g. heat) diffusion. This dependence of the scalar diffusivity on the location of the source was termed by Wyngaard and Weil (1991) "transport asymmetry". They found that the interaction between the skewness and the gradient of the transported scalar flux can induce this asymmetry. Using the kinematic approach, Wyngaard and Weil (1991) derived an expression for the scalar flux which in the case of heat flux has the following form:

$$\overline{w'\theta'} = -K_\theta \left(\frac{\partial \theta}{\partial z} - \frac{1}{2} A S \sigma_w T_L \frac{\partial^2 \theta}{\partial z^2} \right) \tag{3.11}$$

where

$$S = \overline{w'^3}/(\overline{w'^2})^{3/2}$$

is the skewness of w,

$$\sigma_w = (\overline{w'^2})^{1/2}$$

T_L is the Lagrangian integral time scale, and A is a constant. Thus, in order to describe the scalar flux, a term proportional to the second derivative of the mean quantity is added, in the theory suggested by Wyngaard and Weil (1991), to the eddy diffusivity term.

A nonlocal generalization of the Boussinesq hypothesis can be written in the following form:

$$\overline{a'w'} = -\int_0^\infty K(z, z') \frac{\partial a}{\partial z'} dz' \tag{3.12}$$

This idea was first suggested by Berkowicz and Prahm (1979) with application to air pollution studies. Their generalization of the diffusivity theory is based on the so-called spectral turbulent diffusivity concept, according to which the eddy diffusivity coefficient of a single Fourier component of the passive scalar field is treated separately as a function of the wave number k. For each individual mode the Boussinesq closure Eq. (3.4) is used. In the case of spectral diffusivity $K(k) = K_0$ for all k, it follows that $K(z, z') = K_0 \delta(z - z')$, and Eq. (3.12) coincides with formula (3.4). The integral closure of type (3.12) was also used by Fiedler (1984) and Hamba (1995). A similar closure can be derived from a set of equations for the second and third moments applied for modelling nonlocal turbulent transport of momentum in jet-like flows (Lykossov 1992).

3.2.4 The Atmospheric Ekman Layer

It is experimentally known that outside the tropics the mean wind changes direction with height in the transition layer (above the surface layer) and nearly coincides with the free-atmosphere velocity at heights far enough from the underlying surface. At the same time, the momentum fluxes $\overline{u'w'}$ and $\overline{v'w'}$ (as well as the stress components $\rho \overline{u'w'}$ and $\rho \overline{v'w'}$) decrease with height. When horizontal homogeneity is assumed and the viscous stress is neglected, the momentum equations can be written as follows (e.g. Brown 1974):

$$\frac{\partial u}{\partial t} + \frac{\partial \overline{u'w'}}{\partial z} = -\frac{1}{\rho}\frac{\partial p}{\partial x} + fv \tag{3.13}$$

$$\frac{\partial v}{\partial t} + \frac{\partial \overline{v'w'}}{\partial z} = -\frac{1}{\rho}\frac{\partial p}{\partial y} - fu \tag{3.14}$$

where $f = 2|\Omega| \sin (\text{latitude})$ is the Coriolis parameter. Assuming the steady-state conditions, the geostrophic balance between Coriolis and pressure-gradient forces in the free atmosphere can be written

$$\frac{1}{\rho}\frac{\partial p}{\partial x} = fv_g, \qquad \frac{1}{\rho}\frac{\partial p}{\partial y} = -fu_g \tag{3.15}$$

where subscript g indicates the geostrophic wind, which is often taken as the upper boundary condition for the boundary flow.

Let us now consider the steady-state version of Eqs (3.13) and (3.14)

$$\frac{d\overline{u'w'}}{dz} = f(v - v_g) \tag{3.16}$$

$$\frac{d\overline{v'w'}}{dz} = -f(u - u_g) \tag{3.17}$$

where u_g and v_g are substituted for the constant pressure gradients from Eq. (3.15). It is seen from Eqs (3.16) and (3.17) that momentum is generated in the boundary layer by ageostrophic components of the wind velocity and transported down to the surface.

The most idealized model of the wind structure can be derived for this case with the help of the K-theory turbulence closure based on Eq. (3.4). The resulting equations with the constant eddy diffusivity coefficients $K_u = K_v \equiv K$ are known as the Ekman layer equations. When the x-axis is aligned with the geostrophic wind, these equations are written as follows:

$$K\frac{d^2u}{dz^2} + fv = 0$$

$$K\frac{d^2v}{dz^2} - f(u - G) = 0 \tag{3.18}$$

where

$$G = \sqrt{u_g^2 + v_g^2}$$

is the geostrophic wind speed. An analytical solution of these equations for the ocean was derived by Ekman (1905), and for the atmosphere, by Akerblom (1908). In complex notation, these equations take the form

$$\frac{d^2W}{dz^2} - i\frac{f}{K}(W - G) = 0 \tag{3.19}$$

where $W = u + iv$, and $i = \sqrt{-1}$. The solution to Eq. (3.19), subject to the boundary conditions

$$W = 0 \quad \text{at} \quad z = 0 \tag{3.20}$$

$$W \to G \quad \text{as} \quad z \to \infty \tag{3.21}$$

is written as follows:

$$W - G = -G\exp\left(-\frac{z}{h_E}\right)\left[\cos\left(\frac{z}{h_E}\right) - i\sin\left(\frac{z}{h_E}\right)\text{sign}\, f\right] \tag{3.22}$$

where

$$h_E = \sqrt{2K/|f|}$$

For $K = 12.5$ m^2/s and $f = 10^{-4}$/s, the value of $h_E = 500$ m. Note that the quantity πh_E is the lowest height (the Ekman layer depth) where the boundary layer wind is parallel to the geostrophic wind. The u and v components of the solution (3.22) are expressed in the form

$$u = G\left[1 - \exp\left(-\frac{z}{h_E}\right)\cos\left(\frac{z}{h_E}\right)\right]$$

$$v = G\exp\left(-\frac{z}{h_E}\right)\sin\left(\frac{z}{h_E}\right)\mathrm{sign}\,f$$

It is seen from this solution that the wind veers with height, giving the so-called Ekman spiral wind profile, and slightly overshoots the geostrophic value. Since

$$\tan\theta = \lim_{z\to 0}\frac{v}{u} = \lim_{z\to 0}\frac{dv/dz}{du/dz} = \mathrm{sign}\,f \qquad (3.23)$$

the surface wind is parallel to the stress and directed to the left (right) of the free-atmosphere wind in the northern (southern) hemisphere. In this idealized model, the angle θ between the surface and geostrophic wind is equal to 45° and does not depend on geographical location and meteorological situation. However, this is not true for the real atmospheric boundary layer. Observed angles between the geostrophic and surface winds may vary considerably due to the fact that ageostrophic components of the wind and, consequently, the momentum transfer may be influenced by various physical processes. For example, the wind profile is sometimes characterized by the presence of a low level jet in the upper part of the Ekman layer (e.g. Pennel and LeMone 1974; Smedman et al. 1995). A baroclinicity of the free-atmosphere flow, which can be expressed in terms of the height-dependent geostrophic wind, using the thermal wind relationships (Stull 1988)

$$\frac{\partial u_g}{\partial z} = -\frac{g}{fT}\frac{\partial T}{\partial y}$$

$$\frac{\partial v_g}{\partial z} = \frac{g}{fT}\frac{\partial T}{\partial x}$$

may significantly alter the Ekman profile. The constant K assumption is also not valid since K is linearly increasing with height, at least, near the surface. A lot of proposed theoretical distributions of the eddy diffusivity coefficient $K(z)$ are presented in the literature (see, for example, Brown 1974; Holt and SethuRaman 1988 for the review). Nevertheless, it is widely recognized that the modelling of the more or less real dynamics of the boundary layer requires more sophisticated approaches, a brief review of which was given in Section 3.2.2. Whenever modelling the surface winds and stress, cognizance of the nonlinear solution containing organized large eddies is essential as discussed in Chapter 11. The data supporting this solution as the correct geophysical solution for the flow in the planetary boundary layer is found in Chapters 7 and 11.

3.2.5 Surface Layer

In the atmospheric surface layer, typically the lower 10% of the boundary layer, the turbulent fluxes of momentum, water vapour and sensible heat are nearly constant with height. In a surface microlayer less than 1 cm thick, the molecular transport dominates over turbulent transport. Over oceans, diabatic processes and wave motion are dominating in their influence on the wind shear. Customarily, these two effects are treated as independent (e.g. Hasse and Smith 1997).

Orientating the x-axis in the surface stress direction, and following Brown (1974), one can transform Eq. (3.16) to the nondimensional form by the use of an arbitrary characteristic velocity scale V_0 and vertical scale H together with the surface stress τ_0. This produces

$$E\frac{d\tilde{\tau}}{d\tilde{z}} + \tilde{v} - \tilde{v}_g = 0 \tag{3.24}$$

where the tilde indicates nondimensional variables, and $E = \tau_0/(\rho f V_0 H)$ is the Ekman number. For $H \to 0$, the parameter $E \to \infty$, and Eq. (3.24) yields

$$d\tilde{\tau}/d\tilde{z} = 0$$

subject to the boundary condition $\tilde{\tau}(0) = 1$. Integration gives $\tau \equiv \tau_0$. Observations show that close to the surface, where E is large, the layer of nearly constant stress can really be detected. Similarly, it can be shown that for steady-state conditions the buoyancy flux $-(g/\rho)\overline{\rho'w'}$ in the atmospheric surface layer is nearly constant.

Assuming that for neutral conditions in this layer the eddy diffusivity coefficient $K = \kappa u_* z$ (Prandtl 1932), where

$$u_* = \sqrt{\tau_0/\rho}$$

is the so-called friction velocity, one can obtain

$$\frac{\partial u}{\partial z} = \frac{u_*}{\kappa z} \tag{3.25}$$

and on integration the wind profile

$$u(z) = \frac{u_*}{\kappa} \ln\left(\frac{z}{z_0}\right) \tag{3.26}$$

Here z_0 is an integration constant called the aerodynamic roughness length. In some sense, this parameter is an artificial quantity which results from extrapolation of the wind profile to zero wind speed. However, it was experimentally found that z_0 is a parameter, which in the whole characterizes the geometrical properties of the solid underlying surface.

Contrary to the land surface, where the roughness is determined by the roughness elements of fixed geometry, the sea surface should be considered as the interface of two fluids of different density, and that both are in motion and may generate waves. In this case z_0 will not reflect topology of the sea surface but must be obtained, if possible, as a parameter, characterizing dynamics of the interfacial layer. The experimentally derived Charnock (1955) formula for the calculation of z_0 from the friction velocity u_* (see Chapter 2) can be considered as an example of such parametrization. It is necessary to point out that the logarithmic wind profile (3.26) is valid only well above the height z_0 (see, e.g. Monin and

Yaglom 1971). Close to the surface, the viscous stress, which is neglected in Eq. (3.25) should determine the wind profile.

Using an aerodynamic approach, from the surface stress τ, mean wind speed, and density of air ρ can be derived the drag coefficient C_d as

$$\tau/\rho = C_d \mathbf{u}^2 \tag{3.27}$$

Note that since $\mathbf{u}$ is a function of z, this coefficient depends on height. Often the difference in direction between τ and $\mathbf{u}$ is neglected. Typically, it is defined for a reference height of 10 to 25 m above the sea level. This approach is commonly used as a parametrization to describe the air–sea fluxes, the only practicable tool that is available to apply results of empirical investigations. In the neutrally stratified momentum constant flux layer, from the logarithmic profile, z_0 is related to the neutral drag coefficient C_{dN} by

$$C_{dN} = \left(\frac{u_*}{u}\right)^2 = \left[\kappa / \ln\left(\frac{z}{z_0}\right)\right]^2 \tag{3.28}$$

Equation (3.28) shows that there is a one-to-one relation between the drag coefficient and the roughness length so that C_d grows when z_0 increases. At very low winds, no waves are generated and the stress should not be less than for an aerodynamically smooth flow, for which $z_0 = 0.11\nu/u_*$ (Schlichting 1960) where $\nu = 0.14 \times 10^{-4}$ m²/s is the kinematic viscosity of air.[2] This smooth flow constraint causes the drag coefficient to increase with decreasing wind speed below about 3 m/s (e.g. Zilitinkevich 1970; Wippermann 1972; Smith 1988).

For non-neutral conditions, the Monin–Obukhov similarity theory predicts that the dimensionless gradient of the wind velocity can be expressed by a universal function of dimensionless stability z/L only

$$\frac{\kappa z}{u_*}\frac{\partial u}{\partial z} = \phi_M(z/L) \tag{3.29}$$

where $\phi_M(0) = 1$, and the stability parameter L is the Monin–Obukhov length scale

$$L = \frac{\rho u_*^3}{\kappa g \overline{\rho' w'}} \tag{3.30}$$

To calculate the buoyancy flux, the following relation is usually used:

$$\frac{g}{\rho}\overline{\rho' w'} = -\frac{g}{\theta_v}\overline{\theta_v' w'} \tag{3.31}$$

Note that a positive buoyancy flux $\overline{\rho' w'}$ is away from the surface. The flux would be expected to be positive when the virtual potential temperature θ_v increases with height. This is known as stable conditions.

Given the surface roughness, integration of Eq. (3.29) from z_0 to z results in a diabatic wind profile (see e.g. Monin and Yaglom 1971)

---

[2] Generally speaking, ν is dependent on the air temperature, but in the atmosphere, the temperature effect on ν is usually small and can be neglected.

$$u(z) = \frac{u_*}{\kappa}\left[\ln\left(\frac{z}{z_0}\right) - \psi_M\left(\frac{z}{L}\right)\right] \tag{3.32}$$

where

$$\psi_M\left(\frac{z}{L}\right) = \int_0^{z/L} \frac{1 - \phi_M(\zeta)}{\zeta}d\zeta \tag{3.33}$$

is the integrated universal function for velocity. The drag coefficient C_d can be now formally expressed as

$$C_d = \left(\frac{u_*}{u}\right)^2 = \kappa^2 \Big/ \left[\ln\left(\frac{z}{z_0}\right) - \psi_M\left(\frac{z}{L}\right)\right]^2 \tag{3.34}$$

Equation (3.34) shows that in the non-neutral constant flux layer the drag coefficient depends not only on the roughness length z_0 but on the stability parameter L too, which also reflects dynamics of the air–sea interfacial layer. This is especially important for those ocean regions where highly unstable thermal stratification can be formed in the atmospheric surface layer. On the other hand, the z/L dependence of universal functions means that the boundary layer is nearly neutral close to the surface and becomes more non-neutral when the height increases. Note also that when the winds are strong, the u_* value is generally high, and $\psi_M(z/L)/ \ll \ln(z/z_0)$ so that the surface layer can be again considered as neutrally stratified.

Since the stability parameter includes the virtual potential temperature flux, which can be treated as a linear combination of the potential temperature flux and water vapour flux

$$\overline{\theta'_v w'} \approx (1 + 0.61q)\overline{\theta'w'} + 0.61\theta\overline{q'w'} \tag{3.35}$$

it is advisable to give here the corresponding dimensionless fluxes in the form of the Monin–Obukhov universal functions

$$\frac{\kappa z}{\theta_*}\frac{\partial\theta}{\partial z} = \phi_H(z/L) \qquad \frac{\kappa z}{q_*}\frac{\partial q}{\partial z} = \phi_E(z/L) \tag{3.36}$$

and their aerodynamic representations

$$\overline{\theta'w'} \equiv u_*\theta_* = C_H u(\theta_s - \theta) \qquad \overline{q'w'} \equiv u_*q_* = C_E u(q_s - q) \tag{3.37}$$

where $u_*\theta_* = \overline{\theta'w'}_s, u_*q_* = \overline{q'w'}_s$, and subscript s refers to the values at the surface. As above for the drag coefficient C_d, the bulk transfer coefficients C_H (Stanton number) and C_E (Dalton number) are expressed with the help of integrated universal functions

$$C_H = \alpha_H\kappa^2\left[\ln\left(\frac{z}{z_0}\right) - \psi_M(z/L)\right]^{-1}\left[\ln\left(\frac{z}{z_H}\right) - \psi_H(z/L)\right]^{-1}$$

$$C_E = \alpha_E\kappa^2\left[\ln\left(\frac{z}{z_0}\right) - \psi_M(z/L)\right]^{-1}\left[\ln\left(\frac{z}{z_E}\right) - \psi_E(z/L)\right]^{-1} \tag{3.38}$$

where α_H and α_E stand for the ratios of the eddy diffusivities of sensible heat and water vapour to that of momentum, z_H and z_E are the roughness lengths for temperature and specific humidity, respectively.

Similar to the aerodynamic roughness z_0, these quantities are associated with a logarithmic profile and the surface fluxes. The magnitude of z_H and z_E is controlled by transport mechanisms very close to the surface where molecular processes dominate. This is especially important under low-wind, unstable conditions over water (Godfrey and Beljaars 1991). As an example of parametrization for z_H, the following approximation of an experimental data obtained for natural and artificial surfaces (Garratt and Hicks 1973; Hicks 1975; Garratt 1977) is suggested (Kazakov and Lykossov 1982):

$$\ln(z_0/z_H) = \begin{cases} -2.43 & \text{for } Re_* \leq 0.11 \\ 0.83\ln(Re_*) - 0.6 & \text{for } 0.11 \leq Re_* \leq 16.3 \\ 0.49 Re_*^{0.45} & \text{for } Re_* \geq 16.3 \end{cases} \tag{3.39}$$

where $Re_* = u_* z_0/\nu$ is the roughness Reynolds number.

The use of the diabatic wind profile requires knowledge of the stability functions $\phi_M(z/L), \phi_H(z/L)$ and $\phi_E(z/L)$. Variations of stability are more pronounced over land than over sea. Hence it has been found suitable to adopt stability functions determined over land for the use over sea, too. It is usually assumed that $\phi_H = \phi_E$. In the case of stable stratification, the linear type functions are theoretically derived and experimentally supported (see e.g. Monin and Yaglom 1971)

$$\phi_M = \phi_H = 1 + B\zeta$$
$$\psi_M = \psi_H = -B\zeta \tag{3.40}$$

where $\zeta = z/L$, and the parameter B varies, according to observations, from 4.7 to 5.2 (Panofsky and Dutton 1984). For the regions with moderate unstable stratification ($-2\zeta \leq 0$), the Businger–Dyer formulations are widely used (Businger et al. 1971; Dyer 1974)

$$\phi_M = (1 - \alpha\zeta)^{-1/4} \qquad \phi_H = (1 - \alpha\zeta)^{-1/2} \tag{3.41}$$

where values of α ranging from 16 to 28 fitted the data derived from the measurements over oceans (Edson et al. 1991). The corresponding integrated universal functions have the following form (Paulson 1970)

$$\psi_M(\zeta) = \ln\left[\frac{1}{8}\left(1 + \phi_M^{-2}\right)\left(1 + \phi_M^{-1}\right)^2\right] - 2\arctan\phi_M^{-1} + \pi/2$$

$$\psi_H(\zeta) = 2\ln\left[\frac{1}{2}\left(1 + \phi_H^{-1}\right)\right] \tag{3.42}$$

When convection dominates so that ζ large and negative (in particular, in the case of light winds), the universal functions should vary as $(-\zeta)^{-1/3}$, a relation called the free-convection condition (Panofsky and Dutton 1984). Carl et al. (1973) suggested for momentum that

$$\phi_M = (1 - 16\zeta)^{-1/3} \tag{3.43}$$

which satisfies this condition when $-\zeta$ becomes large. The corresponding integrated universal function can be written as follows:

$$\psi_M = \frac{3}{2}\ln\left[\frac{1}{3}(X^2 + X + 1)\right] - \sqrt{3}\left(\arctan\frac{2X+1}{\sqrt{3}} - \frac{\pi}{3}\right) \qquad (3.44)$$

where $X = (1 - 16\zeta)^{1/3}$. A similar $-1/3$ power law dependence is also required for ϕ_H in order to satisfy the theoretical prediction. To combine the Businger–Dyer expressions and free-convection limit, one can use (Kazakov and Lykossov 1982; Large et al. 1994)

$$\phi_a = (b_a - c_a\zeta)^{-1/3} \qquad \text{for } \zeta < \zeta_a \qquad (3.45)$$

where a stands for M or H, and the b_a and c_a are chosen so that both ϕ_a and its first derivative are continuous across the matching value $\zeta = \zeta_a$.

In order to show the importance of air–sea temperature difference, the drag coefficient C_d and transfer coefficients C_H and C_E are sometimes presented in the form that depends on the bulk Richardson number Ri_b

$$Ri_b = \frac{gz(\theta_v - \theta_{vs})}{\theta_v u^2} \qquad (3.46)$$

The variables Ri_b and z/L can be converted into each other, when the stability functions are known (see, for example, Launiainen 1995).

The above presented consideration does not include effects of the sea surface waves. In a wave boundary layer, part of the shear stress is replaced by momentum flux carried by pressure fluctuations to surface waves (Hasse and Smith 1997). On the other hand, if the air flow is modulated by the wavy surface, the mean wind profile may be distorted. For example, Dittmer (1977) derived two average diabatic wind profiles from GATE – for wave heights below 25 cm and between 25 and 75 cm – and found that there was a more pronounced deformation of the wind profile with increased wave height. Moreover, the measurements of wind profiles, carried out during the JONSWAP experiment (Hasselmann et al. 1973), showed that 1) the profile slope depends on wave energy, but not on mean wind speed, and 2) the wave-influence on the profile is confined mainly to the lower heights which are comparable to the wave height (Kruegermeyer et al. 1977; Hasse et al. 1978b).

To explicitly separate the relative influences of mean, wave, and turbulence components of the wind field, one can decompose the instantaneous horizontal and vertical wind velocity as

$$a = \bar{a} + \tilde{a} + a' \qquad a = (u, w) \qquad (3.47)$$

where $\bar{a}$ is the time-average component, $\tilde{a}$ stands for the periodic wave-induced component, and a' is the turbulence component of the motion (e.g. Anis and Moum 1995). Time averages must be performed over time scales much larger than the characteristic wave period. Assuming that the mean, the periodic wave-induced, and the turbulence components of the motion are uncorrelated, one can formulate the constant stress approximation as follows:

$$-\rho(\overline{u'w'} + \overline{\tilde{u}\tilde{w}}) = \text{const} = \tau \qquad (3.48)$$

The problems related to the wave-induced momentum flux $\overline{\tilde{u}\tilde{w}}$ are considered in Chapter 4.

3.2.6   Clouds in the Boundary Layer

Marine low-level clouds cover a large area of the World ocean surface. They are very important for the radiation budget and play a significant role in the surface energy budget and in the water balance of the atmosphere. These clouds determine also the vertical structure of the turbulent fluxes of momentum, moisture and heat. There are two major forms of the boundary layer clouds: stratocumulus clouds and cumulus clouds (Jonas 1993). Solid stratocumulus cover extensive areas of the oceans over the cool water; cumulus clouds are mainly observed in trade-wind regions. A strong correlation between regions of strong radiative cooling and regions of extended stratocumulus cloud cover in the marine boundary layer was derived from the climatological data (Bretherton 1993). Stratocumulus and boundary-layer cumulus have many features in common and very often the transition from one form to another takes place. It is known from observations that there is also a climatological transition from nearly solid relatively shallow subtropical stratocumulus to trade cumulus clouds with lower fractional cloud cover and a deeper boundary layer (Bretherton 1993).

Marine stratocumulus clouds have been intensively studed within frame of recent field programs FIRE (First International Satellite Cloud Climatology Regional Experiment, 1987) and ASTEX (the Atlantic Stratocumulus Transition Experiment, 1992). Selected FIRE and ASTEX results relevant for the development of cloudy boundary layer models are presented by Albrecht et al. (1988), Albrecht (1993), Bretherton (1993) and Tjernström and Rogers (1996). It was found from experimental and model investigations that the list of important physical processes which control the structure and type of stratocumulus includes, in particular, cloud top entrainment instability, diurnal turbulent decoupling, microphysics and drizzle.

Cloud-topped boundary layers are usually capped by warm, dry air and there is the tendency for the cloud to dissipate due to the entrainment of this air into the boundary layer. Negatively buoyant downdrafts produce additional turbulent kinetic energy (TKE) which can enhance mixing and entrainment (Stull 1988). The additionally entrained air can then become unstable and again produce more TKE and cause more entrainment. This positive feedback process can lead to rapid breakup and evaporation of the cloud. On the other hand, one could expect that such a mechanism of top-down convection, which causes enhanced turbulence, might lead also to enhanced momentum transfer.

It is also known (Johnson 1993) that boundary layers capped by stratocumulus clouds demonstrate large diurnal variations. Long wave cooling from the top of the cloud produces TKE that forms large eddies which transport water vapour up from the sea surface to the cloud layer. During the day, solar absorption by the cloud reduces the effect of the long wave cooling and consequently, the size of the vertical eddies. If the surface heat and water vapour fluxes are not sufficiently strong, as it takes place in near-neutral surface conditions, the heating of the lower parts of the cloud may produce a secondary inversion between the cloud base and surface (Tjernström and Rogers 1996). Thus, the surface layer becomes decoupled from the cloud and sub-cloud layer (in the sense that the turbulent fluxes are severed), and this cuts off the moisture supply to the cloud. The entrainment of dry air will tend to thin the cloud and even break up. On the other hand, moisture build up in the well mixed surface layer can lead to conditional instability which produces small cumulus clouds at the top of the surface layer that can grow and penetrate the stratocumulus layer. If the boundary layer becomes deep enough, then it may remain decoupled all the time. In this case, the cumulus clouds will be produced even at night and will then restore a recoupling between the surface layer and the cloud. Such a mechanism may transport enough moisture from the surface layer to the top of the boundary layer to maintain or thicken the stratocumulus. The field experiments, carried out over quite different regions of the ocean, such as FIRE (Moyer and Young 1993), ASTEX (Rogers et al. 1995) and the North Sea stratocumulus study project (Nicholls 1989) demonstrated that decoupling of the surface layer from the cloud and sub-cloud layer is not a very frequent feature of the marine atmospheric boundary layer.

The aerosol characteristics of the boundary layer play an important role in microphysics and radiative transfer of the cloud layer. The cloud condensation nuclei govern the size and number of cloud drops. Drizzle production can be an effective decoupling mechanism due to evaporative cooling and can also limit the cloud liquid water. The presence of drizzle is highly correlated with cumulus–stratocumulus interactions which has a modifying effect on the reflectivity of the stratocumulus (Johnson 1993).

3.3 Oceanic Upper Boundary Layer

3.3.1 Vertical Structure

The structure of the oceanic upper boundary layer can consist of at least three layers (Anis and Moum 1992). The oceanic analogue to the atmospheric viscous sublayer is the cool skin which has an average thickness of a few millimetres and very large temperature gradients (Khundzhua et al. 1977). Dynamics of this layer is mainly driven by the surface wave breaking, Langmuir circulations and shear due to the wind stress and wave drift. Turbulence here is an intermediary in the transfer of momentum, heat and salt between ocean and atmosphere. Below the oceanic surface layer two essentially different layers are observed. First, the upper quasi-homogeneous well-mixed layer includes large scale convective eddies with size of an order of the whole layer. The presence of these coherent structures causes the existence of certain differences in modelling upper ocean dynamics. In particular, the effect of "negative viscosity" may be encountered (Muraviev and Ozmidov 1994). Second, the underlying thermocline is characterized by an abrupt increase of density with depth and, consequently, by a very stable stratification. Turbulence exists in the thermocline and not fully developed but intermittent (Kraus 1977; Monin and Ozmidov 1981; Gargett 1989). The internal waves are very often observed here.

3.3.2 Turbulence

One can note the following major mechanisms of oceanic turbulence production: shear instability, internal wave breaking, double diffusion, and deep convection (Monin and Ozmidov 1981; Large et al. 1994). To calculate the vertical turbulent fluxes of momentum, heat and salt, corresponding to the first three listed processes, the K-theory closure Eq. (3.4) is widely applied. It is assumed (Large et al. 1994) that the eddy diffusivity coefficient can be parametrized as a sum of coefficients characterizing a separate process

$$K_a = K_a^s + K_a^w + K_a^d \qquad (3.49)$$

where a stands for momentum, temperature or salinity. The deep convection mechanism requires a more complicated approach.

1. An instability caused by the vertical gradients of the drift currents velocity. This shear-generated turbulence develops in the whole upper layer of the ocean as the result of direct action of the wind on the sea surface. Since Reynolds numbers for the drift currents are very large (of order 10^7) and significantly exceed the critical Reynolds number (of order 10^3) this kind of turbulence is produced practically everywhere in the World ocean. The turbulent mixing of the oceanic upper layer takes place in the density-stratified sea water when the vertical velocity shear overcomes the stabilizing effect of the buoyancy gradient. This process is characterized by the gradient Richardson number

$$Ri = -\frac{g}{\rho} \frac{\partial \rho / \partial z}{|\partial \mathbf{V} / \partial z|^2}$$

where the vertical coordinate z is positive up, ρ is the water density, and $\mathbf{V}$ is the horizontal current velocity vector with the components u and v. Shear instability occurs when Ri is below some critical value Ri_0. According to oceanic field measurements, Ri_0 is generally higher than the theoretical value of 0.25 and varies from 0.4 to 1 (Large et al. 1994). The eddy diffusivity coefficients K_a^s are often chosen as depending on the gradient Richardson number Ri and are the same for momentum, heat and salt. For example, the parametrization suggested by Large et al. (1994) reads

$$K_a^s = \begin{cases} K_{\max} & \text{for } Ri \leq 0 \\ K_{\max}[1 - (Ri/Ri_0)^2]^3 & \text{for } 0 < Ri < Ri_0 \\ 0 & \text{for } Ri_0 < Ri \end{cases} \qquad (3.50)$$

where $K_{\max} = 50 \times 10^{-4}$ m^2/s and $Ri_0 = 0.7$.

2. A breaking of surface waves and hydrodynamical instability of the wave motions in the oceanic upper layer. The superposition of internal waves increases shear and consequently, the Richardson number decreases. The eddy diffusivity coefficient K_a^w, describing the effect of mixing due to internal wave breaking, seems to be small and depends mainly on the internal wave energy. It is found that the internal wave diffusivity for heat and salt is about 0.1×10^{-4} m^2/s (Ledwell et al. 1993), and for $Ri_0 < R_i$ the wave momentum transfer is expected to be from 7 to 10 times more effective (Peters et al. 1988).

3. Double-diffusive convection. This is a very important process by which the heating (or cooling) and the salting (or freshening) at the sea surface become distributed in the oceanic boundary layer. Such ocean-mixing mechanism is associated with the fact that the density of sea water ρ is determined by the temperature T and the salinity S with large (of two orders) differences in their molecular diffusivities. During this process a statically unstable vertical distribution of one property can be balanced by a distribution of the other. While the resulting density distribution is stable, small-scale instabilities can lead to release of gravitational potential energy from the unstable component. The mixing of momentum from double-diffusive convection is found (e.g. Large et al. 1994) to be the same as for salt $(K_M^d = K_S^d)$, but the temperature diffusivity K_T^d is quite different. Depending upon whether the resulting motions are driven by energy stored in the component of the higher (T) or lower (S) diffusivity, two basic types of convective instabilities occur (the "diffusive" and "finger" forms) which differ in the relative efficiency of heat and salt transport (Monin and Ozmidov 1981; Turner 1985; Gargett 1989). To quantify these two regimes, the following ratio of the density flux due to heat to that due to salt is used:

$$R_f = \frac{bF_T}{aF_S} \tag{3.51}$$

where F_T and F_S are the heat flux and the salt flux, respectively; $a = -\rho^{-1}(\partial\rho/\partial T)$, $b = \rho^{-1}(\partial\rho/\partial S)$. It was found that $R_f < 1$ for the diffusive case and $R_f > 1$ for the fingering case (Gargett 1989). The instability growth rate and the flux ratio R_f are functions of the stability parameter

$$R_\rho = \frac{\beta \partial S/\partial z}{\alpha \partial T/\partial z} \tag{3.52}$$

The diffusive instability occurs in regions where cold, dilute water lies above warm salty water. As the result of this kind of instability, weakly stirred convective layers separated by much thinner interfaces of strong molecular transport are formed. A part of the potential energy released from the heat field during thermal convection is converted to kinetic energy to transport salt upward. It was found that the flux ratio R_f decreases with an increase of the stability parameter and becomes nearly constant $(R_f \approx 0.15)$ for $R_\rho > 2$. Layered structures in T and S are relatively rare in the World ocean. They are mainly observed in polar regions where circumstances are favourable for the diffusive instability (Gargett 1989). For example, the following parametrization is suggested (Fedorov 1988; Large et al. 1994):

$$K_T^d = 0.909\nu_w \exp(4.6\exp[-0.54(R_\rho - 1)])$$

$$K_S^d = \begin{cases} K_T^d(1.85R_\rho^{-1} - 0.85) & \text{for } 1 < R_\rho \leq 2, \\ 0.15K_T^dR_\rho^{-1} & \text{for } R_\rho > 2, \end{cases} \tag{3.53}$$

where ν_w is molecular viscosity of the water.

The salt fingering instability occurs in regions where warm salty water lies above cold, dilute water. Extended thin columns of fluid moving vertically in both directions form the salt finger pattern of the resulting motion. Each upward-moving finger is surrounded by downward-moving fingers, and vice versa. The downgoing fingers lose heat and become more dense, whereas the upgoing fingers gain heat and become less dense. The potential energy is now derived from the salt field. It was experimentally discovered that over a wide range of conditions the flux ratio R_f is nearly constant and close to 0.56 (Turner 1985). Finger activity is strongest at $0.5 < R_\rho < 1$ (Gargett 1989). Large regions of the subtropical and tropical oceans are favourable for the salt fingering process. Mixing due to this process can be parametrized, for instance, in the form (Large et al. 1994)

$$K_S^d = \begin{cases} K_f \left[1 - \left(\dfrac{R_{\rho 0}}{R_\rho} \right)^2 \left(\dfrac{1 - R_\rho}{1 - R_{\rho 0}} \right)^2 \right]^3 & \text{for } R_{\rho 0} < R_\rho < 1 \\ 0 & \text{for } R_\rho \le R_{\rho 0} \end{cases}$$

$$K_T^d = 0.7 K_S^d \tag{3.54}$$

where $K_f = 10 \times 10^{-4}$ m^2/s and $R_{\rho 0} = 0.526$.

4. Convection due to unstable density stratification. Convection in the upper part of the ocean can be caused by cooling of the sea surface in cold seasons or by accumulation of salt due to intensive evaporation of the sea water. This process is most pronounced in the form of open-ocean (deep) convection. During deep convection localized bursts of violent vertical mixing transport and mix water over several hundred metres and short time periods (Aagard and Carmack 1989). Observations indicated that vertical velocities can exceed in this case 0.1 m/s. There are identified three phases associated with deep convection: preconditioning, violent mixing, and sinking and spreading (Alves 1995).

During the preconditioning phase the vertical static stability is reduced so that the formation of convectively unstable layers can begin. The mechanisms which may trigger a deep convection are instabilities of a different kind, e.g. double diffusion, convective instability (heating, cooling, evaporation or brine release), thermobaric instability (caused by depth dependence of the equation of state), baroclinic instability, and Kelvin–Helmholtz instability (Chu 1991). The main areas of deep convection are reviewed by Alves (1995) and include the Labrador, Greenland and Baltic Sea, the eastern Mediterranean, the Adriatic Sea and the Gulf of Lions in the north-western Mediterranean, the Weddel Sea and the Bransfield Strait at the tip of Antarctica, the subpolar oceans near the ice edge and in polynyas.

The violent mixing phase is the period when dense plumes develop and gradually extend deeper into the ocean transporting and mixing the sea water in the vertical. These plumes have vertical scales of 1–2 km and horizontal scales of 100–1000 m. It was found from observations that the plumes form a nearly homogeneous patch of the sea water which is called a chimney. A density-driven rim current of the local Rossby radius of a few kilometres exists between the chimney and the surrounding fluid. Since the vertical static stability is low, the rim current is baroclinically unstable. This leads to the development of baroclinic eddies which advect relatively lighter water from the surroundings into the chimney and limit the convective activity (Madec and Crepon 1991). Thus, the eddies and plumes are responsible for the two vertical mixing processes which determine this stage of deep convection.

The sinking and spreading phase begins when surface forcing turns off. The dense water sinks under gravity, and water from outside the chimney restores the stratification of the near-surface layer (Jones and Marshall 1993). During a few days the chimney breaks up into geostrophically adjusted fragments ("cones") with a spatial size of several kilometres. The possible mechanisms for the breakdown of the chimney are also mixing by internal waves, topography, and shear in the large scale cyclonic circulation (Killworth 1983).

Oceanic deep convection resembles very much convection in the atmospheric boundary layer. In both media, the mean stratification is slightly stable, and a countergradient heat flux may exist. This means that the collective effect of an ensemble of oceanic plumes might be also parametrized on the basis of a nonlocal closure theory.

3.3.3   The Oceanic Ekman Layer

By definition, the ocean drift current is driven only by the wind stress at the sea surface when the pressure gradient in the ocean is neglected. For steady state and horizontally homogeneous conditions, the equations of motion are of the Ekman type Eq. (3.18) but reduced to

$$K\frac{d^2u}{dz^2} + fv = 0$$

$$K\frac{d^2v}{dz^2} - fu = 0 \tag{3.55}$$

It is convenient to orientate the x-axis in the surface stress direction. Assuming that the current vanishes deep in the ocean, and the stress is continuous across the air–sea interface, the boundary conditions subject to Eq. (3.55) read

$$K\frac{du}{dz} = \epsilon^2 u_*^2 \qquad K\frac{dv}{dz} = 0 \quad \text{at} \quad z = 0 \tag{3.56}$$

$$u \to 0 \qquad v \to 0 \quad \text{as} \quad z \to -\infty \tag{3.57}$$

where u_* is the air friction velocity, and ϵ^2 is the ratio of the air density to the water density. The solution to the problem (3.55) to (3.57), obtained first by Ekman (1905), reads

$$u = \frac{\epsilon^2 u_*^2}{\sqrt{Kf}} \exp\left(\frac{z}{h_E}\right) \cos\left(\frac{z}{h_E} - \frac{\pi}{4}\right)$$

$$v = \frac{\epsilon^2 u_*^2}{\sqrt{Kf}} \exp\left(\frac{z}{h_E}\right) \sin\left(\frac{z}{h_E} - \frac{\pi}{4}\right) \text{sign } f \qquad (3.58)$$

where, as for the atmosphere

$$h_E = \sqrt{2K/|f|}$$

but with K applied to ocean values. For $K = 12.5 \times 10^{-4}$ m^2/s and $f = 10^{-4}$/s, the value of $h_E = 5$ m.

As is seen from (3.58), the surface current in the northern (southern) hemisphere is 45° to the right (left) of the surface stress, and consequently, it is parallel to the atmospheric geostrophic wind (compare (3.58) with (3.22) from Section 3.2.4). Taking Eq. (3.27) into account, $u_*^2 = C_D U_a^2$, where C_D and U_a are the drag coefficient and wind speed, say, at the height 10 m, one can calculate from (3.58) the surface current speed

$$U_o = \frac{\epsilon^2 C_d U_a^2}{\sqrt{Kf}} \qquad (3.59)$$

To estimate U_o, let us use $C_D = 10^{-3}$, $\epsilon^2 = 10^3$, $U_a = 10$ m/s, and $Kf = 9 \times 10^{-8}$ m^2/s as a typical values. Their substitution into the expression (3.59) leads to $U_o = 1/3$ m/s. Since $U_a \sim G$, where G is the geostrophic wind speed, this means that roughly the surface drift current is 30 times weaker than the geostrophic wind.

Since

$$\tan \theta = \frac{v}{u} = \tan\left(\frac{z}{h_E} - \frac{\pi}{4}\right) \text{sign } f$$

and z is negative, the drift current turns with depth to the right (left) of the surface current direction in the northern (southern) hemisphere. Thus, large-scale horizontal divergence should be expected in the ocean under atmospheric regions of large-scale horizontal convergence, and vice versa. This means that synoptic low (high) pressure systems in the atmosphere induce upwelling (downwelling) motion in the underlying oceanic domains (e.g. Stull 1988).

3.4 Processes at the Air–Sea Interface

Qualitatively new effects in the air–sea interaction are detected in the case of strong winds (nominally, above 15 m/s). The development of the surface waves leads, at high wind velocities, to significant change of the sea surface roughness length. The air–sea interface is disrupted under stormy conditions, during which intensive spray is detected in the atmospheric surface layer, air bubbles are found in the water, and as a result, can be treated as a two-phase flow. The contribution of spray droplets to the sea surface heat and moisture budgets, as well as to the sea surface aerosol flux, is found to be important (Woodcock 1955; Toba 1965a,b, 1966; Borisenkov 1974; Wu 1979, 1990; Kazakov and Lykossov 1980; Bortkovskii 1987;

Ling 1993; Andreas et al. 1995). Shearing of wave crests by wind, aerodynamic suction at the crests of capillary waves, and bursting of air bubbles at the sea surface are principal mechanisms of sea spray production (Wu 1979). It was shown that bubble bursting is the primary mechanism of producing spray due to a large volume of air entrained by breaking waves. The bubbles in the subsurface plume form a whitecap and produce two distinct types of droplets (film droplets and jet droplets) when they burst (Andreas et al. 1995). The film droplets are generated when the upper, protruding surface of a bubble thins by down-slope drainage and shatters. The jet droplets are produced by a microscopic column of water which forms in the centre of the cavity resulting from the rupture of the bubble film. Jet droplets dominate the spectrum of spray droplet flux in the 1–100 μm radius range with initial vertical velocities of 5–20 m/s, whereas film droplets contribute mainly at radii below 0.1–10 μm (Smith et al. 1996a). At high wind speeds spume drops, mainly larger than 40 μm, are blown off the crests of the spilling waves. It was found that the maximum ejection height for jet droplets is 18 cm (e.g. Wu 1979), but turbulence in the atmospheric boundary layer can carry the small bubble-derived droplets higher (de Leeuw 1986) up to cloud heights, where they contribute to cloud condensation nuclei and form a salt inversion below cloud base (Blanchard and Woodcock 1980).

An analysis of the processes of heat and moisture transfer in spray clouds has resulted in the conclusion (Borisenkov 1974) that it is necessary to solve problems related to the formation and time evolution of the spray cloud and to determination of the thermal regime of an individual droplet. Initially, the droplets have the same properties as the sea surface. After ejection many of them quickly return to the sea but being in the air they attempt to adapt to the air conditions. All spray droplets reach thermal equilibrium within 1 s. The same time is required for the smallest droplets to reach moisture equilibrium, whereas the largest droplets need for this an hour or more (Andreas et al. 1995). The evaporation of droplets contributes to the water vapour flux, decreases liquid water flux, and cools a region where the droplet cloud is formed. In particular, it was estimated (Andreas et al. 1995) that in a 20 m/s wind, with an air temperature of 20°C, a water temperature of 22°C, and a relative humidity of 80%, the sea spray contributes 150 W/m to the latent heat flux and 15 W/m to the sensible heat flux. Additionally, the transfer of radiation between the droplets and the environment can significantly increase the consumption of sensible heat by the droplets. The resulting vertical distribution of droplets and sea salt is one of the factors determining the optical properties of the marine atmosphere. However, from HEXOS field and laboratory experiments it was derived that there is a negative feedback: in a "droplet evaporation layer" close to the surface the evaporating droplets modify the temperature and water vapour gradients and reduce the turbulent sensible and latent heat fluxes (Hasse and Smith 1997).

Effects of sea spray on the wind profile. The sea spray droplets can also influence the density stratification and consequently, the parameters of the surface layer. It was shown (Pielke and Lee 1991) that the water loading effect on the surface-layer wind profile during strong wind conditions can be significant in white-cap sea-spray situations. Thus, the turbulence statistics (especially, the friction velocity and drag coefficient) near the sea surface may be remarkably altered.

The water droplets are embedded into air flow, and if their concentration is large enough, the flow must be considered as multi-phase flow and the drag reduction effects must be included into a parametrization scheme. Assuming that air and droplets form a two-phase fluid, the density ρ of the mixture may be expressed by

$$\rho = \rho_a(1 - S) + \rho_w S = \rho_a(1 + \epsilon_e^2 S) \tag{3.60}$$

where $S(z)$ is the volume concentration of droplets, and ϵ_e^2 indicates the relative excess of the droplet density over the air density:

$$\epsilon_e^2 = (\rho_w - \rho_a)/\rho_a$$

The following expression for the stability parameter L can be derived for these conditions (Wamser and Lykossov 1995):

$$L = \frac{\rho_a(1 + \epsilon_e^2 S)u_*^3}{\kappa g[\rho_a' w'(1 - S) + \rho_a \epsilon_e^2 \overline{S'w'}]} \tag{3.61}$$

In the absence of droplets ($S \equiv 0$ and $\overline{S'w'} \equiv 0$) Eqs (3.31) and (3.61) lead to the expression for the stability parameter L derived by Monin and Obukhov (1954):

$$L = -\frac{\overline{\theta_v} u_*^3}{\kappa g \overline{\theta_v' w'}}$$

In the case of thermally neutral stratification ($\overline{\theta_v' w'} \equiv 0$) the parameter L in the presence of droplets has the form:

$$L = \frac{(1 + \epsilon_e^2 S)u_*^3}{\kappa g \epsilon_e^2 \overline{S'w'}} \tag{3.62}$$

Making an eddy diffusion assumption for the concentration flux of droplets (3.4) and using the balance relation of steady-state sea spray

$$K_S \frac{\partial S}{\partial z} = -w_f S \tag{3.63}$$

where $w_f > 0$ is the falling velocity of droplets, the expression (3.62) can be rewritten as follows:

$$L = \frac{(1 + \epsilon_e^2 S)u_*^3}{\kappa g w_f \epsilon_e^2 S} \tag{3.64}$$

It is seen from this equation that the stability parameter is not constant with height, and, consequently, the Monin–Obukhov similarity theory cannot be applied in its traditional form. A comprehensive review of the modified similarity theory based on the height-dependent stability parameter is presented by Stull (1988).

Since L is positive, the density stratification is stable, and one can use the log–linear universal function (Kondo et al. 1978). Assuming that $\epsilon_e^2 S \ll 1$, Eq. (3.29) reads in this case as

$$\frac{\kappa z}{u_*}\frac{\partial u}{\partial z} = 1 + \frac{\hat{\beta}g\kappa w_f z \epsilon_e^2 S}{u_*^3} \tag{3.65}$$

The value of the parameter $\hat{\beta}$ in this formula is different in this case from that found for scaling with the surface density flux. Kondo et al. (1978) found from observations of the atmospheric boundary layer that the most appropriate value of $\hat{\beta}$ for the height-dependent L equals 7, while a value of $\hat{\beta} = 4.7$ was derived by Businger et al. (1971). The equation for the droplet concentration (3.63) transforms to[3]

$$\frac{\partial S}{\partial z} + \frac{w_f S}{\kappa u_* z}\left(1 + \frac{\hat{\beta}\kappa g w_f z \epsilon_e^2 S}{u_*^3}\right) = 0 \tag{3.66}$$

The solution to this equation, subject to the boundary condition $S = S_r$ at a reference height $z = z_r$, has the following form (Taylor and Dyer 1977):

$$S(z) = \frac{(1-\omega)S_r(z/z_r)^{-\omega}}{1 - \omega + \hat{\alpha}\omega^2[(z/z_r)^{1-\omega} - 1]} \tag{3.67}$$

where

$$\omega = \frac{w_f}{\kappa u_*}, \qquad \hat{\alpha} = \frac{\beta g \kappa^2 z_r \epsilon_e^2 S_r}{u_*^2} \tag{3.68}$$

For $\omega \to 1$, the solution (3.67) converges to

$$S(z) = \frac{S_r(z/z_r)^{-1}}{1 + \hat{\alpha}\ln(z/z_r)} \tag{3.69}$$

Given the concentration profile (3.67), (3.69), integration of Eq. (3.65) from z_r to z results in the following wind profile (Taylor and Dyer 1977):

$$u(z) = u_r + \frac{u_*}{\kappa}\ln\left(\frac{z}{z_r}\right) + \frac{u_*}{\kappa}\begin{cases} \omega^{-1}\ln\left(1 + \frac{\hat{\alpha}\omega^2}{1-\omega}\left[\left(\frac{z}{z_r}\right)^{1-\omega} - 1\right]\right) & \text{for } \omega \neq 1 \\[2ex] \ln\left[1 + \hat{\alpha}\ln\left(\frac{z}{z_r}\right)\right] & \text{for } \omega = 1 \end{cases} \tag{3.70}$$

If $z_r = z_0$, $u_r = 0$, and $\alpha = 0$, Eq. (3.70) coincides with Eq. (3.26). Comparing these equations, one can see that effects of the sea spray on the wind profile are described by the additional logarithmic term. Since $\hat{\alpha}$ and ω are positive, the surface winds should be stronger, for the same friction velocity u_*, in the case of sea spray. Consequently, the drag coefficient should be lower than without spray.

To estimate this effect quantitatively, let us use $u_* = 0.4$ m/s^{-1} as a typical value of the friction velocity, $r = 10\,\mu$m as a typical value of the droplet radius, and

---

[3] Generally speaking, the eddy diffusivity for momentum and droplets may be different, $K \neq K_S$, but any difference are neglected here.

$\epsilon^2 = 10^3$. Applying the Stokes law, $w_f = 2\epsilon^2 gr^2/9\nu$, one can obtain $w_f = 0.016$ m/s and $\omega = 0.1$ from Eq. (3.68). Taking $z_r = 0.18$ m, $z_0 = 10^{-4}$ m, and assuming that the wind profile between z_0 and z_r is not disturbed by sea spray, one can calculate from Eq. (3.70) that for S_r ranging from 10^{-5} to 10^{-4} the wind at 10 m is increased about 4–35% with regard to the neutral case without spray. It is necessary to note, however, that the wave-induced stress, which has an opposite influence on the wind profile, may sufficiently compensate the possible drag reduction effect.

Effects of rain on the surface stress. There are two aspects of this problem. First, when the rain originates in the boundary layer and/or falls through, it interacts with turbulence and may affect the ageostrophic wind component. Second, the rain falling over the ocean may change the sea surface state and exchange momentum with it. We consider here this aspect only.

It was found experimentally (Poon et al. 1992) that the low-frequency waves are attenuated by the rain, and most of the damping occurs in the frequency region of 2–5 Hz. Theoretical studies (e.g. Le Méhauté and Khangaonkar 1990) indicated that the damping rate depends on rain intensity, falling velocity, and inclination of the drops. In particular, observations showed an increase in the damping rate with rain intensity. Reynolds (1900) suggested that the falling raindrops cause the formation of vortex rings which enhance the vertical mixing of the surface water layer. At present time, it is found that under various rain conditions, a mixed layer with depth between 5 and 20 cm is formed with the effective eddy viscosity at least of an order of magnitude greater than the water molecular viscosity (e.g. Katsaros and Buettner 1969; Poon et al. 1992). Van Dorn (1953) found that during a period of moderately heavy rainfall ($\sim$0.5 cm/h) the magnitude of measured stress was significantly larger in comparison with the case without rainfall.

Raindrops move over an ocean at the same velocity as the wind, and when the drops fall and reach the sea surface, they contribute to a rain-induced surface stress. The surface stress τ_r produced by rainfall can be written as follows (Caldwell and Elliott 1971):

$$\tau_r = \rho_w U_s R \qquad (3.71)$$

where ρ_w is the density of rainwater, U_s is the horizontal speed of the rain drop at impact, and R is the rainfall rate. The ratio between the rain-induced and wind stresses at the sea surface was found to be about 7–25% (Poon et al. 1992). The contribution of τ_r into the total stress is higher for lower winds and heavier rain conditions. Contrary to the damping mechanism, the rain-induced stress contributes to the wave growth. Thus, in the presence of wind, the increase in damping with the rain intensity is partially compensated by the increase in the surface stress due to the rain-induced component.

4 Ocean Wave Spectra and Integral Properties

N. E. Huang, Y. Toba, Z. Shen, J. Klinke, B. Jähne and
M. L. Banner

4.1 Introduction

Ocean *wind-waves* play a dominant role in air–sea interaction processes, and, despite much progress, understanding the complex underlying physical processes involved in ocean wave dynamics continues to present a substantial scientific challenge. Two tasks confront us now: first, to establish a viable theory of *wind-wave* generation, and second, to describe the ocean surface quantitatively. Although the emphasis of this chapter is on *wind-wave* spectra, a brief description of *wind-wave* generation theory and the concept of *local equilibrium with the wind* is necessary to appreciate the uncertainty and problems facing us.

There are two existing theories on *wind-wave* generation; both published in 1957. The first one is by Phillips (1957), who proposed resonant interaction between the waves and the pressure field associated with the wind as the generation mechanism. Lacking a feedback mechanism, the original result was only valid for the initial stage of the wave field where the growth is linear. Phillips's theory was subsequently modified by Miles (1959a,b) to include partial feedback and resulted in an exponential growth in the principal stage. The second theory is due to Miles (1957, 1959a,b, 1962, 1967), who proposed the instability of an inviscid shear flow as the main energy transfer mechanism, which causes the waves to grow exponentially. Because this model requires the existence of a wave field at the beginning, it explains more of a *wind-wave* interaction process rather than the actual generation phase (see, for example, Phillips 1977).

Ever since the proposal of these theories, attempts have been made to combine and generalize them. An important development is to establish the initial wave growth as a coupled air–water instability (Valenzuela 1976; Kawai 1979; Blennerhassett and Smith 1987; Caponi et al. 1992; Belcher and Hunt 1993; Belcher et al. 1994), which is shown to be the main mechanism for short *wind-wave* generation and growth. Yet in all these models, the effects of turbulence in the air are mostly neglected except in determining the mean wind profile; therefore, these new developments can be regarded as an extension of the instability theory. All the theories involve a number of assumptions; unfortunately, comparisons between theoretical and experimental results produce only order of magnitude agreements (see, for example, Snyder et al. 1981; Hasselmann and Bösenberg 1991). Possible causes of agreements and discrepancies are many; some of them are listed as follows.

First, all the theories are essentially linear models. The wind is considered to be the energy source to the waves, but the wave motions do not enter into the model in any way other than providing the surface perturbations to the wind field. Laboratory (Mitsuyasu and Honda 1982; Toba 1985) and numerical (Hasselmann and Hasselmann 1985; Young et al. 1987) studies have shown that the spectral growth of the waves is influenced strongly by the nonlinear wave–wave interactions and the turbulence structure in the wind. Although the nonlinear wave–wave interaction process does not generate nor annihilate energy by itself, the interactions can transfer energy in the wave field from one wavenumber to different ones. This alone is critical in maintaining the continuous growth of the wave field. Can the generation mechanism and the nonlinear wave–wave interactions be simply superimposed after the results are computed separately?

Secondly, the ranges of validity for the models are different. As stated, the Phillips model works for the initial range only; one will have to consider the feedback to extend the range of validity. The Miles model, on the other hand, works well only for the range near $c/u_* = 10$. As the wave growth continues, eventually the matched layer of the Miles will shift away from the highest curvature range of the wind profile and render the mechanism ineffective. Under such conditions, what are the most effective mechanisms to transfer energy from the wind to the *wind-waves*? Could it be the resonant interaction mechanism proposed by Phillips? How to account for the full feedback with the full turbulent structure in the air and breaking *wind-waves* of the water surface?

Thirdly, there are also questions on the simplification assumptions in the theoretical models. Turbulence is neglected, other than its effect on the mean wind profile. Theoretical studies by Miles (1967), Townsend (1972), Davis (1969, 1970, 1972, and 1974), McLean (1983), and Al-Zanaidi and Hui (1984) have all shown that the turbulence effect is important, but the computation is sensitive to the turbulence models used. Numerical models produced with super-computers (see, for example, Akylas 1984; Katsis and Akylas 1987) may shed some light, but physical parameters will still come from accurate observations. Should we also consider the effects of an uneven distribution of the surface stress and roughness along the surface in the *wind-wave* generation processes? Numerical models (see, for example, Taylor and Gent 1978; Al-Zanaidi and Hui 1984) seem to suggest that the *wind-wave* growth rate depends critically on whether the flow is aerodynamically smooth or rough. But unfortunately, we do not know for sure what is the roughness distribution relative to the phase of the underlying waves. We will return to this point later. Sharp crested and breaking *wind-waves* will cause more flow separations and, therefore, more turbulence. Are these effects fully accounted for in the mean profile?

Finally, the two-dimensional average used in both the Phillips and Miles theories rules out the case of wind and wave directions that are not exactly parallel, and the genuinely three-dimensional nature of the real wave field. The non-parallel cases turn out to be more common, and they cannot be represented by using a simple cosine wind component.

With this brief summary, we can see that there are uncertainties in how *wind-waves* actually gain their energy. How does the nonlinearity in the *wind-wave* field influence the energy distributed directionally. It turns out that these uncertainties can cause difficulties in quantitative description of the *wind-wave* fields. In the description of the ocean surface, there are two approaches: the probabilistic method and the spectral analysis together with the empirical relationships among the various integrated parameters of the spectra. The probabilistic approach has been summarized recently by Huang et al. (1990a). So we will concentrate here only on the spectral analysis.

The dominant wind seas are characteristically modulated in groups, with the largest *wind-waves* within the group involved in breaking from time to time if the wind forcing is sufficiently strong. A powerful method to describe these dominant *wind-waves* is through the empirical relationship among the various integrated parameters (Toba 1972, 1978; Toba and Koga 1986; and Ebuchi et al. 1992). This part of the spectrum can also be modelled by the Wallops spectral form (Huang et al. 1980). However, the wind driven sea surface also supports a wide range of surface disturbances that have wavelengths and time scales much shorter than the dominant *wind-waves*, and have much wider directionality. These shorter *wind-waves* are believed to be very important in air–sea interaction processes involving, for example, the exchange of momentum between the wind and the sea surface.

The simplest way to include the collection of smaller scales as well as the dominant *wind-waves* is to use a spectral description, i.e. to assume a superposition of Fourier modes, noting that this requires the two horizontal spatial directions to allow for the directionality of the *wind-waves*. Also, it is worth mentioning that the underlying linear character of the Fourier mode decomposition does not imply that the ocean waves themselves are a linear system, and that this description is able to describe nonlinear waves, but with a non-unique dispersion relation. Also, it is noted that the description in terms of wave spectra averages out the local modulational structures, presenting the averaged wave energy content of the *wind-wave* field with respect to scale and direction.

The most commonly used waveheight frequency spectral forms for $F(\omega)$ have been reviewed recently by Huang et al. (1990b). Those are the most readily available spectra as they are calculated from local waveheight records measured by a wave buoy. However, such single point wave data cannot provide information on spatial scales, nor on *wind-wave* directionality. Spatial resolution of the *wind-wave* field requires spatially extensive observations, which have been traditionally difficult to obtain. However, low resolution directional measurements are accessible by using a slope-sensitive pitch and roll buoy to observe the wave slope, a two-dimensional vector field. The importance of the directionality of the *wind-wave* field has been overlooked until fairly recently, where it is now seen to be of fundamental importance in understanding and relating the shapes of commonly used spectral measures such as the (nondirectional) frequency spectrum $F(\omega)$.

The shorter *wind-waves* occupying the high frequency region of the waveheight frequency spectrum have been observed to conform to power law descriptions. This form of organization parallels the spectral structure noted in turbulence measurements, and has generated much interest. As a result, equilibrium spectral subranges have been proposed for the frequencies above the spectral peak. However, until recently, there was incomplete accord between observations and proposed equilibrium range spectral forms for gravity *wind-waves*. Upon close scrutiny of the available data, spectral forms proposed for the time-frequency domain were not consistent with those proposed in the spatial-wavenumber domain.

The resolution of these inconsistencies has recently been shown to depend primarily on the fact that the directionality of the *wind-wave* field changes with wavelength: as the *wind-waves* get shorter, they become more directional. This has provided the underlying reconciliation. As is explained in greater detail below, based on observed wavenumber and frequency spectra for fetch-limited growth of *wind-waves*, an equilibrium gravity range form was proposed for the slice through the directional wavenumber spectrum in the dominant wave direction. This was used in conjunction with a model for the directional spreading function of the wave energy, for which several forms have been proposed by various investigators. Frequency spectra and one-dimensional wavenumber spectra calculated from the present model now conform closely to observed spectra over a wide frequency range, so that this model appears to provide a plausible explanation of the underlying influences on the frequency spectrum for different gravity wave spectral subranges.

4.2   Equilibrium Ranges for Gravity Waves

The quest for a physical model of a gravity equilibrium subrange for deep water waves had its origins in the pioneering work of Phillips (1958), who proposed that the amplitudes of spectral components above the spectral peak were hard-limited by wave breaking, independent of the wind strength, fetch or duration. This hypothesis led to the well known ω^{-5} power law dependence for the equilibrium gravity subrange of the frequency spectrum proposed by Phillips (1958)

$$F(\omega) = \beta g^2 \omega^{-5} \tag{4.1}$$

where ω is the radian frequency, g is the gravitational acceleration and β is a universal constant. Shortly thereafter, Hasselmann (1963a,b) presented the theoretical basis for weakly nonlinear wave–wave interactions in the wave spectrum. Support for the importance of this mechanism in shaping the wave spectrum has strengthened with the increasing availability of reliable data on *wind-wave* evolution and the associated wind input source function. As a result, (4.1) has been shown to be inadequate in accounting for observed frequency spectra, with β found to decrease with fetch (distance downwind from an effective origin for the wave field) and a frequency dependence closer to ω^{-4}, which is the form proposed by Toba (1973) based on experimental observations and dimensional arguments. Toba pointed out that the spectrum could not exist in isolation from its integral properties (cf. Section 4.3). Therefore, from the critical parameters of u_*, g, and ω, he proposed

$$F(\omega) = \alpha g^2 \omega^{-5} \cdot (u_*\omega/g)^n \qquad (4.2)$$

where α is a universal constant. The form above corresponds to $n = 0$; field data, however, support the form with $n = 1$, as proposed by Toba (1973). The spectral level coefficient still shows an unexplained variability between investigators (see, for example, Phillips 1985). This has motivated recent revisions of the Phillips (1958) equilibrium range model by several authors. The new wavenumber spectra often have the form,

$$f(k_i) \propto u_* g^{-1/2} k_i^{-5/2} \qquad [i = 1, 2] \qquad (4.3a)$$

where $f(k_i)$ is defined by

$$f(k_1) = \int F(k_1, k_2)\mathrm{d}k_2$$

$$F(k, \theta) \propto u_* g^{-1/2} k_i^{-7/2} G(\theta) \qquad (4.3b)$$

Here u_* is the wind friction velocity, $\mathbf{k} = (k_1, k_2)$ is the wavenumber vector with magnitude k and polar angle θ, $G(\theta)$ is the directional distribution function. All spectra have the property

$$\int_0^\infty F(k_1, k_2)\mathrm{d}k_1\mathrm{d}k_2 = \overline{\eta^2}$$

The envisaged spectral range of applicability of these models was for gravity wave components with much shorter wavelengths than the spectral peak waves. The same dimensional groups are not appropriate for the waves strongly influenced by surface tension above the phase speed minimum. This range is considered in Section 4.2. On the observational side, there has been a growing body of observed frequency spectra apparently supporting Eq. (4.2) with $n = 1$ (e.g. see Donelan et al. 1985), although many observed spectra become unreliable due to noise coloration above about a few times the spectral peak frequency ω_p. The interesting hurricane wave observations of Forristall (1981), which show supportive ω^{-4} behaviour out to $O(3\omega_p)$ but then transition to ω^{-5} just above this range, remain unexplained by present equilibrium models. At higher frequencies, reliable measurements from several investigators have shown a range of frequency power law dependences from -5 to -3.3 (see, for example, Banner et al. 1989).

Now it has long been recognized that Doppler shifting effects arising from ambient and wind drift currents, as well as orbital motions of the longer wave components, are a potentially important source for influencing the frequency spectrum, particularly for the higher frequency components whose intrinsic phase speeds are lower (Kitaigorodskii et al. 1975). The extent to which these effects influence measured frequency spectra and their correspondence to the proposed equilibrium spectral form Eq. (4.2) are examined in detail in the discussion below. Wavenumber spectra, not subject to these Doppler distortion effects, are also able to resolve wave directionality. These offer a more insightful way to address the question of equilibrium spectral forms and range of applicability, but at the expense of much greater observational complexity. This has limited the body of measured wavenumber spectra to one much more modest than the body of frequency spectra.

4.2.1   Formulation of Wavenumber Spectral Model

From an analysis of published directional wavenumber slope spectra of Jackson et al. (1985) the derived one-dimensional waveheight wavenumber spectra in the dominant wave direction (k_1) and orthogonal direction (k_2) shown in Fig. 4.1 reveal the strong directional anisotropy of the wave energy propagation near the spectral peak, and the subsequent tendency to merge at higher wavenumbers k_i, $[i = 1, 2]$ of $O(10k_p)$, where k_p is the peak wavenumber, to follow a k^{-n} dependence, with $n \sim 3$. One-dimensional spectra for fetch-limited growth reported by other investigators are also plotted in Fig. 4.1. The spectral level appears to be only weakly dependent on windspeed, but this aspect is not resolved clearly by the available data. In this figure, some attenuation may have occurred in the Jackson et al. (1985) data, due to a wind shift several hours earlier. The corresponding equilibrium range

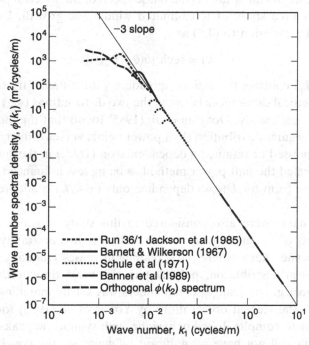

Figure 4.1. One-dimensional wavenumber spectra $f(k_i)$ in the wind direction for fetch limited *wind-wave* growth conditions. The background asymptote shown has a slope of k_i^{-3}. The orthogonal spectrum from Jackson et al. (1985).

model prediction of Eq. (4.3a), i.e. $f(k_1) \sim u_* k_1^{-5/2}$, does not accord with the observed wavenumber dependence. Also, according to the high wavenumber observations of Banner et al. (1989), the observed dependence on the wind speed is much weaker than linear in the observed $O(1\,\mathrm{m})$ wavelength spectral range.

The difference in the two one-dimensional wavenumber spectra near the spectral peak reveal the strong directionality near the spectral peak. Further insight is provided by the directional wavenumber spectrum $F(k, \theta)$ which describes the distribution of wave energy propagation with respect to wavenumber magnitude k and direction θ. Observed wavenumber spectra, if determined from frozen spatial images (such as in stereophotogrammetric methods [e.g. Holthuijsen (1983)]) have an inherent 180° ambiguity, which can be unfolded (as discussed subsequently) using a directional spreading function $D(\theta, k)$. The dependence of the spreading function on k is the crucial element in refining the forms (4.3b) to give a self-consistent picture. A natural form of the spreading function D is defined by

$$D(\theta, k) = F(k, \theta)/F(k, \theta_{\max}) \qquad (4.4)$$

where $\theta_{\max}$ is the dominant wave direction. Several empirical forms have been proposed for $D(\theta, k)$ based on the distribution

$$D(\theta, k) = \cos^{2s}(\theta/2) \qquad (4.5)$$

where the spreading exponent s depends on k/k_p and U/c_p. Here U is the wind-speed at a reference height, and c_p is the phase speed of the spectral peak waves. In a very comprehensive study of fetch-limited wind wave growth, Donelan et al. (1985) proposed a variation to (4.5) as

$$D(\theta, k) = \mathrm{sech}^2[b(\theta - \theta_{\max})] \qquad (4.6)$$

where $b = b(k/k_p)$ controls the rate of spreading with distance from the spectral peak. The geometrical differences between the two distributions (4.5) and (4.6) are not large. More significantly, Donelan et al. (1985) found that the method used to fit the observed angular distribution (half power points versus full angular distribution) either suppressed or retained a dependence on (U/c_p) in the spreading function. They advocated the half power method as being less influenced by noise and on this basis, their form for $D(\theta, k)$ depending only on k/k_p is adopted tentatively as the standard.

Other correlations were also considered in this study in order to assess the relative sensitivity of the findings. An extrapolation to higher gravity range wavenumbers was assumed here, where the spreading increases gradually to an asymptotic broad angular distribution, polarized in the wind direction. The rate of spreading at higher gravity range wavenumbers was set by matching to available radar and wave data. Recent observations by Young et al. (1995) for young wind seas indicate a more complex lobed behaviour, with symmetric peaks off the wind direction, but this will not have a significant influence on the conclusions drawn here. For the present purposes, the form for $D(\theta, k)$ adopted here is shown in Fig. 4.2(a), while Fig. 4.2(b) shows the folded form

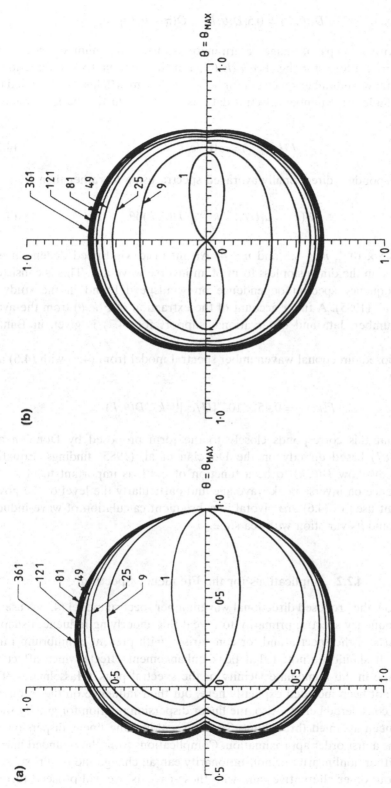

Figure 4.2. (a) Polar plot of the Donelan et al. (1985) directional distribution function $D(k, \theta)$ given by Eq. (4.6) shown for the various values of k/k_p. (b) As in (a) but showing the equivalent distribution function $D(k, \theta) = 0.5[D(k, \theta) + D(k, \theta - \pi)]$ as observed in a purely spatial domain measurement.

$$D_s(\theta; k) = 0.5[D(\theta, k) + D(\pi - \theta, k)]$$

applicable to frozen spatial image determinations of the wavenumber spectrum.

Extraction of a form for the slice $F(k, \theta_{max})$ in the dominant wave direction of the directional wavenumber spectrum for fetch-limited growth has been carried out for the available wavenumber spectral data using (4.6). On this basis it was proposed that

$$F(k, \theta_{max}) = \alpha[U/c_p]^m \, k^{-n} \tag{4.7a}$$

or the corresponding directionally averaged spectra using u_* rather than U

$$F(k) = \alpha[U/c_p]^m k^{-n} \int D(\theta; k) \mathrm{d}\theta \tag{4.7b}$$

with $\alpha \sim 0.45 \times 10^{-4}$, $m \sim 0.5$ and $n \sim 4$. An intrinsic windspeed dependence is included here in the dimensionless form of an inverse wave age. This is consistent with the frequency spectral dependence on windspeed found in the study of Donelan et al. (1985). A fuller account of the extraction of $F(k, \theta)$ from the available wavenumber data and observational support for (4.7) is given in Banner (1990).

The proposed directional wavenumber spectral model from (4.4) with (4.6) and (4.7) is then

$$F(k, \theta) = 0.45 \times 10^{-4}[U/c_p]^{1/2}k^{-4}D(\theta, k) \tag{4.8}$$

We note that this corresponds closely to the form proposed by Donelan and Pierson (1987) based directly on the Donelan et al. (1985) findings. Equation (4.8) does not allow $D(\theta, k)$ to be a function of u_*. It is important to note that the dependence on inverse peak wave age and particularly the level of the power law exponent used in (4.6) are pivotal in subsequent calculation of wave-induced wind stress and its variation with sea state.

4.2.2   Implications for the Frequency Spectrum

In this section, the proposed directional wavenumber spectral model (4.7) is used to calculate frequency spectra, primarily to reveal the underlying influences shaping different parts of the spectra and for comparison with present equilibrium range predictions. It should be noted that peak enhancement effects which affect the spectral shape in the immediate vicinity of the spectral peak (Hasselmann et al. 1973) have not been included explicitly here, but their influence on the frequency spectrum is considered below. Also, the linear dispersion relation for gravity water waves has been assumed throughout. Strictly speaking, the linear dispersion can only serve as a first order approximation. Complications from the bounded harmonics from either nonlinearity or nonstationarity can all change the results substantially. Without other alternatives and with these caveats, we will proceed further.

4.2.3   Frequency Spectral Model

The frequency spectrum is related to $F(k, \theta)$ by

$$F(\omega) = 2g^{-1/2}\left[k^{3/2} \int F(k, \theta)\, d\theta\right]_{k=\omega^2/g}$$

$$= 0.9 \times 10^{-4}[U/c_p]^{1/2}\omega^{-5}\left[\int D(\theta; k)\, d\theta\right]_{k=\omega^2/g} \qquad (4.9)$$

from which it is evident that the departure from ω^{-5} behaviour is predicted only where the spreading function is dependent on k (or ω). From Fig. 4.2(a), this region is largely confined to the spectral subrange below $k/k_p \sim O(10)$ [or $\omega/\omega_p \sim O(3.5)$] and this is where the influence of $D(\theta; k)$ is to be expected. Results for a dominant wavenumber of 0.022 cycles per metre [corresponding to 5.4 s waves] and a wind-speed of 11.22 m/s under fetch-limited growth [$U/c_p \sim 1.3$] are shown in Fig. 4.3. Also shown are results for doubling and halving the windspeed, both at a fixed fetch, with corresponding peak wavenumbers of 0.0104 cycle/m with $U/c_p \sim 1.8$ and 0.0465 cycle/m with $U/c_p \sim 1.0$, based on the empirical nondimensional fetch-limited relationships proposed by Donelan et al. 1985 for the fetch dependence of the peak frequency.

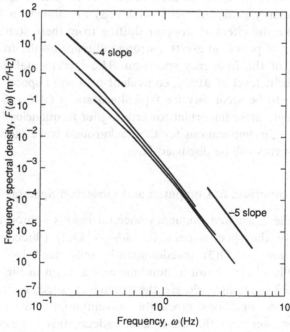

Figure 4.3. Calculated frequency spectra based on the proposed wavenumber spectral form of Eq. 4. For the directional distribution function $D(k, \theta)$ shown in Fig. 4.2. The central solid curve is for a reference spectral peak wavenumber of 0.022 cycle/m and a reference wind speed of 11.2 m/s. The upper and lower solid curves are for peak wavenumbers of 0.014 and 0.0465 cycle/m, respectively, corresponding to doubling and halving the wind speed under fetch-limited conditions.

It is seen that the calculated $F(\omega)$ have a close to ω^{-4} dependence below $\omega/\omega_p \sim O(3)$ and transition to ω^{-5} dependence for higher ω/ω_p. At given frequencies below $\omega/\omega_p \sim O(3)$, the apparent windspeed dependence at a fixed fetch is seen to be very close to linear. Similar calculations were carried out using alternative formulations of $D(\theta; k)$, e.g. Hasselmann et al. (1980). These incorporate a spreading function dependence on U/c_p as well as on k/k_p and result in slightly stronger windspeed dependences than found in the above calculations, but still reproduce a frequency dependence close to ω^{-4} near the spectral peak. Above $\omega/\omega_p \sim O(3)$, the calculated frequency spectra asymptote to ω^{-5} dependence with a very low apparent windspeed dependence, but in view of the potential effects of Doppler shifting discussed in Kitaigorodskii et al. (1975), these effects must be considered before placing too much credence on the calculated spectra.

An analysis was undertaken to examine the typical influence of Doppler shifting effects in order to assess the net modification to the frequency spectrum arising from the assumed model for $F(k, \theta)$ due to (a) local advection by the orbital motions associated with the dominant waves and (b) incremental wind drift current effects. A set of calculations using a two-scale approximation model was carried out, based on the modulation theory described by Phillips (1981) and using an exact Stokes wave model for the dominant wave, a wind drift modulation model (Phillips and Banner 1974) and neglecting local refraction effects (estimated to be of secondary importance). The results of the calculations for different dominant wave slopes (or equivalently peak enhancement levels) are shown in Fig. 4.4. These results indicate that the effect of Doppler shifting from these sources is unimportant near the spectral peak, but exerts a strong influence on the frequency dependence in the tail of the frequency spectrum. The incremental effect of a very substantial wind drift level of $0.05c_p$, equivalent to a wind speed of ~ 30 m/s for 7 s waves, is seen to be secondary for typical sea states ($AK \sim 0.1$), although it becomes increasingly more important towards higher frequencies for lower dominant wave slopes. The implications for the background wave modulation for the capillary-gravity waves will be discussed later.

4.2.4   Comparison of Computed and Observed Spectral Models

A composite of the calculated frequency spectral results appropriate to the distinct ranges above the spectral peak for $\omega/\omega_p < O(3)$ (directional spreading-dominated) and $\omega/\omega_p > O(3)$ (predominantly influenced by orbital motion Doppler shifting) has been shown in nondimensional form in Fig. 4.4. This composite form should reproduce closely the frequency spectra observed for any fetch-limited growth conditions, given the wavenumber and slope AK of the spectral peak, together with the windspeed. Indeed, this has been verified for several examples for both young and mature wind seas which are reported in detail in Banner (1990). The model has also been applied to frequency spectra measured in the recent Surface Waves Dynamics Experiment (SWADE, Weller et al. 1991) conducted off the U.S. East Coast during 1990–91, with a similar close correspondence between observed and calculated frequency spectra. This leads

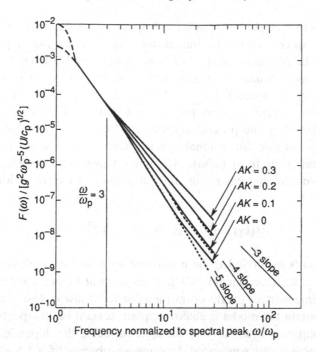

Figure 4.4. Effect of the slope AK of the spectral peak on the frequency spectrum over $< 3\omega/\omega_p < 30$, for the indicated slopes AK; dotted lines, spectral peak orbital motions alone; solid lines after Banner et al. (1989), spectral peak orbital motions and wind drift of $0.05c_p$, representing a wind speed of about 30 m/s for 7 s *wind-waves*. Also shown for $1.3 < \omega/\omega_p < 3$ in nondimensional form, is the computed ω^{-n} dependence, with $n \approx 4$. The nonequilibrium peak enhancement region shown below $\omega/\omega_p \approx 1.3$ is indicated by dashes.

one to conclude that a power law exponent of O(0.5) appears reasonable for the wave-age dependence of the wavenumber slice in the wind direction in (4.7), and has been used in (4.8). However, this issue is not fully resolved yet and awaits further observational support.

One-dimensional wavenumber spectra are given by

$$f(k_1) = \int F(k_1, k_2)\mathrm{d}k_2 \tag{4.10}$$

and similarly for $f(k_2)$, where $F(\mathbf{k}) = F(k_1, k_2) = F(k, \theta)$ are used interchangeably. With the present emphasis of this contribution on the frequency spectrum, we simply note here that calculations of one-dimensional spectra have been carried out for typical ocean wave and windspeed conditions. Full details of these calculations are given in Banner (1990). In brief, both near the spectral peak and towards higher wavenumbers, calculated $f(k_1)$ spectra conformed closely to k^{-3} and $(U/c_p)^{1/2}$ dependence, in agreement with the observations in Fig. 4.1. Calculated $f(k_2)$ spectra reflect the same depletion near the spectral peak as the transverse spectra of Jackson et al. (1985), run 36/1, also shown in Fig. 4.1. At higher wavenumbers above $k/k_p \sim$ O(10), calculated $f(k_2)$ have the same dependences as $f(k_1)$.

Near the spectral peak $[1.5 < \omega/\omega_p < 3.5]$, it is the increasing directional anisotropy of the wave energy with increasing ω/ω_p that has the major influence on the forms of the frequency and of the one-dimensional wavenumber spectra. In addition, the dependence on inverse wave age U/c_p in (4.7) plays a significant role in setting the spectral levels. The one-dimensional wavenumber spectral results are not consistent, near the spectral peak $[k/k_p < O(10)]$, with the form (4.3a) predicted by the present equilibrium theories, as both observed and calculated orthogonal one-dimensional spectra have very different forms in this subrange, each different from (4.3a). At higher wavenumbers, calculated one-dimensional wavenumber spectra in the dominant wave direction conform closely to

$$f(k_1) = 0.6 \times 10^{-4}[U/c_p]^{1/2}k_1^{-3} \tag{4.11}$$

The model results are found to be consistent with the available observational data shown in Fig. 4.1 (see Banner 1990). Thus even at higher wavenumbers the calculated one-dimensional spectra do not appear to conform closely to (4.3a). In the frequency domain the model is able to explain several basic aspects consistent with observed frequency spectra. Near the spectral peak, the dependence on ω^{-n}, with $n \sim 4$, and near-linear windspeed dependence observed for $1.5 < \omega/\omega_p < 3.5$ arises from the strong directional anisotropy near the spectral peak, through its variation with k/k_p. Also contributing to the variation in spectral level is the $[U/c_p]$ dependence. At a fixed fetch, these two effects produce an apparent near-linear dependence on the windspeed under fetch-limited conditions. However, a doubling of the windspeed with the attendant increase in c_p would only cause an increase of 17% in the spectral level: the bulk of the shift in this level arises from the rate at which $D(\theta; k)$ changes with k/k_p, as the spectral peak moves under the action of the increased windspeed. Thus the apparent near-linear windspeed dependence at a given frequency near the spectral peak is seen to arise largely from the variation of directional characteristics with k/k_p as the spectral peak changes. According to Hasselmann et al. (1973), this is controlled largely by nonlinear wave–wave interactions, rather than as a direct response to spectrally local wind-induced wave growth.

Towards higher frequencies, the effects of varying directionality become secondary and Doppler shifting by the dominant wave orbital velocity exerts a major influence on the shape of the frequency spectrum, depending on the dominant wave slope AK, with the wind drift also playing a role when AK is low and towards higher frequencies. Here again, the equilibrium model predictions are not consistent with the observations as they do not address in sufficient detail the underlying physical mechanisms which appear to significantly influence this subrange. Also, there remain questions concerning the applicability of the above analysis to more general wind sea situations, where it is felt that much of the discussion is relevant, but a detailed assessment is left to a future study. In the next section, we will discuss some dynamical implication of the Doppler shift in the capillary–gravity wave range.

4.3 Short *Wind-waves*

4.3.1 Introduction

The lack in understanding of the complex phenomenon of *wind-waves* at the ocean surface, which has been pointed out already is even worse for short *wind-waves* in the short gravity, gravity–capillary, and capillary wave regions. Adding to the complexity is the significant influence on the dynamics of short *wind-waves* on surface active material (Chapter 13) which makes them also dependent on chemical and biological processes at the ocean surface.

Given the importance of short *wind-waves* as roughness elements for the wind drag (Chapter 5), a better knowledge of the spectral densities, the actual shape and its kinematics and dynamics is a significant scientific challenge which requires considerable progress in both experimental techniques and theoretical modelling.

Since the beginning of the century, oceanographers have attempted to measure the small-scale shape of the ocean surface with different optical techniques, including the sun glitter technique (Cox and Munk 1954), the Stilwell technique (Stilwell 1969), stereophotography (Shemdin et al. 1988; Banner et al. 1989), and scanning laser slope gauges (Hara et al. 1994; Hwang et al. 1996). Although the various techniques have been greatly improved over the years, only recently are imaging instruments available (Klinke and Jähne 1995; Zhang personal communication) that provide the required spatial resolution to reveal the spatio-temporal characteristics of the short oceanic *wind-waves*.

This section focuses on the experimental results and discusses the two-dimensional wavenumber spectra available from wind–wave tunnel and field studies. The conditions under which the wavenumber spectra were taken are summarized in Table 4.1, while Table 4.2 gives the spatial resolution of the images and the number range that was covered by the measurements. The largest fraction of the data was obtained in laboratory facilities with a wide range of conditions. In contrast, the wind speed range for which field data are available is still limited. Since the laboratory data have been taken in facilities of quite different geometry and fetches ranging from 5 m to infinity (Table 4.1), it is possible to draw some conclusions about the spectral shape and wind speed dependence of short *wind-waves* in the field.

4.3.2 Wavenumber Slope Spectra

The data about short *wind-waves* were gained by optical techniques. Early techniques of wave imaging relied on the reflection of light at the ocean surface. Jähne et al. (1994) showed that techniques based on light refraction are significantly more suitable. Refraction of light at a wavy water surface has been employed for laser slope gauges for quite some time (Cox 1958; Tober et al. 1973; Hughes et al. 1977; Lange et al. 1982; Jähne and Waas 1989). In an extension of this work, scanning laser slope gauges, used by three research groups recently, are based on the same principle (Lee et al. 1992; Martinsen and Bock 1992; Hwang et al. 1993). The laser

Table 4.1. Summary of the geometrical parameters of the *wind-wave* facilities and the conditions under which data have been taken

Location	Width (m)	Depth (m)	Fetch (m)	Friction velocity	Reference
Linear facility IMST, University of Marseille	2.6	0.70–1.0	5, 29	0.1–0.49	Klinke 1996
Linear facility Delft Hydraulics	8.0	0.80	100	0.073–0.722	Jähne and Riemer 1990
Linear facility Delft Hydraulics	8.0	0.80	25, 50, 75, 100	0.08–0.75	Klinke 1996
Circular facility, Heidelberg Univ.	0.3	0.05–0.3	12[1]	0.08–0.46	Klinke 1991, 1996
Martha's Vineyard sound, 30 m offshore, Nov. 92	–	–	–	0.05 – 0.25	Hara et al. 1994
MBL West Coast Experiment, May 3 1995	–	–	–	0.19	Klinke and Jähne 1995
HiRes Experiment, June 1993	–	–	–	0.03 – 0.21	Hwang et al 1996

[1] Circumference of annular water channel.

Table 4.2. Summary of the image sectors, the spatial resolution, and the corresponding Nyquist wavenumbers. The first figure in the columns with image size and spatial resolution refers to the alongwind direction, the second to the crosswind direction

Facility	Image size (m)	Spatial resolution (mm)	Wavenumber range (rad/m)	Reference
Marseille	0.309–0.234	0.61–0.97	98–3225	Klinke 1996
Delft	0.664–0.473	1.30–1.85	52–1701	Jähne and Riemer 1990
Delft	0.165–0.144	0.32–0.56	170–5592	Klinke 1996
Heidelberg	0.180–0.140	0.35–0.58	164–5389	Klinke 1991, 1996
1996 Martha's Vineyard 30 m offshore, Nov. 92	0.10–0.10 square scan pattern	–	31–990	Hara et al. 1994
MBL West Coast Exp., May 3 1995	0.192–0.151	0.30–0.63	152–4993	Klinke and Jähne 1995; Klinke 1996
MBL West Coast Exp., May 3 1995	0.10–0.04 eight line scan	2.0–5.0	60–1500	Hwang et al. 1996

scanning slope provides spatial information, but the resolution is not sufficient to result in useful two-dimensional images.

As first demonstrated by Keller and Gotwols (1983), light refraction can also be used to take wave slope images. The principle is quite simple, and Jähne (1985) discusses the basic ideas behind this setup. The principles of optical instruments for wave slope or height measurements are discussed in detail as an example of the shape-from-shading paradigm in computer vision in Jähne et al. (1992, 1994). Klinke and Jähne (1995) and Klinke (1996) describe the setup used for wave slope imaging in the field on a freely drifting buoy.

Since the optical techniques to take image sequences of short *wind-waves* measure the slope of the waves, we briefly outline the relation between wave height and slope spectra. The total slope wavenumber spectrum S and the corresponding height wavenumber spectrum F are related in the following way:

$$F(k) = k^{-2}S(k) \qquad S(k) = k^2 F(k) \tag{4.12}$$

The total slope wavenumber spectrum S is defined as the sum of the crosswind spectrum and the alongwind slope spectrum, S_1 and S_2:

$$S_1(k) = k_1^2 F(k) \qquad S_2(k) = k_2^2 F(k) \tag{4.13}$$

These relations allow the height wavenumber spectra to be computed from the slope spectra and vice versa.

In the case of slope measurements, however, it is necessary to measure both wave slope components. For example, if only the along-wind slope spectra is obtained, the total slope spectra can be computed by

$$S(k) = S_1(k)/\cos^2\theta \tag{4.14}$$

where θ is the angle between the wind direction and the propagation direction of the *wind-waves*. From Eq. (4.7a) it is obvious that only the wavenumber spectra in a certain angular cone around the wind direction can be reconstructed reliably. The width of this cone will depend on the noise level. The two-dimensional wavenumber spectra are symmetric. Thus each spectral component contains the sum of upwind travelling waves with wavenumber k and the corresponding downwind component $-k$. All wavenumber spectra presented in this section are half-sided spectra shown as the dimensionless degree of saturation $B(k)$ (Phillips 1985). The degree of saturation is related to the height and the total slope spectra by

$$B(k) = F(k)k^4 = S(k)k^2 \tag{4.15}$$

The omnidirectional saturation wavenumber spectrum is defined as

$$B(k) = \int_{-\pi/2}^{\pi/2} B(k, \theta)\mathrm{d}\theta \tag{4.16}$$

4.3.3   Angular Distribution

The angular profiles in Fig. 4.5 are obtained by averaging the two-dimensional spectra from 200 to 1000 rad/m and normalizing the mean profiles such that the areas under all curves are equal. Figure 4.5 clearly reveals how strongly the angular distribution depends on both the fetch and the geometry of the wind–wave facility. For wind speeds greater than 5 m/s the centrifugal force in the circular facility causes the *wind-waves* to travel slightly offwind. Strangely enough, this effect is only noticeable at short fetch. At infinite fetch the spreading is multimodal at low wind speeds and unimodal at high wind speeds.

When mechanically generated downwind travelling JONSWAP-type waves are superimposed on *wind-waves* a very interesting effect occurs (Fig. 4.5b). Then the angular spreading becomes monomodal and almost independent of the wind speed. Obviously, the angular distribution of short waves is rather forced by irregular yet monodirectional large-scale background waves.

Generally, only one common trend can be identified in all three facilities: the angular spreading of the *wind-waves* tends to broaden as the wind speed increases. Because the angular distribution depends so sensitively on the geometry of the wind wave facility, in the following only omnidirectional spectra, i.e. a profile of the wavenumber spectrum averaged over all angles according to Eq. (4.16) are discussed. The search is for parameters of the wave field which prove to be less dependent on the facility.

4.3.4   Two-Dimensional Wavenumber Spectra

The degree of saturation $B(k)$ is plotted logarithmically in Fig. 4.6 as a function of the wavenumber k, also in a logarithmic scale, and the propagation direction limited to angles θ of the water surface waves relative to the wind direction between $-90°$ and $90°$. The spectra from the Delft facility at 100 m fetch evolve more uniformly with wind speed for all propagation directions (Fig. 4.6). Slowly, the maximum degree of saturation shifts to higher wavenumbers with increasing wind. Contrasting with both the Marseille and Delft results, the spectra from the Heidelberg facility look quite different. For 2.0 m/s wind, the spectrum is mainly contained in three angular bands at $0°$ and $\pm 45°$. In all bands the spectral peaks of the dominant *wind-waves* and the parasitic capillary waves are visible. With higher wind speeds, the angular spreading increases, as does the spectral density at intermediate angles, until at 5.0 m/s all bands have disappeared and maximum spreading is reached at 10.0 m/s. In general, the spectra from the Heidelberg facility show a larger dynamic range and reduced noise levels due to clearer water and a superior imaging technique.

4.3.5   Wavenumber Dependence

At short fetches (Fig. 4.7) and the lowest wind speeds only small gravity waves are generated. For wind speeds greater than 2.0 m/s the spectral density of the spectral

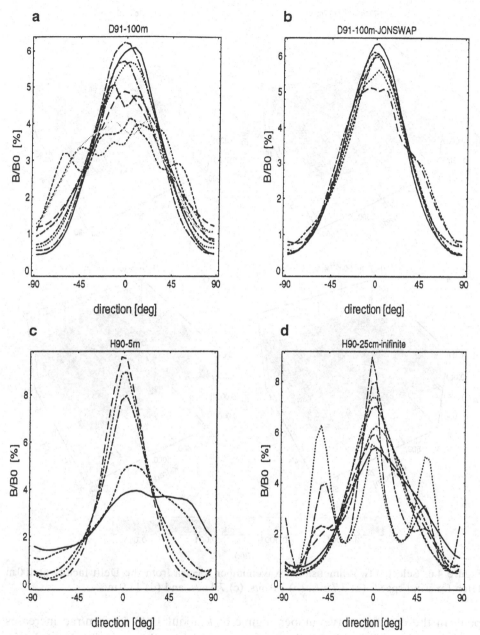

Figure 4.5. Mean angular profiles for (a) Delft 100 m fetch; (b) Delft 100 m fetch with JONSWAP-type background waves; (c) Heidelberg 5 m fetch; and (d) Heidelberg infinite fetch for wind speeds indicated.

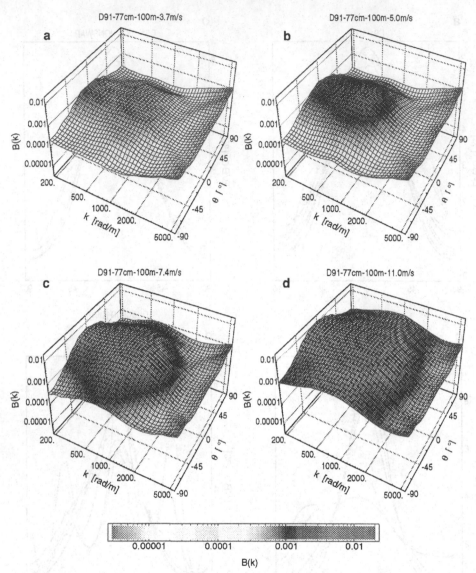

Figure 4.6. Selected two-dimensional wavenumber spectra from the Delft facility at 100 m fetch for wind speeds (a) 3.7 m/s, (b) 5.0 m/s, (c) 7.4 m/s, and (d) 11.0 m/s.

peak in the capillary wavenumber regime at k about 1000–1200 m/rad increases strongly with wind speed. All short fetch spectra show the secondary peak in the capillary wave range and the characteristic dip around the wavenumber for the minimum phase speed $k = 363$ m/rad. In addition to the peak in the capillary wavenumber range, the Marseille spectra contain the spectral peak of the dominant *wind-waves*, due to the slightly larger image sector. As the wind speed increases, the peak of the dominant *wind-wave* shifts towards lower wavenumbers while the peak of the parasitic capillary waves shifts towards higher wavenumbers. The spectral dip is gradually filled up. Only at the highest wind speeds has an equilibrium range been established.

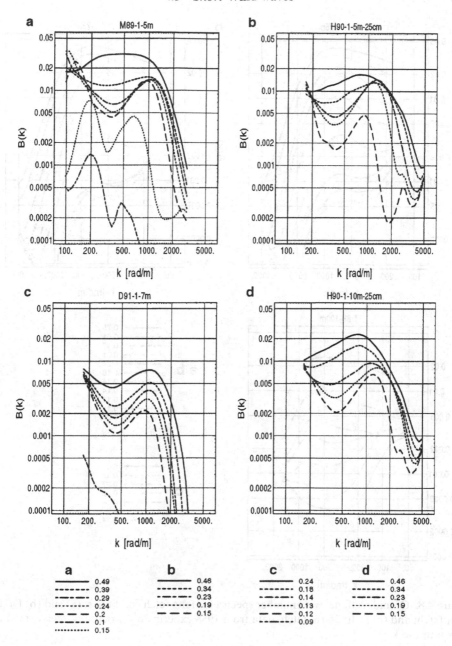

Figure 4.7. Omnidirectional wavenumber spectra for short fetch conditions: (a) Marseille 5 m fetch; (b) Heidelberg 5 m fetch; (c) Delft 7 m fetch; and (d) Heidelberg 10 m fetch. Friction velocity (m/s) is shown in the key.

At higher fetches (Fig. 4.8), the dominant *wind-waves* have continued to grow in wavelength, the lower wind speed spectra are still contaminated by the falloff from the spectral peak of the dominant *wind-waves*. However, at intermediate wind speeds the separation from the dominant spectral peak is already large enough for an equilibrium range to be established. This subrange has a wavenumber dependence of $k^{-0.53}$

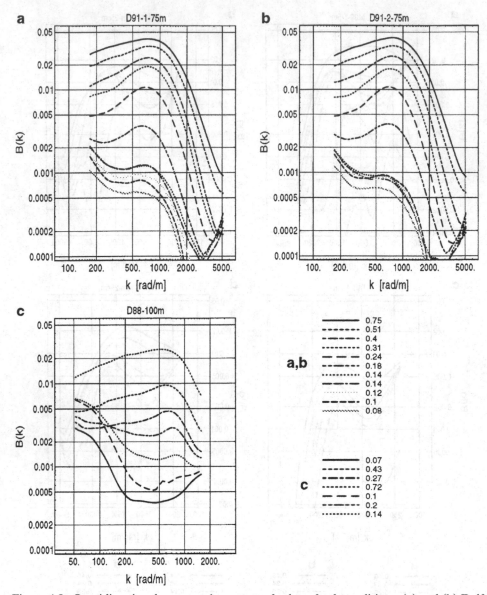

Figure 4.8. Omnidirectional wavenumber spectra for long fetch conditions: (a) and (b) Delft 75 m fetch; and (c) Delft 100 m fetch, data from 1988 experiments. Friction velocity (m/s) is shown in the key.

and extends to the lowest measured wavenumbers. The observed spectral shape is in very good agreement with measurements of Zhang (1995) in the Scripps wind-wave flume at 24 m fetch and friction velocities between 0.4 and 0.66 m/s.

Between 3 m/s and 5 m/s the degree of saturation increases strongly in all facilities. At higher wind speeds an equilibrium range has been established with the spectral densities in the capillary range exceeding the spectral densities at lower wavenumbers. The continuous increase in the degree of saturation extends at higher wind speeds to shorter wavenumbers. At the highest wind speeds this range covers wavenumbers between 125 rad/m ($\lambda \approx 5$ cm) and 1000 rad/m ($\lambda \approx 6$ mm). In this wavenumber range the degree of saturation increases proportional to $k^{0.45}$ for the Delft and $k^{0.5}$ for the Heidelberg data.

It is necessary to match the gravity wave spectra of Eq. (4.7b) to these near-capillary wave spectra. The data of Fig. 4.8 support a $k^{-3.5}$ dependence of the height spectra for wavenumbers near the phase speed minimum, which extends without any change well into the capillary wavenumber range.

Because of the insensitivity of this equilibrium range to the geometry of the facilities (width and fetch) we can infer that oceanic wavenumber spectra should have a similar spectral shape. At higher wind speeds it reaches a $k^{-3.5}$ dependence. This range starts at lower wavenumbers for higher friction velocities and extends down well into the capillary range until the cutoff sets in. The levels of the omnidirectional spectral density at low friction velocities ($u_* < 0.2$ m/s) scatter considerably in wind–wave facilities partly due to various background wave conditions and partly due to surface films. Variations by a factor of four are observed.

At higher wind speeds, however, the omnidirectional wavenumber spectra show a remarkably consistent picture, despite the different geometry and fetch of the facilities. Therefore we expect that also the omnidirectional wavenumber spectra for short *wind-waves* at sea will come close to the laboratory data at high wind speeds. The spectral gap around the minimum phase speed and the secondary peak in the capillary wave range found at low fetches and low-to-medium wind speeds (Fig. 4.7) is generally not relevant for oceanic conditions.

4.3.6 Cutoff at High Wavenumbers

All omnidirectional wavenumber spectra show a common remarkable feature. Within a short wavenumber range, the degree of saturation changes from a slight increase with wavenumber to a steep decrease at least proportional to k^{-3}. In Fig. 4.9, the wavenumbers where the spectral densities of each wind–wave facility start to drop are plotted as a function of wind speed. These so-called cutoff wavenumbers are obtained as the intersection of two straight lines. While one line describes the falloff of the spectral densities towards high wavenumbers, the other line is determined by linearly approximating the wavenumber dependence of the spectral densities in the equilibrium range.

For wind speeds larger than 2.5 m/s and except for the shortest fetches with an overshoot in the spectra caused by parasitic capillary waves, the cutoff wavenumber in each facility is practically independent of the wind speed. The cutoff wavenumber lies between 900 rad/m and 1350 rad/m for all facilities. These results are consistent with the spectra measured by Zhang (1995) in the Scripps wind-wave facility at a fetch of 24 m. He reports a cutoff wavenumber between 700 and 1000 rad/m. In an earlier paper (Jähne and Riemer 1990), a wind-speed independent cutoff wavenumber of 800 rad/m was reported. In light of the new results it appears

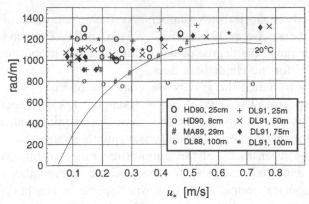

Figure 4.9   Cutoff wavenumbers as a function of friction velocity.

that the earlier results were slightly biased towards lower wavenumbers by the limited wavenumber resolution due to the larger image sector (Table 4.2) and the median filtering of the images.

The new results are also consistent with findings by Trukenmüller (1988) who measured frequency spectra in the small circular facility at Heidelberg University. The cutoff frequencies shown in Fig. 4.10 are practically wind speed-independent. Using the measured phase speeds, the cutoff frequencies correspond to a wavenumber range from 840 rad/m to 1300 rad/m, in excellent agreement with the determined cutoff wavenumbers from much larger facilities.

Even more important, Trukenmüller (1988) found that the average cutoff frequencies do not depend on the kinematic viscosity. By conducting experiments between 5°C and 35°C and adding glycerin he could perform measurements with kinematic viscosities ranging from 0.7 to 2.6×10^{-6} m^2/s. In contrast to these measurements, Kahma and Donelan (1988) found a significant change in the root mean square slope versus water temperature and thus viscosity. They also found that the critical wind speed for the onset of waves decreases with increasing water temperature. This does not necessarily stand in contradiction to our results, since their measurements were conducted under completely different conditions: Kahma and Donelan studied the initial *wind-wave* growth, whereas our measurements were performed with fully developed *wind-waves*. In the initial growth regime, turbulent dissipation is not a dominant factor, while it might very well be in a *wind-wave* field that has reached equilibrium. Thus the cutoff appears to be a universal property of the wind-generated wave field and we expect it to be found in wavenumber spectra obtained from measurements in the open ocean as well.

The wind speed independence of the cutoff strongly rules out the possibility that the balance between wind input to the waves and viscous dissipation, as postulated by the Donelan and Pierson (1987) model, solely determines the spectral density of the short *wind-waves*. Furthermore, viscous damping does not seem to be the dominant dissipation mechanism for short *wind-waves* and, at least at low wind speeds, they are not predominantly generated directly by wind, but by steep short gravity waves. This conclusion also agrees with the early experimental findings of

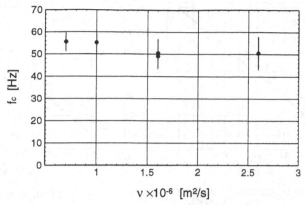

Figure 4.10. Cutoff frequencies extracted from Trukenmüller (1988) for different values of the kinematic viscosity.

Cox (1958), who showed that steep mechanically generated waves also generate parasitic capillary waves in the absence of wind, and the theoretical work by Longuet-Higgins (1992) and Watson and McBride (1993).

4.3.7 Wind Speed Dependence

Generally, the data from the different facilities and fetches give a very consistent picture (Fig. 4.11). The spectral densities scatter only significantly (variation in spectral density by a factor of four) at low friction velocities ($u_* < 0.2$ m/s), while there is a better correlation (scatter is less than a factor two) at higher friction velocities. Given the significant differences of water channel geometries and associated variations in the angular distribution, this is a truly remarkable result.

The u_*-dependence of the spectral densities also varies with wavenumber. First, the scatter of the spectral densities at low wind speeds decreases with decreasing wavelength. Secondly, the exponent of the friction velocity dependence of the spectral densities at intermediate friction velocities increases from approximately 2 for the 3 cm waves to almost 3 for $\lambda = 8$ mm. At higher friction velocities and $\lambda = 8$ mm, the increase in spectral density saturates and varies almost linearly with u_*. At longer fetches, this asymptotic behaviour occurs earlier, i.e. at lower wind speeds. From Fig. 4.11 a mean friction velocity exponent is extracted by fitting the data for $u_* > 0.2$ m/s. The values for the wavelengths $\lambda = 12.5$ cm and $\lambda = 6.3$ cm were obtained from Fig. 10 of Jähne and Riemer (1990).

Figure 4.12 shows a clear trend from rather weak dependence of the spectral densities for short gravity waves (12 cm) proportional to u_* towards a much steeper increase for capillary waves which almost goes with u_*^3 for intermediate friction velocities (0.15–0.3 m/s). For a given wavenumber the exponent is not constant; the increase in the spectral levels varies with friction velocity and has rather an S-shaped form. At high friction velocities (0.4–0.8 m/s), the spectral densities vary almost linearly with friction velocity.

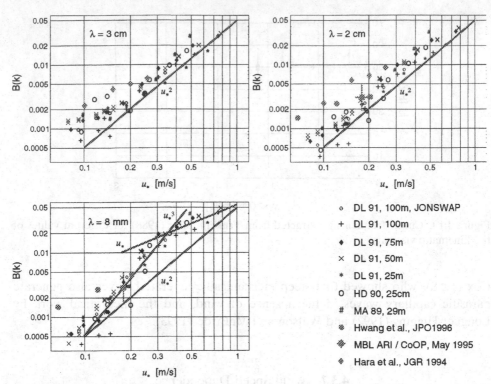

Figure 4.11. Friction velocity dependence of the spectral densities of short *wind-waves* for wavelengths and conditions as indicated.

4.3.8   Comparison with Field Data

In this section the laboratory data are compared with field data obtained in the 1995 MBL cruise (Klinke and Jähne 1995, Tables 4.1 and 4.2) and field data reported by Hara et al. (1994) and Hwang et al. (1996). In addition the *wind-wave* data are compared with radar backscatter data.

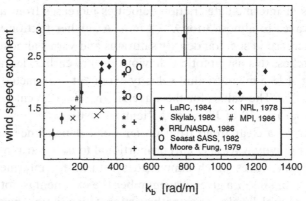

Figure 4.12. Wind speed exponent as function of Bragg wavenumber. Data with error bars are from optical wave measurements, symbols represent radar backscatter measurements taken from Masuko et al. (1986).

An overall similarity in the spectral shape of the lab and field data with regard to the wavenumber dependence exists, except for one noticeable difference (Fig. 4.13). The angular distribution of the field data is much wider and not symmetric with regard to the wind direction. This asymmetry, however, may have been caused by a swell system propagating obliquely to the main wind direction. The spectral cutoff in the omnidirectional spectra, however, occurs around the same wavenumber for the lab and the field data and the spectral levels in the field and laboratory are comparable (Fig. 4.11).

In contrast, Hara et al. (1994) and Hwang et al. (1996) found significantly higher spectral densities at low wind speeds ($u_* = 0.03$–0.21 m/s) than in the laboratory. This comparison was based on their own laboratory data and the early wavenumber spectra from the Delft facility at 100 m fetch (Jähne and Riemer 1990). As one possible explanation they offered the fluctuating component of the wind field, which is high in the field but low in the laboratory. The data collection in Fig. 4.11, however, shows that the 1988 Delft data have the lowest spectral densities.

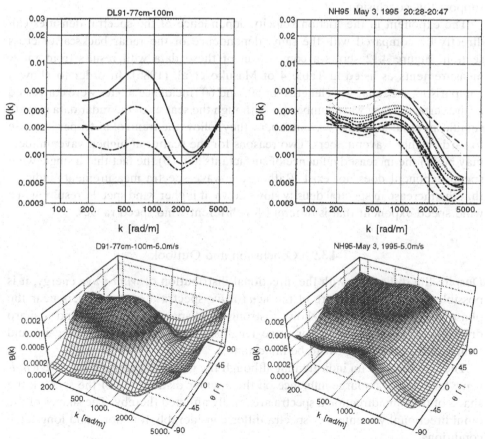

Figure 4.13. Comparison of two-dimensional (top) and omnidirectional (bottom) wavenumber spectra measured in the field during the MBL West Coast Experiment in May 1995 at a friction wind speed of about 5 m/s ($u_* = 0.19$ m/s, Klinke and Jähne 1995) with spectra from the Delft facility at friction velocities of 0.13, 0.19, and 0.23 m/s.

Especially the spectral densities in the Heidelberg facility with unlimited fetch are a factor of 2–4 higher at low wind speeds and thus much closer to the field data.

This suggests another explanation for the higher spectral densities of short *wind-waves* at low wind speed under oceanic conditions. Since the pioneering experiments of Cox (1958) it is known that steep short gravity waves generate parasitic capillary waves without applying wind. We have observed in the Delft wind-wave facility that even at the very low wind speed of 1.5 m/s short gravity waves every now and then become steep enough to generate a train of parasitic capillary waves. Under oceanic conditions, where there are other sources for short gravity waves than the wind at low wind speed, it is natural that parasitic capillary waves should be produced more frequently. This explanation is also supported by laboratory data. When irregular gravity waves with a JONSWAP-type of spectrum were superimposed to *wind-waves*, the spectral densities at low wind speeds are also higher in the laboratory than with pure *wind-waves* (Fig. 4.11). At wind speeds higher than 5–8 m/s ($u_* > 0.2$–0.3 m/s), the influence of background waves seems no longer to be significant. There are, however, no field data yet available to support this conclusion.

The exponent in the friction velocity dependence of the spectral densities can directly be compared with the same dependence of the radar backscatter cross section. Figure 4.12 shows a comparison of these data with results from radar measurements as listed in Table 4 of Masuko et al. (1986). In order to allow a comparison, the radar wavelengths for 50° and 60° incidence angles were converted to the surface Bragg wavenumber k_b. Although the scatter of the radar data is quite significant at larger Bragg wavenumbers, they follow the trend of a steeper increase towards higher wavenumbers. Two reasons for the scatter at higher wavenumbers may be (a) the increased influence of surfactants and (b) the fact that a single wind speed exponent does not exist. Rather, the wave spectra measurements indicate that the increase in spectral density slows down at higher wind speeds, resulting in a wind speed exponent that is different for different wind speed ranges.

4.3.9   Conclusion and Outlook

Despite the difficulties with the directional distribution of *wind-wave* energy, it is possible to present a picture of the *wind-wave* spectra at wavenumbers near the phase speed minimum that shows a strong wind-speed dependence and a sharp drop off of spectral energy for wave lengths shorter than 6 mm. Viscosity and temperature do not appear to be important.

The laboratory data indicate that although the geometry of the respective *wind-wave* facilities bears large influence on the angular distribution of the spectra, the shape of the omnidirectional spectra are hardly affected thereby. The shapes of the omnidirectional wavenumber spectra differ considerably for short- and long-fetch conditions.

The energy balance of small-scale *wind-waves* cannot be determined unambiguously solely by analysing mean wavenumber spectra, nor is it possible to study the geometrical shape and temporal characteristics of the wave field which would be of much significance to the study of the roughness of the ocean surface. Thus the determination and analysis of mean wavenumber spectra can only be regarded as a first step in the evaluation of the variety of pieces of information contained in *wind-wave* slope image sequences.

4.4   The Variability of Gravity–Capillary Waves in Presence of Long Waves

It remains to address the form of the directional wavenumber spectra in the capillary–gravity and capillary wave ranges, where direct observations in the open ocean are even more difficult. These short *wind-waves*, while energetically less significant, make an important contribution to the *wind-wave* drag, and are important in the surface. The low phase and group velocities of these *wind-waves* pre-dispose them to be highly variable. From progress so far, it is felt that the directional aspects of the behaviour of *wind-wave* spectra above the spectral peak need to be considered carefully in future development of spectral models.

As a result of the numerical modelling study (see, for example, Taylor and Gent 1978; Al-Zanaidi and Hui 1984), we have to understand the unsteadiness of the wind stress variation, and to examine the elementary physical processes that are responsible for the unsteadiness. This problem seems to be extremely complicated, and a complete answer is unavailable at the present time. The limited goal of this section is to show that the sea roughness depends not only on the friction wind speed u_* and the *wind-wave* peak frequency ω_p, but also on the wave–current interaction at the range from the dominant wavelength to several times this length. The latter is the controlling term of the in situ fine structures of the sea surface. On the sea surface, this interaction could cause unsteadiness everywhere. Although it has not been precisely defined either theoretically or experimentally, we believe that this is one of the most significant physical processes that make for unsteadiness of the *wind-wave* dependence of the sea roughness.

It is generally accepted that the surface roughness is caused by short gravity–capillary waves. The central idea of gravity–capillary wave variation and the wave blockage phenomenon has been elucidated by Phillips (1981). Phillips has predicted and explained that parasite gravity–capillary waves are relatively prominent on the forward side of the dominant *wind-wave* crests, as a consequence of long wave and short wave interaction. Traditionally, the ocean surface geometry is represented by a Fourier summation of independent free components. The short *wind-wave* components, however, are not really free travelling waves; they are modulated to some degree by the background dominant *wind-waves*. This problem was studied by Unna (1947), Longuet-Higgins and Stewart (1960), Bretherton and Garrett (1968), etc. Their conclusions were that the short *wind-waves* would become both steeper and shorter at the crests of the background long *wind-waves*. These results imply that if we neglect these modulations, the short *wind-wave* train will propagate almost freely on the background long *wind-waves*.

Phillips in his 1981 study, specified the phenomenon in the frame of reference moving with long wave phase velocity c. The riding short *wind-waves* are then those propagating on a slowly varying media with the tangential velocity U induced by the long wave itself. Since the intrinsic dispersion of the short gravity–capillary waves is

$$\omega^2 = g'k + \gamma' k^3 \tag{4.17}$$

where γ is the surface tension coefficient, and g' is the effective gravitational acceleration. Because the tangential surface velocity U is here a function depending only on position x, but not on time t, according to the wave ray theory, the apparent frequency, σ, of the short waves must be invariant, that is

$$\sigma = \omega + kU = \sigma_o = \text{constant} \tag{4.18}$$

For a short wave train initially located at the trough and propagating in the negative direction (corresponding to a negative apparent frequency), there are two solutions k_1, and k_2 ($>k_1$). Phillips designated the solution k_2 as capillary waves (of which the group velocity $U + c_g$ is positive), and the solution k_1 as gravity waves (of which the group velocity is negative). Phillips argued that when a short wave arrived at the point where the two solutions coalesce, it should be reflected and then it will be travelling in the opposed direction as a capillary wave train. Phillips named this coalescence point as the blockage point. So the short wave train under consideration will be blocked, and cannot overpass the long wave crest. Notice that the reflected capillary wave train will conserve the wave action density, and that this capillary wave train will have a much shorter wavelength when propagating forward, so it will be dissipated rapidly by viscosity. As a result, the blocked gravity–capillary waves can only occur at the forward side of the background long wave crests.

This blockage phenomenon predicted first by Phillips has a profound significance to our understanding of the fine structure of the water wave surface, which relates directly to the sea surface roughness. The reason is that this theory suggests that the short waves (characterizing the sea surface roughness) do not satisfy the free wave dispersive relation at the lowest approximation; instead, it may be directly exposed to the influence of the near-surface water currents, which is certainly not characterized only by the two parameters u_*, and σ_p. In the following subsection, we present our theoretical result, which indeed confirms this conjecture mentioned above.

4.4.1   Theoretical Analysis

To simplify mathematical procedures, instead of specifying the problem in the orthogonal curvilinear coordinate system chosen by Shyu and Phillips (1990), we simply use the Cartesian orthogonal coordinate system and we will neglect the influence of the background wave curvature. The reason is directly based on the results of Shyu and Phillips. The curvature and its modification to the gravitational acceleration g do not make any qualitative difference in the solutions as long as the long wave induced modulation has been enclosed in its tangential velocity U on which ride the short waves under consideration.

Let us consider two wave trains travelling on a boundless water surface, one is a gravity–capillary short wave train and the other is a dominant long wave train. They both satisfy the irrotational water wave equation. In absence of the background long wave, the short wave train will be a harmonic wave:

$$\eta = \mathrm{Re}[ae^{i(kx-\sigma_0 t)}] \tag{4.19}$$

The dispersion relation for this short gravity–capillary wave is similar to that given in (4.18). In the presence of the long wave train, it will make the short wave train become a narrow band wave train:

$$\begin{cases} \eta = \mathrm{Re}\{a(x,t)e^{i\theta(x,t)}\} \\ k(x,t) = \theta_x \omega(x,t) = -\theta_t \\ \omega(k;x,t) = \sigma_o(k) + kU(x,t) \end{cases} \tag{4.20}$$

where $U(x,t)$ is the long wave induced horizontal current at $z = 0$, and ω is its visual frequency. $U(x,t)$ is determined by the water wave equation. For a long wave with the surface elevation equal to Re $\{Ae^{i(Kx-\omega_l(t))}\}$, $U(x,t)$ will take the form

$$U(x,t) = \mathrm{Re}\{\omega_l A e^{i(Kx-\omega_l t)}\} \tag{4.21}$$

where A, K, and ω_l are respectively the long wave amplitude, wavenumber, and frequency. The dispersion of this long wave train is given by the linear gravity wave relationship.

The riding short waves specified by Eq. (4.20) are controlled, according to the wave ray theory (e.g. Whitham 1974), by the following system of equations:

$$\begin{cases} \dfrac{dx}{dt} = \dfrac{\partial\omega}{\partial k} \\ \dfrac{dk}{dt} = -\dfrac{\partial\omega}{\partial x} \end{cases} \tag{4.22}$$

Notice that this is a Hamilton system, and the Hamiltonian is the apparent frequency. According to Eqs (4.20) and (4.21), its expression is

$$\omega = \sigma_o(k) + k\omega_l A \cos(Kx - \Omega t) \tag{4.23}$$

Hence

$$\begin{cases} \dfrac{dx}{dt} = \dfrac{\partial\sigma_o(k)}{\partial k} + \omega_l A \cos(Kx - \Omega t) \\ \dfrac{dk}{dt} = k\omega_l KA \sin(Kx - \Omega t) \end{cases} \tag{4.24}$$

We make the following transformation of coordinate:

$$x' = x - \frac{\omega_l}{K}t \tag{4.25}$$

After dropping the prime for x, the Hamiltonian becomes

$$H = \sigma_0 - \frac{\omega_l}{K}k + k\,\omega_l A \cos(Kx) \tag{4.26}$$

Because H does not depend explicitly on time t, it will have to be a constant; therefore, we finally obtain

$$\sqrt{gk + \gamma' k^3} - \frac{\omega_l}{K} k + k \omega_l A \cos(Kx) = H = \text{constant} \qquad (4.27)$$

Here we have chosen the positive branch of $\sigma_0(k)$. For positive k it represents intrinsically forward-travelling short waves. Physically, the short waves are generated both by the forward tilting of the dominant wave crests and by the forward blowing wind, so only the positive branch is physically meaningful. For chosen A and K, the contours in the k–x plan of the above equation are just the trajectories of the short wave trains seen in the frame of coordinates moving with the background long wave; it is also the solution of the transformed Hamiltonian system, which has the equilibrium points in its phase space. It is clear that the equilibrium points must fall on the lines

$$x = \frac{n\pi}{K} \qquad \text{for} \quad n = 0, \pm 1, \pm 2, \ldots \qquad (4.28)$$

and the equation that determines k is

$$\frac{g + 3\gamma' k^2}{2\sqrt{gk + \gamma' k^3}} = \begin{cases} \left(\dfrac{1}{K} - A\right)\omega_l & \text{for } n = 0, \pm 2, \pm 4, \ldots \\[2mm] \left(\dfrac{1}{K} + A\right)\omega_l & \text{for } n = \pm 1, \pm 3, \ldots \end{cases} \qquad (4.29)$$

The first branch of the above equation corresponds to the equilibrium points located on the long wave crests, and the second branch corresponds to those located on the long wave troughs. Notice that for the first branch equation to have solutions, the long wave amplitude must at least satisfy

$$AK = \frac{1}{2\pi} \approx 0.16 \qquad (4.30)$$

In fact, observable gravity waves in deep water rarely have slopes greater than 1/7, so this necessary condition is not always satisfied physically. Higher order analysis would find that locally near the crest the condition might be satisfied. But even if this condition is satisfied, by solving Eq. (4.29), we can see that there exists a critical long wave slope S_c. When the long wave slope is greater than S_c, there will be no equilibrium point on the long wave crests. In such case some new phenomena will appear. Hereafter, we use Case I for the situation with long wave slope less than S_c, and Case II for the situation with long wave greater than S_c.

Figure 4.14 plots one example of Case I, displaying a few selected trajectories of gravity–capillary wave trains. The parameters we chose are $A = 0.025$ m and wavelength at 0.7 m. Figure 4.15 plots an example of Case II, with $A = 0.096$ m and wavelength also at 0.7 m. In Case I, there is a pear-shaped separatrix in the phase space, which passes through the saddle-equilibrium point above each long wave crest. In this case, the short waves can be qualitatively categorized into three groups. The first group is the capillary wave trains corresponding to the ray lines above the separatrix. In Fig. 4.15, we can see that these capillary wave trains would have the wavelength too small to be physical, i.e. they have the wavelength varying more than four times when travelling over one whole long wave period. So these capillary wave trains are hardly observed on the actual sea or experimental wave tank. The second group are the gravity–capillary wave trains corresponding to the

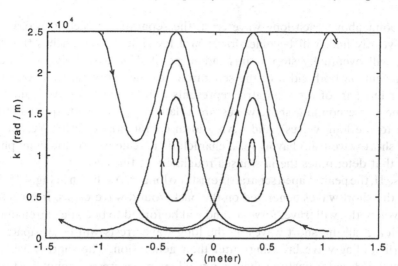

Figure 4.14. Trajectories of gravity–capillary wave trains riding on a long gravity wave with $A = 0.025\,\text{m}$, $\lambda = 0.7\,\text{m}$, and $\gamma' = 0.00007\,\text{m}^3/\text{s}^2$.

closed ray lines inside the separatrix. This group is blocked before each long wave crest. When this wave is generated on the long wave surface, first it will propagate backward to the long wave crest as a gravity wave train, then it will be reflected and turn to propagate forward as a capillary wave train and will be dissipated very rapidly. So the actual ray is not at all a closed line, instead, only the lower left part of those closed lines is realizable. That is just the wave blockage first discovered by Phillips (1981). The third group is the short gravity wave trains corresponding to the ray lines beneath the separatrix.

In fact, the existence of the separatrix in this case is one of the key points that determine the sea roughness. Since the separatrix passes through the saddle-equi-

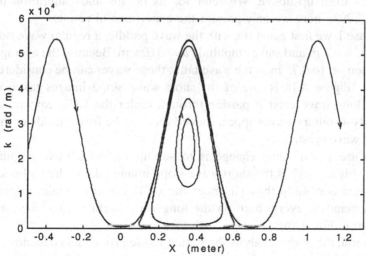

Figure 4.15. Trajectories of the short waves riding on the long wave with $A = 0.096\,\text{m}$, $\lambda = 0.7\,\text{m}$.

librium point above each long wave crest, the separatrix, which is also a special short wave ray line, will become slower and lower as it approaches this saddle point. It will eventually stop at this saddle point. Comparatively the other ray lines adjacent on both sides of the separatrix advance very rapidly. As a result, the lower left part of the separatrix represents a quasi-inertial wave train region, where the corresponding short waves are relatively prominent, quasi-stationary (relative to the long waves), and have a local characteristic scale. We anticipate that this short wave train having local characteristic scale will be the main physical element that determines the in situ sea roughness in this case.

In Case II, the pear-shaped separatrix exists no longer. As shown in Fig. 4.15, in this case, all the short waves generated on the background wave crests, irrespective of their wavelengths, will travel forward, and will be forced to become extremely short waves. Hence all the short waves on the long wave crests will be dissipated very quickly (including wave breaking) after their generation. The short waves represented by the closed trajectories between two long wave crests, cannot exist permanently. This is because the wavelength corresponding to the elliptic equilibrium point is too small to exist on actual water surface. As a result, the blockage phenomenon no longer exists in this case; instead, many irregular short waves of turbulence nature will appear both on the forward and the backward side of the long waves.

4.4.2   Comparisons with Wave Tank Observation

Qualitative verification of the above theoretical analysis in a wind-wave tank has been made possible due to a recently developed technique in the Wallops Wind–Wave–Current Research Facility at Wallops Flight Facility, Wallops Island, Virginia, and the wind-wave tank at the Ocean University of Qingdao. The key idea is to install a linearly decayed light source at the bottom of the tank, which will make the two-dimensional short wave slope patterns automatically visible to a CCD-Camera fixed up above. We refer for its details and calibration to Long and Huang (1993); here we only present the experimental results.

To test Case I, we first generate, with the wave paddle, a regular wave train with wavelength $\lambda = 0.7$ m and wave amplitude $a = 0.025$ m. Because the average water depth has been set to 0.75 m in the wave tank, these waves can be considered deep water waves. Figure 4.16 is one of the short wave slope images taken when a background long wave crest is passing through under the CCD camera. As predicted, gravity–capillary waves appear indeed only on the forward side of the background long wave crest.

To test Case II, the only change is to set up the long wave amplitude to $a = 0.096$ m. Figure 4.17 is the short wave slope image taken when a background long wave crest is passing through under the CCD camera. Again, as predicted, short waves manifest everywhere on the long wave surface, and they are more irregular than in Fig. 4.16.

In conclusion, the above analysis shows that the sea roughness depends crucially on the nature of the near surface water current. It is the different near-surface currents that make a different separatrix, and/or different patterns of ray line. The

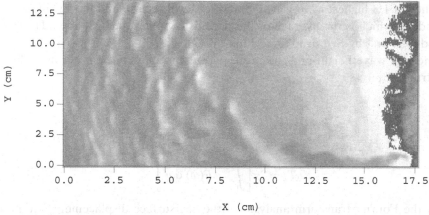

Figure 4.16. A slope image showing short waves on the forward side of one long wave crest: Case I.

near-surface current field could depend on the *wind-wave* field, but the relationship is not unique. Sporadic breakings and organized currents such as Langmuir cells can all generate current fields different from the long wave orbit velocity. Its variation will necessarily produce a corresponding significant variation upon the sea roughness, even if the statistical characteristics of the dominant *wind-wave* field is unchanged.

4.5 Integral Properties and Local Wind–*Wind-wave* Equilibrium

4.5.1 Integral Properties

There are two measures of the sea surface in common use as integral properties. These are the significant wave height H_s, which is the mean of the height to crest

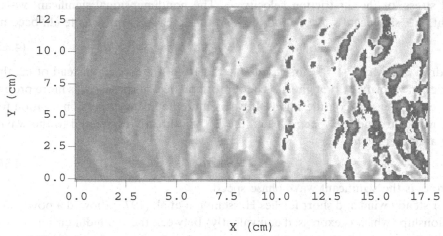

Figure 4.17. A slope image showing short waves everywhere on the long waves: Case II.

distance of 1/3 highest *wind-waves* in a water level record, and the significant wave period T_s, which is the average of the periods of the same 1/3 waves. These were introduced first by Sverdrup and Munk (1947).

The rms wave height and the mean period can be defined in terms of Fourier spectrum, $F(\omega)$, as

$$\overline{\eta^2} = \int_0^\infty F(\omega) \, \mathrm{d}\omega \tag{4.31}$$

and

$$T = 2\pi \left[\int_0^\infty \omega F(\omega) \, \mathrm{d}\omega \right]^{-1} \tag{4.32}$$

With the Fourier transform analyses of the sea surface displacement, the natural measure is the characteristic frequency of the displacement spectrum $\omega F(\omega)$, or the spectral peak frequency, ω_p. Although significant wave measures were developed before the advent of computers able to produce Fourier transforms of the sea surface displacement, there are relationships actually between the spectral measures and the significant wave measures, such as

$$E = \int_0^\infty F(\omega)\mathrm{d}\omega = H_s^2/16 \tag{4.33}$$

after Longuet-Higgins (1952) and

$$\omega_p = 2\pi(1.05T_s)^{-1} \tag{4.34}$$

after, e.g., Mitsuyasu (1968) and Toba (1973). The physical bases of these relationships are in the similarity structure found in windsea, as will be described in this section, and which stems from the wind–*wind-wave* equilibrium as given in the title.

It is convenient to look for relationships between waveheight, period and fetch in terms of nondimensional groups, since we can reduce the number of variables by proper nondimensionalization, and can make data analyses much simpler. The obvious external variables are the acceleration of gravity g, and the kinematic wind stress, or the air friction velocity u_*. The nondimensional significant waveheight, H^*, significant wave period, T^*, and the fetch X^*, with x as fetch, become

$$H^* = gH_s/u_*^2 \qquad T^* = gT_s/u_* \qquad X^* = gx/u_*^2 \tag{4.35}$$

Sverdrup and Munk (1947) first used U_{10}, the 10 m wind speed, instead of u_*, the air friction velocity. Although U_{10} is better for practical purposes, u_* is more proper in cases for physical considerations. If we assume the linear dispersion relation for *wind-waves* around the spectral peak, the T^* can directly be related to the wave-age parameter, c_s/u_*, by

$$T^* = 2\pi c_s/u_* \tag{4.36}$$

where c_s is the significant wave phase speed.

For steady wind and short fetches Hasselmann et al. (1973) showed a power law relationship (which is expressed equivalently) between these nondimensional variables as

$$H^* = 0.041X^{*1/2} \qquad T^* = 0.8X^{*1/3} \tag{4.37}$$

These relationships are generally consistent with the empirical relationship proposed by Wilson (1965) and also Mitsuyasu et al. (1971) for short fetches. For long fetches Wilson (1965) proposed more complicated expressions as the windsea tends to saturation.

4.5.2 The 3/2-Power Law

For steady wind of some duration, there exists a conspicuous similarity law which can be expressed as the nondimensional significant wave height, H^*, proportional to the 3/2-power of the nondimensional period, T^*:

$$H^* = BT^{*3/2} \qquad B = 0.062 \tag{4.38}$$

where B is an empirical constant (Toba 1972). An alternative expression in a dimensional form is

$$H_s = B(gu_*)^{1/2} T_s^{3/2} \tag{4.39}$$

This relation is alternatively recognized as indicating that the significant wave steepness is statistically limited, as a kind of equilibrium state of *wind-waves* under the action of the wind. This relation is expressed by

$$\frac{H_s k_s}{2\pi} = (2\pi)^{1/2} B \left(\frac{c_s}{u_*}\right)^{-1/2} \left\{1 + u_s^* \left(\frac{c_s}{u_*}\right)^{-1}\right\}^{-2/3} \qquad u_s^* \equiv \frac{u_s}{u_*} = 0.206 \tag{4.40}$$

where k_s is the significant wavenumber and c_s is the phase speed of significant waves (Bailey et al. 1991). The wave age of significant waves, c_s/u_*, in the second term appears as the consequence of an averaged local wind drift, of which the velocity is very large in the windward face of each small scale individual *wind-wave* and small or even of negative on the lee side for small *wind-waves* (Okuda et al. 1977). This relation can be used, as Eqs (4.38) and (4.39), for conditions of growing windsea fields effectively free of swells.

This power law was first derived by Toba (1972) by eliminating the nondimensional fetch from complicated empirical formulas for the growth of *wind-waves* by Wilson (1965), together with his experimental data. The Mitsuyasu et al. (1971) observation formula and the JONSWAP formula by Hasselmann et al. (1973) for short fetches (4.37) gave essentially the same relation. However, it is not appropriate to consider that Eq. (4.38) is a consequence of fetch laws for H_s and T_s, since the relation holds robustly in the ocean where the fetch may not necessarily be defined by virtue of intrinsic variation of the wind field.

Figure 4.18 shows Eq. (4.38) with a data set, including the conversion from the above-mentioned fetch graph formulas, together with wind-tank experiment data by Toba (1972) and the tower station data by Kawai et al. (1977). It is noted that the value of B for laboratory *wind-waves* was smaller than that for *wind-waves* in the sea, e.g. $B = 0.043$ in Tokuda and Toba (1981). The points of laboratory *wind-wave* data of Fig. 4.18 were entered first by using H_s values estimated by $H_s = 1.6$

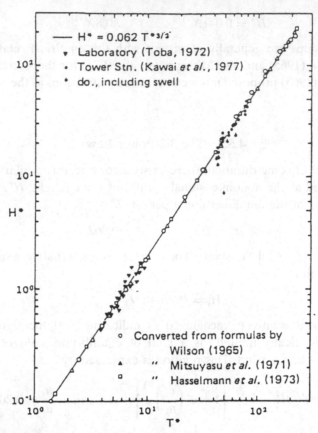

Figure 4.18. The 3/2-power law expressed by Eq. (4.38) with a composite data set, converted from empirical fetch graph formulas, experimental data by Toba (1972), and observational data at a tower station at sea. For the laboratory data, $H_s = 1.6\,\overline{H}$ was used, where $\overline{H}$ is the average wave height (cited from Kawai et al. 1977).

$\overline{H}$, where $\overline{H}$ is the average wave height. Figure 4.19 shows Eq. (4.40) with some observation data for significant steepness relation from Bailey et al. (1991).

Some data giving the observational values in the sea are available where u_* was measured. Kawai et al. (1977) showed an average value of 0.062 ± 0.010. Toba et al. (1990) gave the observed scattering of the measured values as $B = 0.062 \pm 20\%$ by synthesizing the data including Kawai et al. (1977), Donelan (1979), and Merzi and Graf (1985). Later, Ebuchi et al. (1992) examined all the data collected by Ocean Data Buoy stations around Japan during some period, containing swells, and showed that the 3/2-power law, with the above mentioned value of B, holds for cases without significant swells. The perturbation of the 3/2-power law will be mentioned in Chapter 9.

Komatsu and Masuda (1996) developed a new numerical scheme of nonlinear energy transfer among *wind-waves* called the RIAM Method, and reported that Toba's constant B of the 3/2-power law is attained numerically, by using the RIAM method as well as by using the WAM method (The WAMDI Group 1988), with the standard value of $B = 0.06$.

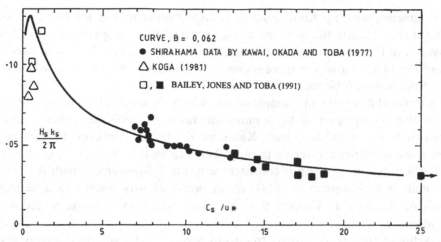

Figure 4.19. Steepness of significant waves as a function of wave age, Eq. (4.40). (Cited from Bailey et al. 1991.)

4.5.3   One-Dimensional Wave Spectral Form Consistent with the 3/2-Power Law

The frequency spectrum does not exist in isolation from its integral properties. Dimensional forms that involve g, u_* and f are

$$F(\omega) = \alpha g^2 \omega^5 (\omega u_*/g)^n \tag{4.41}$$

Phillips (1958) initially suggested arguments for $n = 0$ over a broad range of frequencies less than the most energetic frequency. Later evidence supported the value of $n = 1$ proposed by Toba (1973) from wind-wave tank data. As a matter of fact, he proposed this as the spectral form which is consistent with the 3/2-power law Eq. (4.38), that is

$$F(\omega) = \alpha_s g_* u_* \omega^{-4} \qquad \omega > \omega_p \tag{4.42}$$

where α_s is a constant, ω is the angular frequency, ω_p the peak angular frequency, of *wind-waves*, and g_* the acceleration of gravity expanded to include the effect of surface tension, γ, i.e.

$$g_* = g(1 + \gamma k^2/\rho_w g) \tag{4.43}$$

where k is the wavenumber, ρ_w the density of water. The spectral form proportional to $gu_*\omega^{-4}$ was also consistent with later observational data (e.g., Kawai et al. 1977; Mitsuyasu et al. 1980; Kahma 1981; Forristall 1981; Donelan et al. 1985; Battjes et al. 1987; Liu 1989), and Phillips (1985) re-proposed this spectral form theoretically, by assuming that energy input, nonlinear interactions and dissipation are of the same order.

In observed *wind-waves*, we can see this form of spectra in some main part of the high frequency side called the equilibrium range. The existence of the form of Eq. (4.42) with g instead of g_*, in some ranges of the windsea spectra, was first predicted dimensionally by Kitaigorodskii (1961). Zakharov and Filonenko (1966) predicted theoretically the one-dimensional spectral form proportional to ω^{-4} for the system of the gravity wave interactions. Kitaigorodskii (1983) derived also the form of Eq. (4.42). However, there seems to be no purely theoretical article which derived the u_*-proportionality.

It is worthwhile to cite an example of data which shows observed u_*-proportionality of the energy level of the equilibrium range of one-dimensional windsea spectra. It is Fig. 4.20, cited from Kawai et al. (1977). However, the observed value of the coefficient α_s seems to range between (6 to 12) $\times$ 10^{-2} (e.g. Phillips 1985). The value of α_s which is consistent with the 3/2-power law with $B = 0.062$ was obtained, by Joseph et al. (1981) for the self-similarity spectra, as $\alpha_s = 0.096$. As will be discussed in Chapter 9, α_s varies especially in response to the wind fluctuation.

In actual ocean waves there frequently exists a characteristic called peak enhancement, as the JONSWAP spectral form (Hasselmann et al. 1973) indicates. Also, some data shows that the coefficient α_s of Eq. (4.42) may be a function of the significant slope as proposed by Huang et al. (1981a), and described by Huang et al. (1990a,b). Also, laboratory short fetch *wind-waves* do not have self-similar spectra, but have steeper peaks. However, there is evidence to show that the main constituent of individual *wind-waves* in the laboratory wind-wave tank also obey the 3/2-power law (Tokuda and Toba 1981), though the value of B is about 0.04 for fetches between 4 m to 15 m. Consequently, the 3/2-power law, being a relationship of

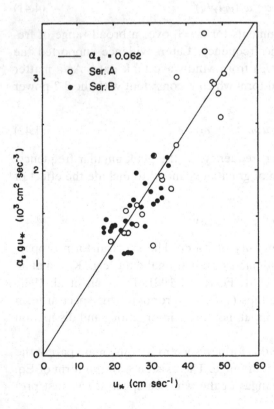

Figure 4.20. An example of observed u_*-proportionality of the energy level of the equilibrium range of one-dimensional windsea spectra. The ordinate corresponds to the spectral density normalized by ω^{-4}. (Cited from Kawai et al. 1977.)

integral properties, seems more robust than the delicate variation of the spectral form of *wind-waves*.

For short *wind-waves* in the capillary–gravity range of 10 to 50 Hz, the spectral level is proportional not to u_*, but to, say, $u_*^{2.5}$, as found earlier by Mitsuyasu and Honda (1974). Recent studies of two-dimensional spectra for these frequencies by use of optical techniques clearly show this (e.g. Jähne and Riemer 1990). Readers are referred back to Section 4.2 for more details.

4.5.4   Physical Implication of the 3/2-Power Law: The Local Wind–Windsea Equilibrium Concept

The 3/2-power law (4.38) claims that the relation between H_s and T_s is given only by consultation of u_* (or the wind speed U_{10}), at least under growing windsea conditions, since Eq. (4.38) is conceptually expressed as

$$f(H_s, T_s, u_*) = 0 \qquad (4.44)$$

The similarity laws, Eqs. (4.38) or (4.42), are results of a series of strongly non-linear processes, such as air separation from the surface of individual *wind-waves*, reattachment of the high shear layer to the windward face of individual *wind-waves* (Kawamura and Toba 1988), the consequent wind stress distribution (both pressure and shear stress) along the individual *wind-wave* surfaces (Okuda et al. 1977; Banner 1990), consequent local shear in water side, in particular under crests of individual *wind-waves* (Okuda 1982a,b; Banner and Peirson 1998). Corresponding to the flow separation and ordered motion in the air side of the interface, there are breaking waves. Besides visible wave breaking which carries air bubbles into the water, there are breaking waves that are not accompanied by air entrainment. Banner and Phillips (1974) named these "incipient breaking". It is not visible by eye, but is recognized by using hydrogen bubble lines, and it is intrinsic in windsea (Toba et al. 1975; Okuda et al. 1976). All of these types of wave breaking cause turbulence. Toba and Kawamura (1996) showed that this special turbulent layer produced by windsea (downward-bursting boundary layer) is of the order of $5H_s$ thick. The windsea is thus a special strongly nonlinear phenomenon, which is generated at a sheared interface between the air and water, and which connects two turbulent boundary layers of air and water.

It seems that some self-adjustment mechanisms exist for the windsea field through these strongly nonlinear processes. We believe that the similarity law, the 3/2-second power law, and the similarity structure of energy spectrum of *wind-waves* are the consequence of these self-adjustment processes. Also, the continuities between velocities, turbulent structures and momentum fluxes in the two turbulent boundary layers of the air and water should be achieved by these self-adjustment processes.

Toba (1972) called this situation the local balance (equilibrium) in the air–sea boundary process. This concept has been investigated intensively by wind–wave tank experiments as reviewed by Toba (1985) and further discussions were given in Kusaba and Masuda (1988). Toba (1988, 1998) presents some reviews. Tulin (1996) proposed to explain the 2/3-power law by simultaneous consideration of *wind-wave* energy and wave momentum change rates resulting from wave breaking, keeping the wave momentum and energy relation: $M = E/C$. This leads to the necessary downshifting of *wind-wave* frequency to satisfy the 2/3-power law, keeping a balance between the wind input and dissipation by wave breaking.

Perturbations of the local equilibrium between wind and windsea will be discussed, with respect to wind unsteadiness, in Chapter 9.

4.6   *Wind-wave* Modelling

Having presented the spectral form, we should make a few remarks on *wind-wave* modelling as reported by Komen et al. (1994). W*ind-wave* prediction models have certainly served useful purposes, and need to be further developed. But in their present form, they would be of limited use as a tool to study *wind-wave* generation processes. To begin with, the de-coupling of the various source functions into independent terms is not defensible from a dynamical point of view; this decomposition totally rules out the fully interactive processes which will call for coupled source functions. Even considering just the decoupled source functions, we have to admit that not enough is known about all of them. The best known one is the nonlinear term. Other than that, the input term is quite uncertain. There are the linear growth function by Snyder et al. (1981) and the quadratic growth by Plant (1982); obviously, they cannot both be right. The generation mechanisms also need to be clarified.

Of all the source functions, the most elusive one is the dissipation term, which is, by consensus, due mainly to the wave breaking. The expression adopted now was based on the analysis by Hasselmann (1974), and some tuning. With the necessity of patching and tuning, we should be aware of the limitations when using the prediction model as a research tool to understand the physics and dynamics of the *wind-wave* generation process. The usefulness of the *wind-wave* prediction model should be diagnostic: to check the physics in the source terms without tuning.

Until some idealized cases can be predicted more closely, the extension to more complex situations, such as turning winds and refraction by horizontally sheared currents, while possible, leaves questions as to the reliability of the predictions, especially as observational support for such calculations is not widely available. Large scale ocean experiments are required with well-defined wind fields, and the recent Surface Waves Dynamics Experiment (SWADE) conducted off the U.S. East Coast during 1990–91 (Weller et al. 1991), which provided such data, is serving as a very valuable basis for testing *wind-wave* models in complex wind fields and current systems.

4.7   Concluding Remarks

A picture of the sea surface roughness, the *wind-waves*, has been presented in terms of the Fourier power spectra. These spectra and their integral properties of significant wave height and period obey laws in steady conditions that are reasonably understood. A recent development of representing a nonlinear and nonstationary process in terms of Hilbert spectrum (Huang et al. 1996, 1998) provides a new approach of the wave field representation, which will eliminate the unnecessary complication of separating waves into free and forced categories. The implication of such an approach is to be explored in the future. Considerable progress has been made on establishing the form of the directionally averaged wave spectrum between the long *wind-wave* peak value and the short *wind-wave* cut off. Wind speed, fetch and duration play a role which is illustrated earlier in Fig 1.8b. The long *wind-waves* are concentrated in the wind direction near the spectral peak but become more uniformly distributed in angle for the shorter gravity waves. The still shorter capillary waves again have a strong directional preference for the wind direction. Figures 4.2 and 4.5 show this quantitatively.

Questions still remain that have a strong impact on our understanding of the drag over the ocean. In particular, it has been observed that *wind-wave* energy growth with fetch is different in confined bodies of water than in the open ocean, a subject discussed in Chapter 15. The differences have been attributed to finite basin geometry, but the details of the mechanism(s) responsible need to be identified. The time and manner that a wave field returns to be "in local equilibrium with the wind" is still uncertain as is the influence of surface tension.

Another major issue is the value of the power law exponent of the wave age dependence of the gravity wave region below the spectral peak. The wave age is set by the wind friction velocity and the peak wavelength. These long *wind-waves* influence the rest of the spectra. In the above discussion it has been proposed that influence declines as the *wind-waves* become shorter, being unimportant in the capillary range. The experimental evidence is inadequate to be confident.

5 Drag Generation Mechanisms

G. T. Csanady

5.1 Introduction

Disequilibrium between the fast air stream and sluggish water drives the irreversible process of air–sea momentum transfer, much as temperature difference drives heat transfer. The rate of transfer per unit area and unit time, the momentum flux, equals the tangential force per unit area applied to the air–sea interface, "drag" for short. In laminar flow viscous shear would be the instrument of momentum transfer, an air-side boundary layer the principal resistance, as a solution of the Navier–Stokes equations reveals. If we could magically contrive to sustain laminar flow above and below the sea surface, we would witness momentum transfer rates some two orders of magnitude lower than we actually find. Hydrodynamic instability of laminar shear flow makes this impossible and brings two important ingredients to the air–sea momentum transfer process: Reynolds flux of momentum, which becomes the main route of momentum transport to and from the air–sea interface, and *wind-waves* on the interface, which play a prominent role in the momentum handover process from air to water.

5.2 Pathways of Air–Sea Momentum Transfer

Turbulence in the air and in the water sustains Reynolds fluxes. Anything that tends to suppress turbulence increases the resistance to momentum transfer, as the example of drag-reduction chemicals shows: introduction of high-polymer substances into the viscous sublayer of a streamlined object (such as a submarine) dissipates turbulence energy, thickens the viscous sublayer, and reduces drag. In this connection it is interesting to note that the mobility of the air–sea interface allows the transfer of turbulence energy from one side to the other, as fluctuations of shear stress and pressure "work" the interface. Anomalously high dissipation of turbulence energy on the water side points to vigorous transfer from air to water

(Craig and Banner 1994), the drag-reducing effect of which should be substantial, although it remains to be documented or quantified. We can only speculate how much greater the drag would be without this drain on air-side turbulence energy. It partly explains why hydrodynamic roughness inferred from the mean velocity distribution should be so much higher on the water side than the air side (Csanady 1984), as illustrated earlier in Fig. 1.6.

As they propagate on a water surface, waves transport momentum horizontally, the more so the higher the wave. Pressure forces on their inclined surfaces may cause waves to grow, increasing their momentum transport. The divergence of this transport equals a portion of the total downward momentum flux from the air, which is often described as the fraction going into "wave momentum" (actually wave momentum transport). Where and when the waves decay, this fraction is converted into momentum of horizontal shear flow, just as any momentum transferred from the air via viscous shear stress. Even while wind-generated waves are growing, they also continuously lose momentum, to some extent through viscous (and eddy-viscous) drag on orbital motions, but mostly through "breaking", a complex overturning motion.

The classical model of a free (unforced) surface wave is a sinusoidal, small amplitude undulation propagating at a celerity depending only on wavelength, surface tension, and the force of gravity. The *wind-wave*, using the term collectively for all the surface perturbations on a wind-blown water surface, is a much more complex phenomenon. A passage from Toba (1988) puts it as follows:

> The *wind-wave* is a special water wave as it is generated by the action of wind at the air–water interface. Since the *wind-wave* is associated with air and water flows above and below the waves, its characteristics are determined by the coupling process between the boundary layers in air and water. The important elements in the *wind-wave* are the surface wave motion, the local wind-drift and turbulence in the air and water boundary layers.

In other words, the *wind-wave* could not exist in the absence of the accompanying air–water shear flow and turbulence, being just one manifestation of a set of tightly coupled processes. It is certainly not a superposition of small amplitude sinusoidal free surface waves. It is a gross conceptual error to suppose that a Fourier resolution of the ephemeral instantaneous shape of the sea surface yields such an assembly. The stochastic average properties of the *wind-wave* are, however, stable, and allow us to regard separate ranges of the wavenumber or frequency spectrum of surface elevation as signatures of physically existing short or long waves. They define the "characteristic wave" as the one with a frequency and wavenumber at the peak of the energy spectrum, the narrow band around this frequency with high spectral density as the "energy containing band", and allow us to refer to short waves, meaning those at the high wavenumber end. The characteristic waves behave more or less as small amplitude sinusoidal waves, obeying, in particular, the theoretical gravity-wave relationship between wavenumber and celerity. Short waves of the spectrum, on the other hand, arise spontaneously on account of hydrodynamic instability of the shear flow, grow and steepen rapidly and break much more readily than longer waves, promptly transferring their momentum to the water-side shear flow. Much circumstantial evidence suggests that they mediate a large fraction of the total momentum transfer.

A useful idealization of the *wind-wave* concept postulates constant wind, absence of "swell", i.e. of waves propagating from some remote generating region into the area of ocean of interest, and describes the evolution of the waves as a function of "fetch", that is the distance from an upwind shore perpendicular to the wind. At constant fetch and wind speed, wave properties are statistically steady, so are a special case of *wind-waves* that are in *local equilibrium with the wind*. The height, wavelength and period of the characteristic wave all increase with fetch, while its steepness drops slowly. A convenient nondimensional variable characterizing wave growth is c_p/u_*, wave celerity at the peak of the spectrum, divided by friction velocity in air, $u_* = \sqrt{\tau/\rho}$, where τ is shear stress, ρ air density. It is often referred to as "wave age", although if one thinks of X/U, fetch over wind speed as a more appropriate measure of age, c_p/u_* is approximately proportional to the third root of X/U. Similarity laws connect the properties of the idealized *wind-wave* to wave age: nondimensional fetch, wave height and period are expressible as simple power laws of wave age. From these it follows that the fraction of air–sea momentum flux supporting the growth of wave momentum transport is 6% or less, declining with wave age (Toba 1976). The characteristic wave of very "young" *wind-waves* can be very short, their wavelength in laboratory flumes often no more than 10 cm or so.

Although breaking waves and associated separated flow regions disrupt the air flow over the *wind-wave*, away from the disruptions the air–water shear flow sustains viscous shear stress on the interface, much as on a smooth surface. Especially intense is viscous momentum transfer in regions of flow reattachment behind separation bubbles. A portion of the total momentum transfer therefore remains garden variety viscous stress.

To sum up this introductory discussion, we may think of air–sea momentum transfer in the open ocean (at fairly long fetch) as splitting above the sea surface into three streams: one stream going to the long waves of the energy containing band, causing these waves to grow with fetch while losing little momentum to shear flow; a second, stronger stream to ephemeral short waves which transfer their momentum through breaking promptly to shear flow, and a third stream going via viscous stress directly to shear flow (Fig. 5.1). Below the surface the streams unite again into Reynolds stress, as they do above the surface.

While we do not have firm quantitative estimates of the fraction going into each individual stream of momentum transfer, laboratory investigations and circumstantial evidence suggests the orders of magnitude. As already mentioned, the growth of the energy containing waves absorbs about 6% the total momentum transfer, or less. An additional fraction goes into maintaining these waves against the action of viscosity, turbulence and wave breaking: just how large a fraction, is not entirely clear. It is inherently not likely that they dissipate substantially more energy than they retain for growth: short waves do this, with the result that their lifetime is short. The waves of the energy containing band, on the other hand, travel long

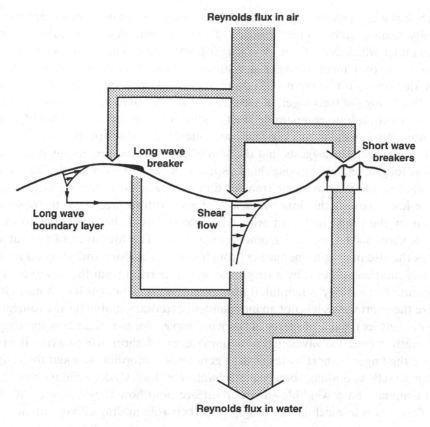

Figure 5.1.   Pathways of air–sea momentum transfer.

distances out of their generation region before decaying. The rapid growth and decay of short waves is largely due to air flow separation at their sharp crests. By contrast, no flow separation accompanies the energy containing waves of low steepness, propagating with a celerity not much less than wind speed: a study by Jones and Kawamura in 1990 showed the large eddies tied to sharp crested short waves to be absent over long wind waves. Another point is that the so-called sand-grain roughness length of the sea surface (comparable to the size of protrusions in the case of rough solid surfaces, see Section 1.3) is much smaller than the heights of the energy containing waves, the former typically a few centimetres, the latter a few metres. This strongly suggests that small-scale disturbances are the primary transferers of momentum, not the house-tall long waves. All in all, it is reasonable to suppose that the total momentum gain from the air by the energy containing waves remains of the same order as needed for their growth, i.e. 6% or so of the total, except at a very young wave age, when the characteristic waves are quite short.

For the fraction going directly into viscous shear, we have the empirical fact that wind stress on the sea surface is not that much greater than on a comparable smooth surface, typically, say, twice or three times as great, in moderate winds. Viscosity remains active in the presence of waves: it is difficult to imagine circumstances under which fast air flow over sluggish water would not be accompanied by viscous shear over much of the water surface. Laboratory experiments, at least, clearly demonstrate this point (Okuda et al. 1977; Banner and Peirson 1996; see Fig. 1.7). Putting the facts together we arrive at the estimate that viscous shear and short waves contribute comparable fractions to the total momentum transfer, the short wave route gaining market share with increasing wind speed.

This argument also suggests that the sum of the two streams, energy containing waves at long fetch, and viscous shear, is more or less predictable, the long-wave one constant, the viscous shear transfer diminishing with increasing wind speed, more or less following the drag coefficient of a smooth surface. Thus the observed increase of the drag coefficient with wind speed has to be attributed to more vigorous short-wave mediated momentum transfer. The literature of the subject ascribes the rise in drag to increasing "roughness", in analogy with drag on a solid boundary, and quantifies it by a roughness length derived from the observed drag coefficient. The analogy is helpful, if not taken too far: whatever length may characterize the short waves it is not an independent external variable (as the roughness of a solid surface), but a property of the *wind-wave*. Nor is it clear how the roughness length relates to wavelengths or amplitudes of short *wind-waves*. Beyond pointing the finger at short waves as drag generators, information about the roughness length tells us nothing about the mechanism of drag. Understanding how short waves originate on a wind-blown water surface, and how they interact with the shear flow, leads to much better insight into their role in drag enhancement.

5.3 The Origin of Short *Wind-waves*

How do short waves arise on a wind-blown water surface in the first place? In a laboratory flume (wind tunnel over a water channel) waves develop on an initially smooth surface within seconds upon turning on the air flow. In a seminal paper on this subject, Kawai (1979) sums up the salient observed facts as follows:

> Now we describe, as an introduction, an outline of the time sequence of the phenomena appearing after the abrupt start of the wind on the water surface. A shear flow first starts and grows in the uppermost thin layer of the water, and then the appearance of waves follows several seconds later. The waves which appear initially are long-crested and regular, and so they are distinguished from those appearing later, which are short-crested, irregular and accompanied by forced convection.

By forced convection Kawai means turbulence in the water; short-crested, irregular short waves characterize the *wind-wave*. The initial period during which the waves are long-crested and regular, is of order 10 s. The observed scenario suggests hydrodynamic instability of the initial shear flow, similar to the instability of wall-bound shear flow, which is the origin of turbulence in pipes and boundary layers. Understanding the fluid mechanics of instability required the concentrated efforts of a host of outstanding theoreticians in the first half of this century, until the experimentum crucium of Schubauer and Skramstad (1947) brilliantly verified the theory for wall-bound shear flow: as the theory predicted, the Tollmien–Schlichting waves of highest growth rate were the initial disturbances observed in laminar flow, painstakingly made clear from other disturbances. Memoirs of Wuest (1949), Lock (1954), Valenzuela (1976) and others developed the application of the now classical instability theory to shear flow on the two sides of the air–sea interface. A small sinusoidal perturbation is imposed on this flow, and its growth or decay explored by solving the Orr–Sommerfeld equation. Essential for the success of the theory is to satisfy all four boundary conditions at the interface, expressing continuity of two velocity components, pressure and shear stress. Perturbations within a certain wavelength range turn out to be unstable, a particular wavelength the most unstable. Selective amplification of the small perturbations which inevitably occur in the flow is then supposed to cause the most unstable wave to become observable. This is exactly how turbulence in wall-bound shear flow originates, see Schlichting (1960) or Landau and Lifshitz (1963).

Kawai (1979) has summarized the instability theory as applied to the air–sea interface, carefully chosen realistic mean velocity profiles in air and water and solved the equations with the four interface boundary conditions. He then also carried out the equivalent of the Schubauer and Skramstad experiment in a laboratory flume, and showed "that the initial wavelets are the waves whose growth rate by the instability mechanism is maximum". He further noted that the widely quoted theory of wave generation due to Miles (1957, 1962), and Brooke Benjamin (1959), while identifying a possible mechanism of energy transfer from the air-side shear flow to a wavy boundary, does not account for the origin of the waves he observed. The Miles theory simplifies the water-side response, and relies on approximate interface conditions.

Further calculations have elucidated the physical nature of the instability waves on the two sides of the interface. The e-folding depth of the initial laminar flow in water, L_s, is the characteristic length scale of these waves, the wavenumber of the most unstable wave being typically $k_m = 1/L_s$ (Wheless and Csanady 1993; Fig. 5.2). The Reynolds number Re formed with this length scale, the surface velocity of the laminar shear flow, and water viscosity, is typically $Re = 500$, similar to other shear flows on the verge of instability. Within the short period when regular waves occur on the interface, the length scale L_s is of order 3 mm, putting k_m near the wavenumber of minimum celerity of free waves, where gravity and capillarity are about equally important.

This seemed at first mysterious, and prompted the investigation of Wheless and Csanady. The pressure boundary condition apparently controls the celerity of the instability waves: it is nearly the same as of free waves at the same wavenumber. Not so for some other properties: the flow structure differs dramatically, because the viscous solution of the Orr–Sommerfeld equation dominates it on both sides of the interface (Fig. 5.3). On the water side, for example, the streamfunction varies vertically on a scale shorter than inverse wavenumber typically by a factor of eight or so. Thus the surface velocity is much higher than in a free wave of the same

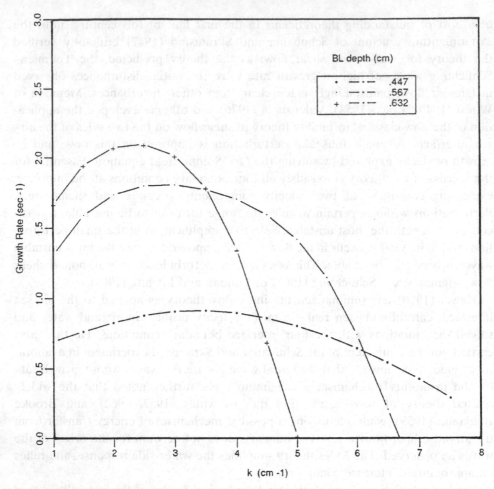

Figure 5.2. Calculated growth rate of instability waves versus wavenumber, at varying water side boundary layer thickness scale: $L_s = 3.2$ mm (dash-dot); $L_s = 4.0$ mm (dashed); $L_s = 4.5$ mm (solid). Inverse wavenumber at maximum growth rate, k_m^{-1}, approximately equals L_s. The friction velocity $U_{*a} = 15$ cm/s.

wavenumber and amplitude. The waves also grow rapidly, and their surface velocity soon approaches their celerity, invalidating linear theory, but suggesting that the waves break at quite small amplitude. While water-side parameters determine most of the wave properties, they gain most of their energy from the air flow, on account of the high perturbation velocities in air, which increase rapidly with height in a viscous wave-boundary layer.

Although Kawai's focus was the initial occurrence of short regular waves on the water surface, perturbations of the laminar shear flow on the water side accompany the waves, as Fig. 5.3 shows. They are equivalent to the Tollmien–Schlichting waves of wall-bound shear flows, subject only to more complex boundary conditions at the free surface. When the surface becomes chaotic, the water side interior flow breaks down simultaneously into turbulence. Photographs of Okuda et al. (1976) illustrate this point graphically. The instability waves thus do double duty, as gen-

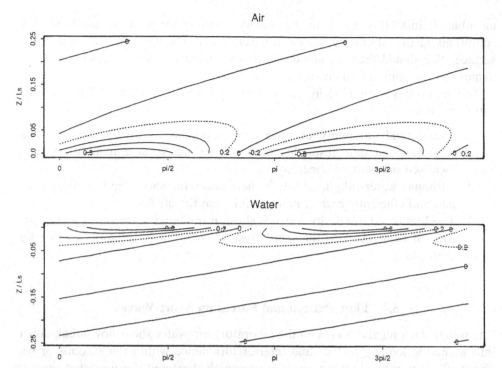

Figure 5.3. Calculated nondimensional stream-function amplitude at maximum growth rate, on the air and water sides of the interface. Viscous solution perturbations reach to about 1/20 of boundary layer thickness scale, L_s.

erators of short *wind-waves* on the surface, and of water-side turbulence below. Or, from another point of view, short *wind-waves* and water-side turbulence are two parallel outcomes of the same breakdown process of the free surface laminar shear flow. Adding that energy supply to the waves comes from the air-side wave-boundary layer, we arrive at Toba's vision of the *wind-wave*, as a phenomenon encompassing the air–water shear flow, turbulence, and chaotic surface waves, all evolving together. The air to water energy flow through growing instability waves seems to be the beginning of turbulence energy transfer in the same direction.

While the laboratory experiments in which instability waves arise on a quiescent water surface under sudden wind seem remote from oceanic reality, something very similar does occur under a wind gust. The "cat's paws" of everyday experience are patches of sea surface covered by capillary–gravity waves, looking for all the world like the initial waves in Kawai's laboratory flume. It is eminently reasonable to suppose that they arise from the same causes, wind-driven surface shear and hydrodynamic instability. Within seconds they evolve into the more chaotic typical *wind-wave* surface containing also somewhat longer waves. As the gust passes (after a typical duration of perhaps 100 s) the shortest waves die out until the next such episode. Each gust may be supposed to generate a new internal boundary layer at the surface, thin and strongly sheared, setting the stage for new instability waves to grow. This is not mere speculation: there can be no doubt that a wind gust greatly intensifies surface shear, nor that such a shear layer is hydrodynamically unstable. If instability waves in the gravity–capillary range occur under similar conditions in the laboratory, they can hardly fail to show up on a natural water surface. Nor should there be any doubt that turbulence generation below accompanies their evolution into chaotic surface waves.

To sum up this section briefly, from theory aided by laboratory and field observation we infer:

1. The natural variability of wind creates conditions for the spontaneous appearance of short instability waves on a water surface, namely episodic, localized thin surface shear layers.
2. Although water-side instability is their cause, the short waves rapidly steepen and efficiently extract momentum from the air flow.
3. The lifetime of instability waves is short, terminated by breaking and handover of momentum to the shear flow, as longer, more chaotic waves take their place.

5.4   Flow Pattern and Forces on Short Waves

The initial short regular waves on the laboratory air–water shear flow break down into a chaotic looking surface and interior turbulence within ten seconds or so. Eventually a statistically steady-state ensues, with short and sharp-crested characteristic waves similar to those seen on small lakes and ponds under moderate to strong winds. This is what the *wind-wave* looks like at a young age. How precisely the short characteristic waves of this stage of wave growth extract momentum from the air flow, is a fascinating story rich in unsuspected detail, unravelled only recently after many careful and difficult laboratory investigations of the complex flow phenomena involved.

In a coordinate system moving with short (order 10 cm wavelength) laboratory waves, the air flow smoothly follows the undulating surface of waves of moderate steepness. Wave celerity being of order 0.5 m/s, air flow in a wave-following frame of reference flows downwind everywhere except in a thin layer (below the critical level) above the water surface, where it moves upwind, along with the water. At the sharp crest of steeper waves, however, the air flow separates (Kawai 1981, 1982; Kawamura and Toba 1988; Banner and Peirson 1996). A separation bubble forms downwind of the crest, separated from the main air flow by a high-shear layer (Fig. 5.4). The flow reattaches ahead of the next crest, the length of the bubble being thus not much less than a full wavelength, the exact length varying from wave to wave. The reattaching streamlines bend upward sharply where they reach the water surface as may be seen in a photograph of Banner and Peirson (1996). See Fig. 5.5a.

On the water side, in the same wave-following frame of reference, the flow is also smooth under waves of moderate steepness, but the crest of steeper waves carries a small separated flow region or roller. In this frame of reference, a mean flow streamline separates the highly turbulent roller from the underlying fluid, connecting two stagnation points on the surface. In a stationary frame of reference the stagnation points move with wave celerity, and the roller moves bodily with

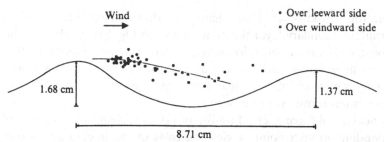

Figure 5.4. Location of high shear layer at the top of separation bubbles on laboratory waves, after Kawamura and Toba (1988).

them. Presence of a roller signifies wave breaking, the technical criterion of which is that fluid velocity reaches wave celerity somewhere on the surface, and short waves with small rollers are called microscale breakers. The front of the roller is steep, similar to the front of a much larger-scale bore sometimes seen on tidal rivers. Unlike the separation bubble on the air-side of a breaker, the roller is much shorter than a wavelength. It serves as the trigger point for air flow separation, however: as several authors have pointed out (see e.g. Fig 5.5b), the bubble

(a)

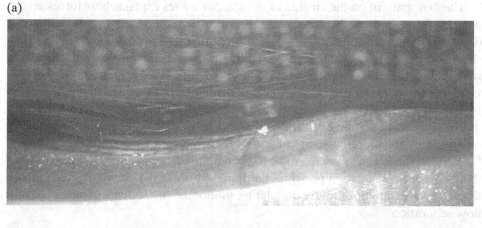

(b)

Figure 5.5. (a) Reattachment in the vicinity of the downwind crest, after Banner and Peirson (1996). (b) Visualization of water particles in the air flow travelling from left to right.

starts at the top of the roller. Underneath and downstream (i.e. upwind) of the roller a turbulent boundary layer forms, as shown in Fig. 5.6, much as in the wake of a solid obstacle, as tracer injection reveals (Peregrine and Svendsen 1978). The surface of less steep waves carries a viscous boundary layer, its thickness and shear determined by variations of the shear stress on the surface, and by downward viscous momentum transport in the water.

The net horizontal force exerted on the interface, driving the water or braking the air depending on your point of view, consists of the integrated viscous shear stress and pressure forces acting on the inclined surfaces of the waves. The proportion, net viscous to pressure force, depends on wind speed: at low wind speeds all of the force (alias momentum transfer) comes from viscous shear, in stronger winds only a fraction. A pioneering study by Okuda et al. (1977) concluded that viscous shear is very unevenly distributed, most of the viscous shear force acting on the windward face of the waves. Banner and Peirson (1996) confirmed this, although they found a more even distribution, significant shear force also on the leeward face. The latter authors also emphasized the importance of high pressure on the windward face, and estimated the magnitude of the resultant net pressure force at 0 to 70% of the total drag, increasing from low to high wind speeds.

The flow pattern on the air side of the steeper waves explains how forces on the windward face come to be as large as they are: where the air flow reattaches at the downwind end of a separation bubble, both shear and pressure are intense. The shear stress is high because the boundary layer is very thin near the point of reattachment, and the velocity outside the boundary layer is the high free stream velocity. The pressure force is high because it has to redirect a high-speed jet (the free shear layer) through an angle of the order of wave steepness. Note that this mechanism of separation–reattachment works contrary to the idea of sheltering of one wave by the preceding one: it exposes the windward face of a wave to the full fury of the wind. The pressures to be expected at reattachment are on the order of the free stream stagnation pressure multiplied by wave steepness, or much higher than pressure variations in the undulating motion of the less steep waves without flow separation.

5.5 Handover of Momentum Transport

Momentum transport crosses the air–sea interface as shear stress and pressure force. Unlike a boundary between two miscible fluids, the sea surface cannot support Reynolds stress: water and air don't mix, although spray in the air at very high wind speeds may create something like a two-phase fluid. Disregarding such

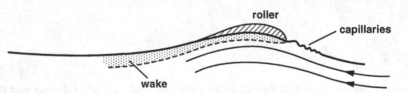

Figure 5.6. Wave to wake momentum conversion in laboratory waves.

extremes, how do interface forces on laboratory waves hand over the momentum transport to fluid motions, to reappear as Reynolds stress some distance above and below the water surface?

The flow patterns described above allow some obvious deductions. On the air side, the interface viscous shear stress slows down fluid in a boundary layer, growing in thickness and in momentum deficiency, hence in (negative) momentum transport. On the steeper waves, high shear stress over the windward face generates strongly sheared fluid. Flow separation transfers this fluid to the shear layer over the bubble, with the slowest fluid presumably forming the interior of the bubble. On waves of lesser steepness, modulation of the shear stress by the wave motion still gives rise to overlying internal boundary layers of greater and smaller momentum defect, an inherently highly unstable flow pattern.

One usually thinks of the pressure force on the interface as affecting the water side, but of course the same force in reverse acts on the air flow, i.e. on the impinging shear layer. A simple conceptual model is a two-dimensional jet, of a thickness comparable to wave amplitude, impinging on an inclined solid surface (the water being so much more massive than the air, this is permissible). A pressure force normal to the surface deflects the jet through an angle θ, of the order of wave steepness (refer to Fig. 4.19). The wave surface being inclined, the pressure force has a horizontal component of order $\theta^2 a U^2$ times air density, where a is wave amplitude and U is wind speed. The horizontal momentum transport of the air flow is diminished by this amount, in each wave with air flow separation. Divided by wave length a/θ, the contribution to the drag coefficient is of order $\theta^3 U^2$, or typically of order of drag over a smooth surface.

Comprehensive observations of Kawamura and Toba (1988) fill in much detail: eddies of wavelength size span the turbulent boundary layer in the air, attached to the waves on the surface. As in wall-bound turbulent flows, at a fixed point the eddies generate bursts of low speed fluid upward, alternating with sweeps of high speed fluid downward. Over the wavy surface, however, the bursts are statistically tied to the windward face of the waves, the sweeps to the leeward face (Fig. 5.7). By way of explanation, Kawamura and Toba offer the conceptual model of a separation bubble drained by the flow, as the steep wave originating it decays, Fig. 5.8. The low speed fluid in the bubble leaves over the downwind crest, the void behind it filled by a sweep of high speed fluid. This, and the impinging jet conceptual model go a long way to explain the big bursts of low speed fluid which the authors described. They also reported, however, that less intense bursts accompany most, if not all, waves, a finding not surprising in view of the uneven shear and pressure force distribution over the waves. At any rate, the interaction of the waves with the air-side shear flow gives rise to ordered eddy motion, its length scale the wavelength, extending throughout the boundary layer. The burst and sweep cycles of the ordered motion add up to Reynolds stress carrying the entire momentum transport, i.e. windward momentum deficiency upward, or surplus downward.

On the water side, where the flow is upwind in the wave-following frame, the boundary layer on the windward face, where the shear is high, grows in thickness and momentum defect, as a remarkable illustration of Okuda et al. (1977) demon-

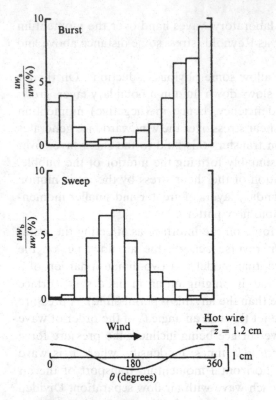

Figure 5.7. Frequency of burst and sweep events versus wave phase, after Kawamura and Toba (1988). The results were obtained for young waves in a laboratory tank.

strates, Fig. 5.9. Under the separation bubble, i.e. over most of the rest of the wave, the shear stress vanishes or even changes sign, remaining small, however. Accumulations of boundary layer fluid at the crests of waves suggest that the slowed down fluid makes its way from one windward face to the next (upwind) crest. The unevenness of the shear stress distribution over the wave has negligible

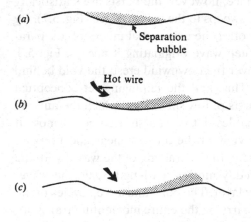

Figure 5.8. Big Burst model of Kawamura and Toba (1988): stagnant fluid in separation bubble being swept out into outer boundary layer.

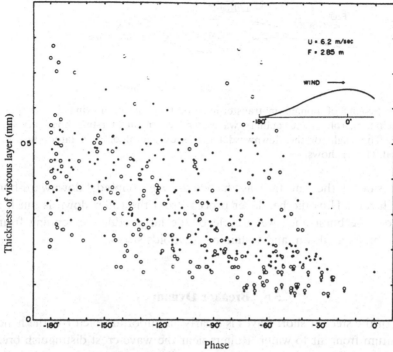

Figure 5.9. Growth of water-side boundary layer on upwind face of laboratory wave, under high shear stress, after Okuda et al. (1977).

wave-generating effect (e.g. Csanady 1991), and all of the momentum transfer via shear stress goes into momentum of the surface shear layer.

The pressure force on the windward face acts very differently: in the wave-following frame, the force presses down on the free surface of a fluid in motion, and its primary effect is wave generation. In the absence of other influences, one would expect the net horizontal pressure force to appear as horizontal wave momentum transport. This turns into shear flow momentum transport in a turbulent boundary layer at the surface, through wave breaking, as discussed in the next section.

The combined surface shear flow sustained by viscous shear and wave breaking transfers its momentum downward again through large eddy motions coupled to the waves. Toba et al. (1975) reported forced convection relative to the crest of short waves, in the first of a remarkable series of laboratory studies at Tohoku University. They found that the "surface converges from the crest, making a downward flow there". In the light of later studies we may place the convergence at the front of the roller on a microscale breaker. According to Ebuchi et al. (1993), "From a point close to the crest on the leeward face of individual waves, a parcel of water with high velocity (burst) goes downward through a particular route relative to the wave phase." They showed the particular route by an illuminating illustration, Fig. 5.10: starting at the toe of the roller, the burst dives down before moving upwind. In further extension of these studies, Yoshikawa et al. (1988) and Toba and Kawamura (1996) have referred to the resulting turbulent flow regime on

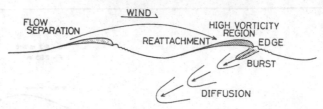

Figure 5.10. Schema of momentum transfer in laboratory waves, showing air flow separation and reattachment, rollers on capillary waves, and downwardbursting boundary layer on water side. The angle of the downward burst is exaggerated, as a later illustration of Ebuchi et al. (1993) shows.

the water side of the interface as the *wind-wave* coupled "downward-bursting boundary layer". Thus the handover process from the roller down is much as on the air side, the burst and sweep cycle of the large eddies extending from the microscale breakers down, and sustaining Reynolds stress.

5.6   Breaker Dynamics

Breaking of the steeper short waves is clearly an important factor in the handover of momentum from air to water. Rollers near the wave-crest distinguish breakers from other waves. Three questions arise regarding breaker dynamics:

1. What triggers the formation of a roller, turning a wave into a breaker?
2. By what process does a breaker lose wave momentum?
3. How does it transfer the same to the underlying shear flow?

Banner and Melville (1976) pointed out that where the particle velocity of the fluid equals wave celerity, in a wave-following frame the particle is stationary, i.e. the flow has a stagnation point. The kinematics of the fluid dictates such a singularity at wave surface both at the front and the rear of a roller, where both the air and the water move with wave celerity. In the absence of a surface shear layer, particle velocity at the wave crest is a function of steepness, and when it reaches wave celerity at high enough steepness, the wave forms a roller. Banner and Phillips (1974) noted further that a shear layer reduces the steepness necessary for breaking. Wave momentum transport increases with wave steepness, so that one may also say that a wave breaks when its momentum transport is higher than some threshold, which depends on the intensity of the surface shear flow, if present.

In two important memoirs Duncan (1981, 1983) answered the second and third questions posed above. Recall that in the wave-following frame the water side is moving upwind, at deep enough levels with a speed equal to wave celerity. An easily studied laboratory analogue is a steady current in a laboratory flume, with a streamlined obstacle (a hydrofoil) submerged at shallow depth, generating a train of standing waves in its wake. These may also be thought of as progressive waves in a frame of reference moving with the current, propagating upstream with a celerity equal to the speed of the current. Their wavelength corresponds to this celerity, and their momentum transport equals the horizontal pressure force the hydrofoil exerts on the flow, known as wave-making resistance. This, at least, is the case as long as the force on the obstacle remains suitably small, either due to slow enough flow or to deep enough placing of the obstacle. Increasing the speed of the current (say) beyond some threshold one finds that the first wave breaks, and forms a roller just upstream of its crest, which is quite turbulent, but statistically steady. The trailing wave train, now of reduced amplitude, transports much less momentum than before breaking.

Duncan (1981, 1983) has elucidated in some detail the effect of the roller on the underlying flow and wave motion. Figure 5.11 shows a schematic sketch of his apparatus and of the control volume for his momentum balances. The key hypothesis of his approach is that the roller on the breaker is an entity separate from the rest of the fluid, interacting with the wave motion as well as with the flow. In justification, recall that a mean streamline separates the roller from the underlying fluid, so that it could be replaced by a solid obstacle, as in many other problems of fluid mechanics. Duncan found that the roller "imparts a shearing force along the forward slope [of the first wave] equal to the component of its [the roller's] weight in that direction. The force produces a turbulent, momentum-deficient wake similar to the wake of a towed, two-dimensional body in an infinite fluid." The weight also presses down on the inclined surface of the fluid below. The horizontal component of this pressure force balances part or all of the force on the submerged foil, while the vertical component generates a phase-shifted wave, reducing the momentum transport of the wave train behind to the magnitude of the unbalanced force. The overall momentum balance is now: wave-making resistance equals wake momentum deficit plus remaining wave momentum transport. The roller has converted most of the wave momentum transport to wake momentum deficit, i.e. generated a shear layer on the surface of the steady current, as if the roller had

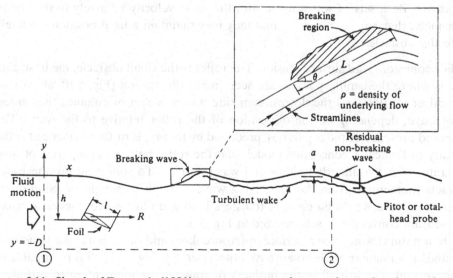

Figure 5.11. Sketch of Duncan's (1991) apparatus generating a stationary breaker and wave train on a flowing stream, also showing control volume (dashed) used in calculating momentum balances.

been a blunt obstacle placed in the fluid. The pressure force of its weight extracted momentum from the wave, and the turbulent shear force on its bottom boundary transferred it to a shear flow below.

There is every reason to believe that rollers on microbreakers of a wind-blown surface act exactly in the way Duncan found: his equations describing the momentum balances apply with few modifications to short wind-blown microbreakers (Csanady 1990). The net horizontal pressure force of the wind acting on a single short wave increases wave momentum transport in the first instance, unless or until a roller forms. In Duncan's experiments, without a surface shear layer, waves with a nondimensional momentum transport gM/c^4 of 0.02 or higher formed rollers (M is kinematic momentum transport, i.e. u^2 depth-integrated, c is wave celerity, i.e. stream velocity). With a shear layer present microscale breakers form at much lower momentum transport, the viscous boundary layers on the interface supplying the fluid in the roller.

The roller extracts momentum from the underlying *wind-wave* if its centre of gravity is over the leeward face, where the wave motion of the surface is upward. In this case its surface pressure on the moving fluid generates a wave of the same wavelength as the underlying wave, but with a phase shift such as to partly cancel the latter. The hydrodynamical problem is the same as of the fishing line, described in Lamb (1957) by a quote attributed to Russell:

> When a small obstacle, such as a fishing line, is moved forward slowly through still water, or (which of course comes to the same thing) is held stationary in moving water, the surface is covered with a beautiful wave-pattern, fixed relatively to the obstacle. On the up-stream side the wave-length is short, and as Thomson has shewn, the force governing the vibrations is principally cohesion. On the down-stream side the waves are longer and are governed principally by gravity. Both sets of waves move with the same velocity relatively to the water; namely, that required in order that they may maintain a fixed position relatively to the obstacle.

For cohesion, read surface tension. The roller is the small obstacle, the upstream side is where the capillary waves are seen in the illustration (Fig. 5.10, above), of Ebuchi et al. 1993 and the downstream side waves cancel or enhance the underlying wave, depending on the disposition of the roller relative to the wave. The observed presence of the capillaries, predicted by theory, is further evidence for the validity of Duncan's conceptual model, that the roller behaves as a parcel of fluid separate from the underlying flow and wave motion. To sum it all up, the roller extracts momentum from the underlying wave through its weight pressing down where the wave would go up, and transfers it to a trailing wake, "wave to wake momentum conversion", schematized in Fig. 5.6.

On a natural wind-blown surface, microbreakers and other bore-like structures abound, as a number of observers have noted (see e.g. Fig. 5.5b). It is reasonable to suppose that they are all in the business of transferring momentum to turbulent surface shear layers, having gained momentum from pressure and shear forces on the interface one way or another. We may lump all their contributions to momentum transfer via the short wave route, all of them contributing to the high wave-number end of the wave spectrum.

5.7   Concluding Remarks

Two remaining questions are, how precisely the energy containing waves extract the momentum from the air flow that they transfer to the water, a minor fraction of the total that may be, and how or to what extent the same long waves influence momentum transfer via the short waves riding on their backs? Other chapters of this volume address these questions, which are still in the forefront of current research.

6 Coupling Mechanisms

J. Bye, V. Makin, A. Jenkins and N. E. Huang

6.1 Introduction

Traditionally, the wind has been considered as the driving force for all ocean dynamics phenomena. Thus, we have the classical works on wind generated ocean circulation (Robinson 1963), and wind generated waves (Kinsman 1965). Almost half a century has elapsed, yet the prevailing thinking in the ocean community remains unchanged: Scientists engaged in ocean model development still fall into two categories – ocean circulation modellers who produce General Circulation Models (CGMs) and *wind-wave* modellers who have constructed, for example, Wave Modelling (WAM). In each respective endeavour, wind is considered as given and unchanged. This viewpoint is now changing. Research seems to be moving in the direction of treating the atmosphere and the oceans as a single system.

Wind-generated waves and currents are fundamental features of the world oceans. As the wind starts to blow over a resting ocean surface, it first generates small-scale *wind-waves*. These *wind-waves* extract momentum and energy from the wind field and modify the effective momentum flux into currents and also influence the wind field itself. The momentum flux will generate the drift currents, which in turn begin to influence the amplitude and directionality of the surface *wind-wave* field. The general ocean circulation pattern will also transport heat from one region to another to modify the global atmospheric dynamics. Thus, they form a closely knitted interacting trinity. The interactions among the wind, *wind-wave*, and current are the essence of the world ocean dynamics. In storm systems and in the coastal regions, the wave field never becomes uniform, so that the atmospheric forcing should also remain spatially inhomogeneous. The key to understanding lies in the detailed processes of *wind-wave* interactions, for that is the controlling mechanism of the energy and momentum fluxes.

Ian S. F. Jones and Yoshiaki Toba, *Wind Stress over the Ocean*. © 2001 Cambridge University Press. All rights reserved. Printed in the United States of America.

   The ultimate aim is to produce an interactive model in which the specification of the surface geostrophic wind and current as an evolving field can be used to obtain the surface wave field. In this process interpolation algorithms for approaching the sea surface from the top of the atmospheric boundary layer, and from the bottom of the oceanic boundary have to be developed. In general these algorithms must take account of the stratification in the two fluids, however the most important consideration is the influence of the *wind-waves* themselves on the dynamics of the boundary layers. In particular, the surface shearing stress, τ, without doubt, cannot be specified from the free stream velocities of the two fluids in isolation from the wave field.

   It is salutory to remember however that even if a fully interactive scheme is developed in which τ is adequately predicted, for the purposes of the local wave-turbulence dynamics, this exercise may not reveal everything about the momentum exchange between the two fluids. Important exchange processes may also occur on the larger scales, ultimately on the scale of the ocean basin, which contains the wave field. The link between the local dynamics and mesoscale transfer is explored in Chapter 11. In this chapter, we focus almost exclusively on only one aspect of the local wave-turbulence problem, namely the influence of the wave field on the specification of the local surface shearing stress.

   Some of the fundamental questions, that need to be addressed, are: What are the momentum and energy fluxes across the air–sea interface? How are the fluxes partitioned among *wind-waves*, currents, and turbulence? What are the feed back mechanisms? At the outset however, we briefly summarize the *wind-wave* model, which depends to a large extent on the validity of the specification of τ^{w} which is that part of the shearing stress τ which enters the wave field, and is termed the "wave-induced stress".

6.2   The *Wind-wave* Model

The *wind-wave* model which is becoming extensively used in coupled ocean–atmosphere modelling is the third-generation ocean *wind-wave* prediction model, known as the WAM model, which was developed by a group of mainly European scientists who called themselves the WAM Group (WAMDI 1988). In the WAM model, the air–sea interface is modelled using the energy transport equation,

$$\frac{\partial \Phi}{\partial t} + \left(\mathbf{u} + \mathbf{c}_{\mathrm{g}}\right) \bullet \nabla \Phi = s_{\mathrm{in}} + s_{\mathrm{nl}} - s_{\mathrm{ds}} \tag{6.1}$$

in which $\Phi(k, \theta)$ is the directional *wind-wave* spectrum, where

$$\overline{\eta^2} = \int \Phi(k)\mathrm{d}\mathbf{k}$$

is the mean square surface elevation, s_{in} is the input source function, s_{nl} is the nonlinear interactions source function, s_{ds} is the dissipation source function, $\mathbf{c}_{\mathrm{g}}$ is

the group velocity and **u** is the current velocity. If we integrate the above equation
with respect to the wavenumber, **k**, then we have

$$\frac{\partial E}{\partial t} + (\mathbf{u} + \mathbf{c}_g) \bullet \nabla E = S_{in} - S_{ds} \qquad (6.2)$$

in which,

$$E = \rho \int \frac{\omega^2}{k} \Phi(\mathbf{k}) d\mathbf{k}$$

is the wave energy, and

$$S_{in} \equiv \left. \frac{\partial E}{\partial t} \right|_{in} = \rho \int \frac{\omega^2}{k} s_{in} d\mathbf{k}$$

$$S_{ds} \equiv \left. \frac{\partial E}{\partial t} \right|_{ds} = \rho \int \frac{\omega^2}{k} s_{ds} d\mathbf{k}$$

are respectively the total energy flux to the wave field by the wind, and from the
wave field through dissipation due to whitecapping, and the advection term has
been approximated using the group velocity of the dominant part of the wave
spectrum (Weber 1994). Therefore, the total input to the wave field is the sum
of *wind-wave* energy growth and the loss. In any storm, the sea state can have three
different stages: the growth, the equilibrium, and the decay stages, which are characterized by S_{in} greater than, equal to, or less than S_{ds}. The dynamics will be
different in each stage as indicated by laboratory and field measurements (Hsu
et al. 1982) and modelled by Papadimitrakis et al. (1987), and the controlling
source functions (s_{in}, s_{nl} and s_{ds}) can be expressed in terms of the directional
spectrum $\Phi(\mathbf{k})$. Details of these relations can be found in Phillips (1985),
WAMDI (1988) and Komen et al. (1994). For our purposes the essential transfor-
mations which relate s_{in} and s_{ds} to the wave momentum,

$$\mathbf{P} = \rho \int \frac{\omega}{k} \Phi(\mathbf{k}) \mathbf{k} d\mathbf{k}$$

and which correspond with the energy fluxes, $(\partial E/\partial t)_{in}$ and $(\partial E/\partial t)_{ds}$ are respectively,

$$\tau^w \equiv \left. \frac{\partial P}{\partial t} \right|_{in} = \rho \int \frac{\omega}{k} s_{in} \mathbf{k} d\mathbf{k}$$

and

$$\tau^o \equiv \left. \frac{\partial P}{\partial t} \right|_{ds} = \rho \int \frac{\omega}{k} s_{ds} \mathbf{k} d\mathbf{k}$$

τ^w is commonly called the *wind-wave* induced stress, and τ^o is a dissipation stress,
i.e. rate of loss of momentum from the wave field to the ocean.

Using these definitions, and integrating Eq. (6.1) over wavenumber space, we
have, corresponding to the energy equation (6.2), the wave momentum budget
equation

$$\frac{\partial \mathbf{P}}{\partial t} + (\mathbf{u} + \mathbf{c_g}) \bullet \frac{\partial \mathbf{P}}{\partial x} = \tau^w - \tau^o \tag{6.3}$$

The relative importance of τ^w (or of the turbulent stress, $\tau^t = \tau - \tau^w$) to the total shearing stress τ is a measure of the *wind-wave* effect on the surface stress, the study of which has motivated this monograph.

We can divide the open ocean sea state into two classes:

(1) Duration limited in which there is an actively growing wave field, and,
(2) Oversaturated in which a developed wave field turns to swell, which is unsupported by the local wind.

State (1) is characterized by a wind-wave stress τ^w which is a substantial proportion of the shearing stress τ, and so it will be assumed that $\tau^w \sim \tau^o$ such that the majority of the shearing stress is transmitted to the ocean locally through the wave field. This assumption is sharply different from the view of Chapter 5. The mechanisms of these processes are discussed in Sections 6.3 and 6.4. In State (2), the situation is more complicated and there is the possibility of momentum transfer in the opposite direction from the ocean to the atmosphere through the swell. This process is explored theoretically in Section 6.5, and observations of negative stress are discussed separately in Chapter 7.

The equilibrium in both states also depends on local and advective changes, see Eq. (6.3), however a quasi-steady balance can be considered as a first approximation. In other words, we can confine ourselves to a horizontally homogeneous balance calculation for both turbulent and wave momentum and energy. Under these assumptions one expects the stress near the water surface to be constant with height as illustrated in Fig. 6.1.

6.3 Momentum Transfer Below the Sea Surface

Near the surface, the downward flux of momentum is partly carried by the wave orbital motions $(\tilde{u}, \tilde{w})$. In a fixed rectangular coordinate system, this part of the momentum flux is equal to

$$\tau = -\rho \overline{\tilde{u}\tilde{w}}$$

where ρ is the water density, and is zero if the wave motion is irrotational (extra terms appear in a curvilinear coordinate system, and also when using Lagrangian hydrodynamic equations or a generalized Lagrangian mean formulation although the physical situation is unchanged (Jenkins 1992)). In order to evaluate, τ, it is instructive to consider the case where we represent all viscous stresses and turbulent motions by employing a viscosity or eddy viscosity which depends only on a vertical coordinate (curvilinear or Lagrangian) in which the air–sea interface is a coordinate surface and perform a perturbation expansion of the wave field to $O\{(ka)^2\}$, a being the wave amplitude. For a constant (eddy) viscosity (Longuet-Higgins 1953 and 1960; Weber 1983; Jenkins 1986), the loss of momentum from waves due to wave damping provides a drag force τ^o on the water column which is exerted from a thin boundary layer near the surface of thickness

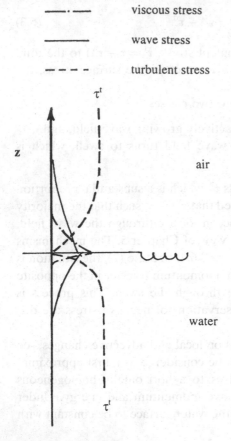

——·——   viscous stress

————   wave stress

— — — —   turbulent stress

Figure 6.1. Conceptual model for the partitioning of wave stress, viscous stress and turbulent stress assuming a region of constant total stress each side of the interface. The problem is assumed to be two dimensional.

$(2\nu/\omega)^{1/2}$ where ν is the kinematic viscosity and ω is the *wind-wave* angular frequency. The approach can be generalized by including the Coriolis force, computed from the Lagrangian mean current, including the wave-induced Stokes drift, not just the Eulerian mean current (Hasselmann 1970; Jenkins 1986; Weber 1990; Huang 1979; Weber and Melsom 1993): and an eddy viscosity which varies with depth. This model (in which part of the force associated with *wind-wave* damping appears as a body force with a magnitude proportional to the eddy viscosity gradient (Jenkins 1986)) has been used by Jenkins (1989) to compute near-surface currents using the wave field from the spectral evolution model WAMDI (1988). A necessary condition is that the *wind-wave* model treats the wave momentum balance correctly, i.e. the nonlinear transfer term should conserve momentum when integrated over all wavenumbers. The momentum transferred from the atmosphere to the wave field, τ^w, as well as directly to the surface current τ^t should also be correctly specified, as in Janssen (1989) and Jenkins (1992, 1994).

It is however, unrealistic to describe turbulence and *wind-wave* dissipation using just an eddy viscosity. The short time scales of wave motions will reduce the effective eddy viscosity acting upon them, and this effect can be quantified by rapid-distortion turbulence theory. Systematic treatments of the interaction between the mean flow, wave orbital motion, and turbulent velocity fluctuations have been presented by various authors such as Thais and Magnaudet (1995). Where breaking *wind-wave* crests are present, forward momentum will be injected into the water column from the plunging jet and/or the turbulent wake behind the crest (Csanady 1985, 1990; Huang 1986; Jenkins 1994; Longuet-Higgins 1973, 1983, 1990, 1992). Breaking crests also act as a significant source of turbulent kinetic energy in the water column, and increase the effective roughness length of the surface when seen from below, to values very much in excess of the corresponding values for a solid wall supporting the same shear stress (for example see Fig. 1.6 and Alekseenko and Egorov 1991; Anis and Moum 1995; Newell and Zakharov 1992; and Yegorov 1996). In fact, the scale of the turbulent structures near the surface appears to be so large that it may be better to describe them as coherent structures (Soloviev 1990), even when *wind-wave* breaking is weak or absent. The span-wise vortex rolls of Langmuir circulation can also provide significant vertical transport of momentum, mass etc. (Leibovich 1983 or see Chapter 11).

These effects also feed back on the wave field, since the turbulent mean part of the near-surface current adds a rotational component to the surface waves. This reduces the maximum wave amplitude required for breaking (Banner and Phillips 1974; Phillips and Banner 1974), and will thus increase the drag coefficient for the airflow (Banner 1990). The presence of surface films, whether monomolecular films of surfactants or finite-thickness films of viscous fluids such as heavy oil, on the other hand will cause wave damping and a corresponding drag reduction, see Chapter 13.

Finally, we note that although τ^o can be evaluated from the wave field, momentum transfers below the sea surface due to $(\tau - \tau^o)$ and τ^o are so inextricably linked that they are treated as a single process.

6.4   Coupling *Wind-waves* with the Atmosphere

The momentum flux above the sea surface depends both on the wind speed and the sea state. A theory that takes into account the sea state to calculate the momentum flux was introduced by Janssen (1989) and later by Chalikov and Makin (1991) in which the state of the sea which was characterized by the directional *wind-wave* spectrum $F(k, \theta)$ is explicitly taken into account to describe the exchange of momentum.

The approach is based on the conservation of momentum in the marine atmospheric surface layer, which implies that the total stress is independent of height, so that in horizontally homogeneous conditions the momentum flux at a given height is equal to the momentum flux at the sea surface. It is also assumed that the lateral stress is zero at all heights. By this assumption the total momentum flux equals the square of the friction velocity u_*

$$\frac{\tau(z)}{\rho_a} = u_*^2$$

The total stress τ is supported by both turbulent motions of the air τ^t and by the organized wave-induced motions due to the presence of *wind-waves* τ^w:

$$\tau = \tau^t(z) + \tau^w(z) \tag{6.4}$$

Figure 6.1 shows this with the addition of the viscous stresses. The turbulent flux is parametrized in terms of the mixing-length theory:

$$\frac{\tau^t}{\rho_a} = (\kappa z)^2 \frac{du}{dz} \left| \frac{du}{dz} \right| \tag{6.5}$$

At the sea surface ($z = 0$) the momentum flux is supported by viscous stresses and

$$\tau^w \equiv \tau^w(0)$$

where,

$$\frac{\tau^w}{\rho_a} = \int_o^\infty \int_{-\pi}^\pi \omega^2 F(k,\theta) \Gamma \cos\theta k dk d\theta \tag{6.6}$$

in which the wavenumber k is assumed to satisfy the dispersion relation $\omega^2 = gk$ and ω is the *wind-wave* frequency. The energy flux from the atmosphere to waves s_{in}, defined in terms of the growth rate parameter $\Gamma(k,\theta)$ is $s_{in} = \omega\Gamma F$. The *wind-wave* induced flux $\tau^w(z)$ decays rapidly with height, and Janssen (1989) found that the height dependence is such, that the wind profile following from the mixing length theory is:

$$u(z) = \frac{u_*}{\kappa} \ln \frac{z + z_0 - z_b}{z_0} \tag{6.7}$$

with a derivative

$$\frac{du}{dz} = \frac{u_*}{\kappa} \left(\frac{1}{z + z_0 - z_b} \right)$$

Substituting Eq. (6.7) into Eq. (6.5) and evaluating Eq. (6.4) at $z = z_b$ to relate z_b to the sea roughness, the following expression is obtained:

$$z_0 = \frac{z_b}{\sqrt{1 - \psi}} \tag{6.8}$$

where $\psi = \tau^w/\tau$ is the coupling parameter and the 'background' roughness z_b is assumed to be given by the Charnock relation:

$$z_b = \hat{z}_b \frac{u_*^2}{g} \tag{6.9}$$

where $\hat{z}_b$ is a constant. The fact that the background roughness is parametrized with the Charnock relation suggests that here also gravity waves play an important role, which appears to be inconsistent with the assumption that the gravity waves are already accounted for in τ^w defined by Eq. (6.6).

This problem was resolved by Chalikov and Makin (1991) and Chalikov and Belevich (1993) who used detailed calculations of the turbulent boundary layer above *wind-waves* of finite amplitude (Makin 1989) to parametrize the vertical distribution of the *wind-wave* induced flux $\tau^w(z)$. They assume that the background roughness (z_b) is proportional to the height (h_T) of the short gravity waves, which scales with u_*^2/g, and using the Phillips spectrum Eq. (4.1), is related to the Phillips parameter (β). Hence the Charnock type relation Eq. (6.9) can be used to parametrize the background roughness. For seas which are old, the use of the Charnock type relations ensures the proper values of the drag coefficients calculated from the models. An alternative approach to fix this inconsistency has been proposed by Caudal (1993) who attempted to bypass the problem by assuming that the total surface stress is supported only by *wind-waves*, i.e. $\tau = \tau^w$. This assumption is inconsistent with the picture in Chapter 5 and is shown to be deficient by Makin et al. (1995).

A consistent theory to calculate the sea drag which accounts for the balance between the *wind-wave* induced and the turbulent stress at the surface and which avoids the use of the Charnock type relation for the background roughness was introduced by Makin et al. (1995). They assume that the instantaneous water surface can be treated as a smooth one. That assumption allows them to relate the local roughness length to the scale of the molecular sublayer; such that

$$z_b = 0.1 \frac{\nu}{u_*^t} \tag{6.10}$$

where ν is the kinematic viscosity of the air and $u_*^t = \left(\tau^t/\rho_a\right)^{1/2}$. Thus z_b is independent of the *wind-wave* stress. Once this assumption has been made the sea roughness can be derived as a result of the balance between the *wind-wave* induced and the turbulent stress. The contribution of *wind-waves* to the total wave induced stress (τ^w) obtained by Makin et al. (1995) is shown in Fig. 6.2.

The weakness of all these models is that the momentum flux to *wind-waves* calculated from (6.6) is very sensitive to the form of the high frequency part of a *wind-wave* spectrum. Makin et al. (1995) use the Donelan et al. (1985), Donelan and Pierson (1987) model of a *wind-wave* spectrum. In that model the *wind-wave* spectrum consists of two parts. The energy containing part is proportional to the inverse wave age parameter $\left(U_{10}/c_p\right)^{0.55}$, where U_{10} is the wind speed at 10 m height and c_p is the wave phase speed at the peak of the spectrum. The high frequency part, which is patched to the former at the wavenumber which is about 10 times the peak wavenumber of the fully developed spectrum, has no wave age dependence. Nor was it wind speed dependent in the manner of Section 4.2. They found that about 90% of the *wind-wave* induced stress at the surface is supported by *wind-waves* shorter than 7 m and less than 5% is supported by short gravity and gravity–capillary waves with lengths shorter than 0.05 m. The drag which resulted from that model was virtually independent of wave age. This is the consequence of the fact that the *wind-wave* induced stress (τ^w) was supported by the high frequency part of the spectrum which is wave age independent. Both Janssen (1989), Chalikov and Makin (1991) found a pronounced wave age dependence of the drag, when the wave age dependence of the JONSWAP spectrum was taken to be $\beta \sim \left(U_{10}/c_p\right)^{3/2}$, which diminished if it was assumed that $\beta \sim \left(U_{10}/c_p\right)^{2/3}$. The actual dependence on wave age of the high wavenumber

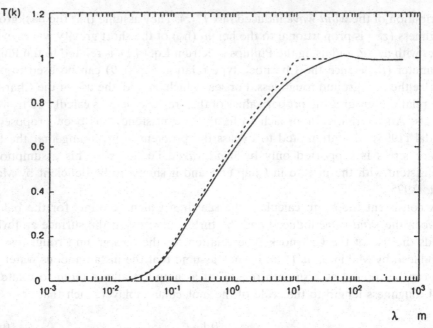

Figure 6.2. Cumulative contribution of waves to the total wave-induced stress $T(k) = \tau^w(k)/\tau^w$ with increasing wavelength component. Solid line, $U_{10}/c_p = 0.83$; dashed line $U_{10}/c_p = 3$ where c_p is the phase speed of the peak of the wave spectrum, and $U_{10} = 10$ m/s (from Makin et al. 1995).

portion of the spectrum is difficult to determine from the frequency spectrum (see Section 4.2.3). This issue is still in need of experimental verification.

In summary the *wind-wave* coupling model indicates that Charnock's parameter should be modified by the factor $(1 - \psi)^{-1/2}$ when the momentum flux to the *wind-wave* spectrum is included. Janssen and Viterbo (1996) applied this two-way interaction scheme to the wave model WAM with $\hat{z}_b = 0.01$, and found that significant changes in *wind-wave* climate were produced in comparison with simulations using a constant Charnock parameter ($\psi = 0$).

6.5 Flow Similarity in the *Wind-wave* Boundary Layer

The discussions of Sections 6.3 and 6.4 may suggest that the *wind-wave* boundary is such a complex entity that it is not possible to describe it, even conceptually, using generalized arguments, such as similarity which have been applied as models for wall boundary layer dynamics.

Observations of drift profiles under the water surface (Bye 1965, 1988; Chang 1969; Lin and Gad-el-Hak 1984; Shemdin 1972) however yield the result that the velocity profile can be also represented approximately by a logarithmic variation in depth in a similar manner to the logarithmic variation with height in the atmosphere. Also Jones and Kenny (1977) found that some distance below the surface the turbulence behaved as in a constant stress wall layer. Hence although the physical processes operating in the two fluids are different, e.g. there is a critical layer mechanism embedded in the turbulence over the *wind-waves*, whereas under the *wind-waves* there is a significant shear due to the irrotational wave dynamics, it would appear that the statistical equilibrium in the constant stress layer in both fluids can be represented using similarity profiles.

The roughness lengths obtained from the logarithmic profiles however cannot be defined without introducing a reference velocity in each fluid. This reference velocity is not prescribed a priori as in the wall boundary layer, where it is zero. It is possible however to choose the two reference velocities so that the roughness lengths in the two fluids are equal, in contrast with the discussion in Chapter 5. This model recognizes that the two fluids are coupled by a shared interface, notwithstanding that this interface separates two very different dynamical systems.

The similarity velocity profiles for this model, assuming that the flow is two-dimensional, are (Bye 1988) in air,

$$u(z) - u_R = \chi u_* + \frac{u_*}{\kappa} \ln \frac{z}{z_R} \tag{6.11}$$

and in water

$$u(-z) - u_R = \chi w_* - \frac{w_*}{\kappa} \ln \frac{(-z)}{z_R}$$

in which $z = 0$ is the mean interface, $u(z)$ is the fluid velocity, u_R is the mean reference velocity common to both fluids, u_* is the friction velocity in air, and w_* is the friction velocity in water, and χ is a similarity parameter. This arrangement is illustrated in Fig. 6.3.

The profiles share a common roughness length (z_R) which occurs at a velocity $\chi u_* + u_R$ in air, and $\chi w_* + u_R$ in water. In the absence of a shearing stress, both

Figure 6.3. Schematic picture of the coupling process described in Section 6.4.

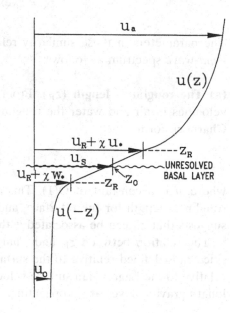

fluids travel at the reference velocity (u_R). The aerodynamic roughness length (z_0) used by Charnock and others can be recovered from these similarity relations by formally extrapolating each profile into the "unresolved basal layer" (Wilson and Flesch 1993), to the level at which the velocities are equal.

By this procedure (which can be used in the field when simultaneous velocity profiles in air and water are available) we obtain the intersection (u_d, z_0) where

$$u_d = u(z_0) = u(-z_0)$$

is the drift velocity, which yields the alternate pair of similarity profiles,

$$u(z) - u_d = \frac{u_*}{\kappa} \ln \frac{z}{z_0} \tag{6.12}$$

(from which we recover a definition of z_0 which may be different to that used previously), and in water

$$u(-z) - u_d = -\frac{w_*}{\kappa} \ln \frac{(-z)}{z_0}$$

which represent the fluid motion without explicit reference to u_R, and following Kraus (1977), where $\varepsilon = (\rho_a / \rho)^{1/2}$ and $u_* = \varepsilon w_*$

$$u_d = \frac{\varepsilon u(z) + u(-z)}{1 + \varepsilon} \tag{6.13}$$

and also from Eq. (6.11) and Eq. (6.12),

$$z_R = z_0 \exp\left(\frac{\chi(1 - \varepsilon)\kappa}{(1 + \varepsilon)}\right) \tag{6.14}$$

The parameters in these similarity relations are related to the properties of the *wind-wave* spectrum as follows:

(a) The roughness length (z_R). From simultaneous observations of logarithmic velocities in air and water the roughness length (z_R) was observed to be of the Charnock form,

$$z_R = a u_*^2 / g \tag{6.15}$$

where, a is a constant of O(1). This result is consistent with z_R being the true roughness length for the interface, and for a moderate breeze $z_R \sim 1$ cm, which suggests that z_R can be associated with the shorter wavelength gravity waves.

The relation between z_R and Charnock's roughness length (z_0) is kinematic, since z_0 is defined relative to the surface drift velocity (u_s) whereas z_R is defined relative to the Lagrangian surface velocity ($u_R + \chi w_*$), on which the shorter wavelength gravity waves are propagating.

(b) The similarity parameter (χ). The simultaneous profile observations also indicated that $\chi \sim 12$, such that from Eq. (6.14) $z_R \sim 10^2 z_0$.

(c) The water similarity velocity (χw_*). It was also found (Bye 1988) that the water similarity velocity (χw_*) could be estimated by the spectrally integrated Stokes surface velocity (εu_L) where from Kenyon (1969).

$$\varepsilon u_L = \int_o^\infty \int_{-\pi}^\pi 2F(k)g^{1/2}k^{3/2} \cos \theta k d\theta \, dk \qquad (6.16)$$

(d) The air similarity velocity (χu_*). More recently, the air similarity velocity (χu_*) has been estimated by the spectrally integrated wave phase velocity (u_L) where,

$$u_L = \int_o^\infty \int_{-\pi}^\pi cF \cos \theta k d\theta dk / \overline{\eta^2} \qquad (6.17)$$

in which $c = (g/k)^{1/2}$, and $\overline{\eta^2}$ is the mean square surface elevation. Equations (6.16) and (6.17) have the property that they are satisfied by the Phillips spectrum (Eq. 4.1) with $\beta = 2\varepsilon/5$ which predicts $\beta = 1.36 \times 10^{-2}$ for the air–water system in remarkable agreement with experimental observations (Phillips 1977), and also $u_L < 0.8c_p$ where $c_p = g/\omega_p$ in which ω_p is the peak frequency of the spectrum. On substituting for χu_* the phase velocity (u_L), Eq. (6.14) can be expressed in the form

$$z_R = z_0 \exp\left(\frac{\kappa u_L}{u_*}\right) \qquad \varepsilon \to 0$$

which, since $u_L < 0.8c_p$, identifies z_R as a low estimate of the matched height (z_M) of Phillips (1977) and Belcher and Hunt (1993), at which level the momentum input to the *wind-waves* through the critical layer mechanism is centred. This relation suggests that Charnock's parameter is controlled by two factors; the nondimensional matched height, a, and the nondimensional wave phase velocity, χ. The predicted increase in Charnock's parameter for a decrease of χ, i.e. a reduced wave state, is consistent in sign with the increase in Charnock's parameter due to *wind-wave* stress (τ^w), see Section 6.3.

This discussion illustrates that the similarity formulation is able to articulate both the turbulent and *wind-wave* aspects of the boundary layer. The full significance however was only realized later, when the collinearity of the *wind-wave* and turbulent components was relaxed, and the similarity relations (6.11) were expressed using the drag coefficient K_R relative to z_R,

$$K_R(|z|) = \left(\frac{1}{\kappa}\ln\frac{|z|}{z_R}\right)^{-2}$$

On eliminating $\mathbf{u}_L$ between the generalized two-dimensional relations it was found (Bye 1995) that

$$\tau = \rho_a K \left|\mathbf{u}(z) - \mathbf{u}_R - \frac{1}{\varepsilon}(\mathbf{u}(-z) - \mathbf{u}_R)\right| \bullet \left(\mathbf{u}(z) - \mathbf{u}_R - \frac{1}{\varepsilon}(\mathbf{u}(-z) - \mathbf{u}_R)\right) \qquad (6.18)$$

where $K = K_{R/4}$ or alternatively on eliminating u_R, that

$$\tau = \frac{\rho_a K_R}{(1+\varepsilon)^2} |u(z) - u(-z) - (1 - \varepsilon)u_L| \bullet (u(z) - u(-z) - (1 - \varepsilon)u_L) \qquad (6.19)$$

The remarkable feature of the inertial coupling relation (6.18) is that the ocean shear is weighted by a factor $(1/\varepsilon \sim 30)$ relative to the atmospheric shear. The alternative relation (6.19) illustrates that this effect is due to the presence of the wavefield. Implications for ocean general circulation modelling are discussed in Bye and Wolff (1999).

On applying these results at the edge of the wave boundary layer $(z = \pm z_B)$ at which $u(z) = u_a$ and $u(-z) = u_0$ where u_a is the surface wind, and u_0 is the surface current, the following conclusion can be drawn: In commonly occurring seastates the surface current (u_0) is usually much smaller than either the surface wind (u_a) or the spectrally integrated phase velocity (u_L) and hence Eq. (6.19) reduces to the approximate bulk aerodynamic relation,

$$\tau \sim \rho_a K_R |u_a - u_L|(u_a - u_L) \qquad \varepsilon \to 0 \qquad (6.20)$$

which indicates that the surface shearing stress only lies along the direction of the surface wind u_a if the wave field (u_L) is aligned with the wind. In changing weather conditions significant deviations would occur as discussed in the observations of Chapter 9. In the absence of wind

$$\tau_s \sim -\rho_a K_R |u_L| u_L \qquad (6.21)$$

which implies that a negative stress occurs, due to the propagation of swell. Observations of this process are discussed in Chapter 7.

The generality (including the possible dependence of z_R on wave parameters) of the predictive set of equations (6.18–6.20) for τ, which take into account the *wind-wave*, swell and current fields, rather than relying solely on a relation between roughness and wave age (Chapter 10), remains to be explored.

In summary, progress has been made towards an understanding of the coupling process between the ocean and the atmosphere, which is beginning to model the physics which occurs at the fluid interface; however it is likely to be some time before a comprehensive theoretical treatment will be generally accepted and implemented.

7 The Measurement of Surface Stress

S. E. Larsen, M. Yelland, P. Taylor, Ian S. F. Jones, L. Hasse and
R. A. Brown

7.1 Introduction

While the previous chapters have described the models of drag coefficient and the physics of momentum transfer at the sea surface, it is still necessary to make some direct measurement of the stress. There are unanswered questions about the value of the drag coefficient in unsteady and non-ideal conditions. One needs stress measurement to resolve these issues. In some upper ocean experiments the exact flux of momentum is important enough to justify the added work in making stress measurements rather than the simpler wind measurements needed to use with the drag coefficient. There is also the possibility that from ships and other complex structures that induce large flow distortions it may be easier to measure the stress accurately than to measure the wind speed accurately.

7.2 The Measurement of Surface Stress

Over solid surfaces the stress over a portion of the surface can be obtained by relatively straightforward methods such as measuring the force on a drag plate. The sea surface offers no such opportunity. The breaking waves that dominate the surface under conditions of strong forcing make it difficult to imagine a direct measurement technique. Thus we rely on remote observation from which we infer the drag force.

This chapter outlines the three common ways of estimating the surface stress. They are known as the Reynolds stress method, dissipation method and profile method. A fourth method of observing the setup of a closed body of water has much more restricted application. A fifth way, using radar remote sensing, is gaining popularity with the introduction of satellites carrying such instruments. Such an observing technology is described in Section 7.8.

Atmospheric stability has an important influence on the velocity profile so it is common to introduce the concept of an equivalent neutral stability drag coefficient The wind profile can be approximated as

$$U(z) = \frac{u_*}{\kappa}\left(\ln\frac{z}{z_0} - \Psi\right) \tag{7.1}$$

where Ψ is a stratification function. Now, defining the equivalent neutral wind, $U_n(z)$ as:

$$U_n(z) = \frac{u_*}{\kappa}\ln\frac{z}{z_0} \tag{7.2}$$

Equations (7.1) and (7.2) can be combined to give:

$$U_n(z) = U(z) + \Psi u_*/k \tag{7.3}$$

The stratification function is found empirically for moderate instability to be:

$$\Psi = 2\ln\left(\frac{1 + \Phi_m^{-1}}{2}\right) + \ln\left(\frac{1 + \Phi_m^{-2}}{2}\right) - 2\tan^{-1}\left(\Phi_m^{-1}\right) + \pi/2 \qquad z/L < 0 \tag{7.4}$$

where $\Phi_m = (1 - c_1 z/L)^{-1/4}$; and L is the Monin–Obukhov length (see Eq. 3.30), while for stable conditions

$$\Psi = -c_2 z/L \qquad z/L \geq 0$$

where c_1 is about 16 to 18 and c_2 is about 4 or 5.

Thus, using the measured wind speed, wind friction velocity and Monin–Obukhov length, the equivalent neutral wind speed at a given elevation is obtained and from this velocity the equivalent neutral drag coefficient determined.

7.3 Reynolds Stress Measurements

In Chapter 3 it was shown how the flux of momentum to the sea surface could be approximated by the Reynolds stress $\langle u'w'\rangle$ and $\langle v'w'\rangle$ (where w' is the fluctuating component of the vertical velocity and $\langle\ \rangle$ indicates a mean quality) some distance above the surface. Some attention must be paid in dividing the velocity into its mean parts U and V and its fluctuating components u' and v'. Even the vertical velocity W cannot be assumed to be zero for averages over short times.

If the stress is measured some distance above the surface, an estimate of the surface stress can be made. The simplest assumption is that the stress is uniform and aligned with the wind at the measurement height. Atmospheric stability and unsteadiness can severely compromise this assumption. For stable to moderately unstable atmospheric conditions a reasonable expression for the stress profile is (e.g. Sorbjan 1989)

$$\frac{\tau(z)}{\rho u_*^2} = \left(1 - \frac{z}{h}\right)^n \tag{7.5}$$

where n is in the range 1–2, and h is the boundary layer height. For neutral and unstable conditions h will usually be of the order of several hundred metres, meaning that for a measuring height of the order of several tens of metres z/h can easily amount to 0.1 or so. Even larger variation can occur for stable atmospheric boundary layers, where h will be of the order of 150 m or less. For such layers the difference between the surface stress and the stress measured at an elevated position can be quite large. In addition, over *wind-waves* some of the stress is supported near the surface by pressure forces and so the Reynolds stress is only part of the flux of momentum.

The direct measurement of the atmospheric momentum flux is known as the eddy correlation method.

7.3.1 The Reynolds Stress

A typical Reynolds stress record is shown in Fig. 7.1 where the product $u'w'$ is plotted after the 15 minute mean of the wind U and W have been removed. In this instance, while $u'w'$ averaged over 15 minutes was -0.189 m²/s², the orthogonal Reynolds stress $v'w'$ has the value of -0.005 m²/s², significantly different from zero. Thus stress was not aligned with the 15 minute average wind, even though the coordinate system was chosen so that $V = 0$ at the height of the observation.

The $u'w'$ record is obviously not normally distributed. It contains more spikes than a Gaussian record. These bursts of momentum have been frequently studied but in many cases are no different than would be expected from the product of a pair of normally distributed variables. Sreenivasan et al. (1978) provide an analysis. This can be illustrated by plotting the probability distribution of the record shown above as has been done in Fig. 7.2 and comparing it with a joint normal distribution with the same expected $\langle u'w' \rangle$ value.

The high kurtosis of such a signal has implication for the length of record that must be used to obtain a reliable estimate of the mean of $u'w'$. A single large value of $u'w'$ in a short record can influence its mean.

On this day the value of the 15 minute mean U and W were not constant but varied from period to period. The Reynolds stress comes from writing $(U + u')(W + w')$ as can be seen in Chapter 3. Averaging the four terms gives, $\langle u'w' \rangle$, in addition to $\langle UW \rangle$, $\langle Uw' \rangle$ and $\langle u'W \rangle$, all of which are zero for long term averages in steady (horizontally uniform) flow. However the terms $\langle UW \rangle$ can carry

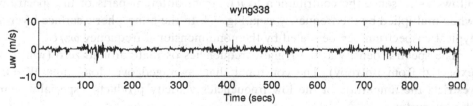

Figure 7.1. The time series of u' and w' from a 15 minutes record taken at a height of 53 m over the sea in Bass Strait. Details are in Jones and Negus (1996). The positive w' is defined as flow away from the sea surface.

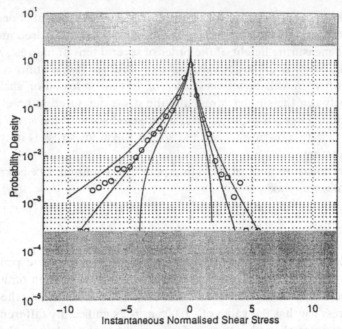

Figure 7.2. The probability distribution of the record in Fig. 7.1 compared with the joint normal distribution shown as a line. The 95% confidence limits are also shown for comparison. Normalization is by the standard deviation of $u'w'$. Correlation coefficient $= -0.46$.

significant momentum when the averaging is for short periods. This is a difficulty that is not easily solved because the planetary boundary layer is adjusting to mesoscale weather systems.

7.3.2 Measurement of Reynolds Stress Spectra

The co-spectrum of the Reynolds stress has the property

$$\overline{u'w'} = \int CO_{uw}(\omega)\mathrm{d}\omega \qquad (7.6)$$

Only the real part of the complex cross-spectrum $CO_{uw}(\omega)$ is used since the imaginary part is required to integrate to zero. The right hand side of this expression allows us to sense the contribution to $u'w'$ from different parts of the spectrum when multiplied by the frequency as in Fig. 7.3. In the atmospheric surface layer a typical cospectrum can be scaled by the nondimensional frequency $\omega z/U$.

The spectral density at the highest frequencies oscillate around zero (which is expected from isotropy). The constraint that, averaged over long time $W = 0$, restricts the amplitude of the low frequencies velocity products, especially near the sea surface.

Figure 7.3 taken from 53 m above the sea in an experiment described in Jones and Negus (1996) shows a commonly observed form of cospectrum where the

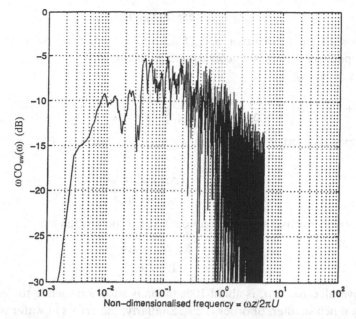

Figure 7.3. The cospectrum between u' and w' for a period of sustained wind of order 10 m/s over the sea in Bass Strait. The significant wave height was 2.5 m while the stability was about $z/L = 0.05$. The record was 15 minutes long. No pre-whitening nor smoothening was applied. There are 20 degrees of freedom. The plot of ωCO_{uw} is spectrum weighted by ω, the frequency in radians per second.

greatest contribution to the Reynolds stress comes from frequencies around $\omega z/2\pi U = 0.1$ with an energy containing range between $\omega z/2\pi U = 0.01$ and 1.0. As the number of degrees of freedom are increased, the curve becomes smoother. A reasonable length of record might be 10 cycles of the lowest energetic frequency, i.e. $T = 1000z/u_*$. Care needs to be taken in applying this result to situations where there are structures present in the boundary layer that do not scale on the height z above the sea surface. Atmospheric rolls discussed in Chapter 11 are such a phenomenon. Difficulties also arise when the flow is unsteady.

Further discussion of the relation between the statistical accuracy and the averaging time can be found in Wyngaard (1973), Larsen (1993) and Edson et al. (1991).

7.3.3   Confidence Limits and Errors

For statistical quantities such as the Reynolds stress, it is usual to estimate the confidence one has that the estimate of an average is close to the average of a very long record. In geophysical problems the concept of an expected value has some difficulty as conditions are not steady on time scales of interest; a problem alluded to in Section 7.3.1.

Strongly related to the uncertainty of the u_* estimate is the accuracy of the roughness length that is obtained from field measurements. Recall from Eq. (7.2) that the sea roughness, z_0 can be calculated in terms of the 10 m neutral wind as:

$$z_0 = 10/\exp(\kappa U_{10n}/u_*) \tag{7.7}$$

Now, suppose there is an error in κ, U_{10n} and u_*, given as $\Delta\kappa/\kappa$, $\Delta U_{10n}/U_{10n}$, $\Delta u_*/u_*$ respectively, then the corresponding error in z_0, $(\Delta z_0/z_0)$, is given as (Johnson et al. 1998):

$$\frac{\Delta z_0}{z_0} + 1 = \exp\left\{\frac{-\kappa/\sqrt{C_{dn}}}{(1 + \Delta u_*/u_*)}\left[\frac{\Delta U_{10n}}{U_{10n}} + \frac{\Delta\kappa}{\kappa} + \frac{\Delta\kappa}{\kappa}\cdot\frac{\Delta U_{10n}}{U_{10n}} - \frac{\Delta u_*}{u_*}\right]\right\} \tag{7.8}$$

The corresponding error in the nondimensional roughness $(\Delta z_* = \Delta z_0 g/u_*^2)$ is given as:

$$\frac{\Delta z_*}{z_*} + 1 = \frac{(\Delta z_0/z_0) + 1}{1 + \Delta u_*/u_*^2} \tag{7.9}$$

Lastly, the error in wave age (see Section 1.5) can be written as:

$$\frac{\Delta(c_p/u_*)}{c_p/u_*} = \frac{\Delta c_p/c_p - \Delta u_*/u_*}{1 + \Delta u_*/u_*} \tag{7.10}$$

Now, the overall error in u_* is about 10% in many situations, while the error in U_{10} is generally much smaller, of order $< 2\%$. Similarly, the errors in water depth, peak frequency and thus phase celerity, c_p, are usually small. Assuming κ equals 0.4 and C_{dn} is 1.5×10^{-3}, and negligible errors in all other parameters except u_*, then an error of $\pm 10\%$ in u_* implies that z_0 can vary between 0.39 to 2.13 of the true value. This calculation is illustrated by the results of Johnson et al. (1998). Assuming the mean value of all the data is the true value, then one can plot the corresponding error bars for $\pm 10\%$ error in u_* as shown in Fig. 7.4. The corresponding error in c_p/u_* is approximately $\pm 9\%$. Since both axes in Fig. 7.4 contain u_*, an inaccuracy that leads to larger value of u_* will lead to an increased wave age and a decrease in nondimensional roughness. Such a trend, known as a spurious correlation, is shown in Fig. 7.4 and has some similarity to the data presented.

Since many of the data points in Fig. 7.4 lie within the $\pm 10\%$ error band for u_*, the apparent trend in the data cannot be accepted with confidence. In this situation, one cannot talk of a trend in the data. These large errors go some way to explain the scatter in aerodynamic roughness diagrams such as Fig. 2.3.

7.4   Inertial Dissipation Method

7.4.1   Formulation

With the exception of Smith (1980), the major data sets of open ocean wind stress estimates (Large and Pond 1981, 1982; Anderson 1993; Yelland and Taylor 1996) were obtained using the inertial dissipation method. This technique, which was suggested by Hicks and Dyer (1972) and developed and applied by Pond et al. (1979) and Large (1979), has been more recently evaluated in some detail by Edson et al. (1991). It may be formulated in terms of the structure function (Fairall and Larsen 1986) or power spectral density of the turbulence; here we will follow Yelland and Taylor (1996) in summarizing the latter method.

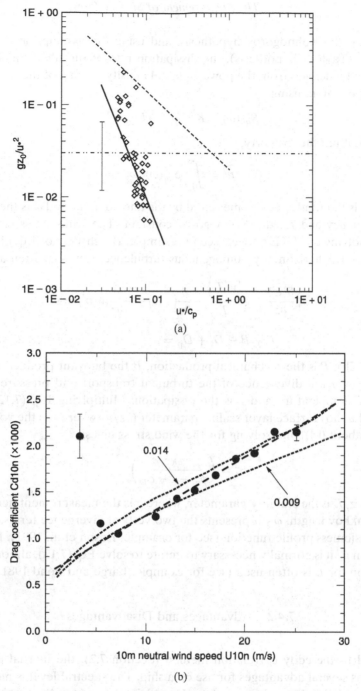

Figure 7.4. The nondimensional roughness length for a small range of wave ages. The error bar is for $\pm\,10\%$ in friction velocity and the broken line is Charnock's expression, $z_0 g/u_*^2$ equals a constant. The solid line is a spurious correlation trend. Adapted from Johnson et al. (1998). (b) Values of the drag coefficient plotted against the wind speed (corrected to 10 m height and neutral stability). Data from Yelland and Taylor (1996) have been corrected for the disturbance of the airflow over the RRS Discovery (Yelland et al. 1998). Also shown is the Smith (1980) relationship (dashed line) and values given by the Charnock (1955) formula with constants of 0.009 and 0.014 (dotted lines).

Based on the Kolmogorov hypothesis, and using the assumption of "frozen" turbulence (Taylor's hypothesis), the dissipation rate of turbulent kinetic energy (TKE), can be derived from the power spectral density, $S_{uu}(\omega)$, of the down-stream wind component, u, using:

$$S_{uu}(\omega) = K_\varepsilon^{2/3} \omega^{-5/3} (U_{\text{rel}})^{2/3} \tag{7.11}$$

where $S_{uu}(\omega)$ has the property,

$$\overline{u^2} = \int_0^\infty S_{uu}(\omega) \mathrm{d}\omega$$

Here, U_{rel} is the wind speed as measured by the anemometer and ω is the measurement frequency and K_ε the Kolmogorov constant. The wind stress can then be found by solving the TKE budget (see for example, Busch 1972 or Eq. (3.3)) which for steady state, horizontally homogeneous turbulence, can be written as

$$u_*^2 \frac{\partial \langle U \rangle}{\partial z} + g \frac{\langle w'T' \rangle}{T} - \frac{\partial}{\partial z}\langle w'e' \rangle + \frac{1}{\rho}\frac{\partial}{\partial z}\langle w'p' \rangle = \varepsilon \tag{7.12a}$$

$$P + B - D_t + D_p = \varepsilon \tag{7.12b}$$

In Eq. (7.12b), P is the mechanical production, B the buoyant production, D_t and D_p are the vertical divergence of the turbulent transport and pressure transport terms, T is temperature and ε is the dissipation. Multiplying Eq. (7.12a) by the Monin–Obukhov surface layer scaling parameter ($\kappa z/u_*^3$ where κ is the von Karman constant; about 0.4), and solving for the wind stress gives:

$$u_*^2 = \left(\frac{\varepsilon \kappa z}{\phi_m - \zeta - \phi_D}\right)^{2/3} \tag{7.13}$$

where $\zeta = z/L$ is the stability parameter, where z is the measurement height and L is the Obukhov length, ϕ_D represents the two vertical divergence terms, and ϕ_m is the dimensionless profile function (see for example, Edson et al. 1991). Because L depends on u_* it is normally necessary to iterate to solve Eq. (7.13), although a bulk formulation for L is often used (see for example, Large and Pond 1981, 1982).

7.4.2   Advantages and Disadvantages

Compared to the eddy correlation method (Section 7.2), the inertial dissipation method has several advantages for use on a ship. The spectral level is measured at higher frequencies, typically around $\omega z/U = 18$, implying a smaller spatial scale for the turbulence and less error due to structures near to the anemometer such as the supporting mast or boom (Edson et al. 1991). The measurement frequency is above the range contaminated by ship motion and there is no need to attempt to determine the fluctuations of the vertical wind component in a situation where "vertical" is rarely well defined. The method can be implemented with automatic instrumentation (Fairall et al. 1990; Large and Businger 1988; Yelland and

Taylor 1996) allowing the assembly of a large data set. The disadvantage of the inertial dissipation method is that it depends on the validity of various assumptions; those inherent in the Kolmogorov hypothesis (Eq. 7.11), and in the application of Eq. (7.13). For these reasons, where data has been available from a stable platform, detailed comparisons have been made between the inertial dissipation and the eddy correlation derived fluxes. Reasonable agreement was found by Large and Pond (1981) and Edson et al. (1991) whereas, in lower wind speed conditions, Geernaert (1988) found cases of significant difference. Here we shall discuss three possible sources of error.

Firstly, concerning the existence of an inertial sub-range, most authors have quality controlled their data using the slope of the spectrum (Eq. 7.11). Recently Neugum (1996) has suggested that a more critical test is to ensure that the cross-wind and along-wind spectra have the 4/3 relationship expected of isotropic flow. Although the constancy of this ratio for turbulence over the sea has been questioned (see the review of Schmitt et al. 1978), Dupuis et al. (1995) confirmed a 4/3 ratio even for low wind speeds (<5 m/s). Secondly, in Eq. (7.13), the vertical divergence term, ϕ_D represents the imbalance between local production and dissipation of turbulent kinetic energy. Although the land-based results of Wyngaard and Cote (1971) had suggested that dissipation exceeded local production, Large (1979) argued that, within the scatter of the results, this term was negligible. However later authors (Schacher et al. 1981; Edson et al. 1991; Frenzen and Vogel 1992; Kader 1992; Fairall and Edson 1994) have either reinterpreted the Wyngaard and Cote (1971) results and/or analysed new data. The various suggested formulations for ϕ_D differ both in sign and in their stability dependence, and the implied uncertainty in the stress estimation is significantly greater than that due to the uncertainty in ϕ_m (discussed by Frenzen and Vogel 1994). Recently, Dupuis et al. (1995) and Yelland and Taylor (1996) have empirically determined ϕ_D by comparing wind stress estimates from neutral and diabatic conditions. The assumption is either that production and dissipation are balanced for $\xi = 0$ or that any imbalance has been allowed for in the value used for the Kolmogorov constant (Deacon 1988). The Yelland and Taylor (1996) formulation for ϕ_D depends on both ξ and wind speed; if confirmed this may explain the disparate results of other authors. In an alternative approach, Janssen (1999) has proposed a method to evaluate ϕ_D over a developing wind sea.

Lastly, the effects of measurement error, instrument noise and the uncertainty in the calculation of the true wind speed were evaluated by Yelland et al. (1994) by comparing inertial dissipation estimates from four anemometers mounted on the foremast of the RRS *Charles Darwin*. The major error source was the airflow distortion due to the presence of the ship (see Section 7.6). With regard to inertial dissipation derived wind stress data this has two effects, both of which vary with relative wind direction. At a typical foremast anemometer site, and for flow directly over the bows of the ship, the airflow is decelerated and lifted compared to the free stream conditions (Yelland et al. 1998). The combined effects of the correction for airflow distortion and for the imbalance term, ϕ_D, will be discussed in the next section using the large set of open ocean wind stress estimates obtained by Yelland

and Taylor (1996) during three cruises of the RRS *Discovery* in the Southern Ocean. The results reduce the uncertainty in the open ocean drag coefficient to wind speed relationship, and quantify the maximum magnitude of the variations in the wind stress estimates that might be ascribed to varying sea states.

7.4.3   Wind Stress Results Obtained Using the Inertial Dissipation Method

By including the imbalance term ϕ_D (or limiting their data to conditions very close to neutral, $|\zeta| < 0.02$), Yelland and Taylor (1996) obtained values for the neutral stability, 10 m height, mean drag coefficient C_{D10n} which, for a given wind speed, were about 5% higher than the results of Smith (1980). However after correcting the Discovery Southern Ocean data for airflow distortion effects, Yelland et al. (1998) obtained the relationship:

$$1000C_{D10n} = 0.50 + 0.071U_{10n} \qquad (7.14)$$

which is similar to that suggested by Anderson (1993). The *Discovery* Southern Ocean data points were now also well fitted by the Smith (1980) formula (Fig. 7.5) which was based on eddy correlation results from a taught moored spar buoy for which flow distortion effects would be small. However this agreement may well have been fortuitous since the scatter of the eddy correlation results was large. Indeed, whereas Smith (1988) was able to represent the Smith (1980) data using a Charnock (1955) relationship:

$$\beta = z_0 g/u_*^2 \qquad (7.15)$$

with the Charnock "constant", or nondimensional roughness length, $\beta = 0.011$, a linear relationship is a significantly better fit to the *Discovery* Southern Ocean data. This implies that the effective value of α must increase with wind speed. At wind speeds of about 10 m/s, $\beta = 0.009$ fits the data whereas $\beta = 0.014$ at around 20 m/s.

It is tempting to ascribe this observed increase in the nondimensional roughness length to the effect of varying sea state but Fig. 7.5 suggests that this was not the

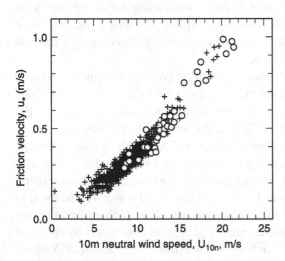

Figure 7.5. Values of the friction velocity data of Yelland and Taylor (1996) plotted against the 10 m neutral wind speed. The symbols represent the ratio of the observed significant wave height to that for a fully developed sea (Bouws 1988): + > 1.25 (i.e. apparently overdeveloped); O < 0.5 (i.e. underdeveloped sea state).

case. The significant wave height, H_s, has been compared to that for fully developed waves based on a Pierson–Moscowitz spectrum (Bouws 1988), H_s. Comparing cases where the sea-state was apparently overdeveloped (due to the presence of swell) to those where the sea-state was not fully developed, no significant difference in the wind stress to wind speed relationship could be found. The same result was obtained irrespective of whether H_s was estimated using data from the ship-borne wave recorder on RRS *Discovery* or from the ECMWF wave model. This implies either that the wind sea was always under-developed (unlikely) or that the sea-state induced effects were small. Wind stress anomalies due to varying wave age (e.g. Smith et al. 1992; Juszko et al. 1995) do not appear to have been present. Donelan et al. (1993) found that any relationship between drag coefficient and significant wave height, observed for a purely wind induced sea, was obscured when the wave field was dominated by swell; possibly this is generally the case in the open ocean.

Examination of the *Discovery* Southern Ocean data did not identify the anomalies in the drag coefficient (of order of 20% over a period of a few hours) which some previous studies (e.g. Geernaert et al. 1986; Denman and Miyake 1973; Large and Pond 1981) had explained by the effects of varying sea-state. For the u_* estimates, 97% were within 0.075 m/s of the mean u_* value for a given wind speed and the rms scatter was less than 10%. This scatter can easily be explained by the measurement errors identified by Yelland et al. (1994). Following the results of Yelland and Taylor (1996) and Yelland et al. (1998) one might speculate that at least some of the anomalies found in previous studies were caused by errors in applying the inertial dissipation technique. With the passage of an atmospheric front, changes occur both in the atmospheric stability, and hence in the imbalance term, and also in the wave field. With a changed wave field a research ship is likely to change its mean orientation to the wind, thus also altering the required airflow disturbance correction (Section 7.6). The combined effect of airflow disturbance and imbalance term could easily account for the reported wind stress anomalies.

Further, we note that any relationship between nondimensional roughness length and wave age in the Yelland et al. (1998) data needs careful consideration. Older waves predominantly occur in light winds when C_D and hence z_0 is small. At higher wind speeds the waves are younger and the roughness greater. However, at a given wind speed, younger wave ages are not correlated with higher roughness length, or older waves with lower roughness length. This does not corroborate the relationship suggested by Fig. 1.15 for data collected from the open ocean.

7.5   The Profile Method to Measure Momentum Fluxes at the Sea Surface

7.5.1   The Flux–gradient Relationship

The basis for the profile method is the assumption that turbulent mixing near the boundary results in a nearly logarithmic wind profile. The gradient of the wind profile is related to the momentum transferred. A flux–gradient relationship is necessary to determine fluxes from profiles. For this, the semi-empirical Monin–Obukhov similarity theory can be seen as the diabatic extension of the Prandtl theory of the logarithmic wind profile. This theory is detailed in Chapter 3 of this book and outlined in Eqs (7.2)–(7.4).

The logarithmic wind profile can be used only under strictly neutral conditions to infer stress. Since the influence of stability on the profile changes considerably with small deviations from neutral conditions, knowledge of the flux–gradient relationship, also called the stability function, is the core of the profile method. The range of stability variations is typically larger over land than over sea and stability functions are easier to obtain over land. From various field experiments, for unstable conditions the 'KEYPS' formulation (Panofsky 1963) was obtained, and the log–linear formulation for stable conditions (see Eq. 7.4). These together with the von Karman constant of 0.40 have been verified for use over sea in an indirect way. From measured profiles, using these stability functions, the flux at the sea surface was determined and expressed as a drag coefficient at the given stability. Using the same stability functions, the drag coefficient was reduced to neutral stability. The neutral drag coefficients should be independent of stability. If we used a faulty stability function, a significant variation of the neutral transfer coefficient would result. This test has been applied to wind profile measurements at low wind speeds (Dittmer 1977), where the stability effects are most felt, and where influences from wind waves should be negligible. The results, Fig. 7.6, verify the applicability of the stability function. A similar proof for the latent heat flux is given by Hasse et al. (1978a). It should be noted that the diabatic influence is felt more in the gradient than by the curvature of the profile. For the limited height range that is accessible the curved diabatic profiles still look logarithmic (Fig. 7.7)! In the early use of the profile method neglect of diabatic influences ("stability was

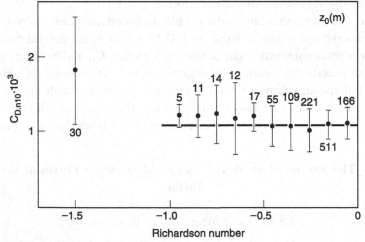

Figure 7.6. Verification of applicability of the stability function at the ocean: neutral drag coefficients are independent of stability (from Dittmer 1977).

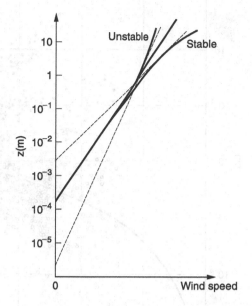

Figure 7.7. Influence of stability on the wind-profile. The sketch explains that in the typical height range of profile measurements at sea the variation of the mean gradient is more pronounced than the curvature. In fact, within the inevitable scatter of measurements, it is not really feasible to determine the curvature with recourse to theory (from Hasse 1993).

not measured, but profiles were fairly logarithmic and have been treated as neutral") has led to disastrous results, when estimating z_0, as has been convincingly shown by Brocks and Kruegermeyer (1971), Fig. 7.8.

7.5.2   Waves and the Wind Profile

Wind and waves have different velocities. One would expect that the air flow is accelerated above the crests, and decelerated above the troughs, and that the deformation of the streamlines decays with distance from the waves. This will result in two effects, namely wave coherent pressure fields, and a deformation of the wind profile. The wave coherent pressures have been measured by fine static pressure heads by Snyder et al. (1981) and Hasselmann and Bösenberg (1991) and found to transfer momentum to the waves. The wind profile above the waves has been measured with a surface following buoy (Hasse et al. 1978b). In a surface following coordinate system, the wind profile is deformed and shows a systematic deviation towards lower velocities in the lowest layers, Fig. 7.9. Both observations together demonstrate that the definition of a roughness length from the logarithmic (or diabatic) wind profile in the presence of waves has no direct physical meaning. From a set of detailed wind profile measurements in the trade wind regime it was possible to separate the effects of wind speed and waves by multiple regression. Figure 7.10 shows that the influence of waves, feigns an increase of the drag. This effect disappears if instrument levels above the crest of the waves only are used. For the determination of stress by the profile method these observations resulted in the advice to have the lowest anemometer level at least higher than the wave height (double amplitude) above the mean water level. This requirement is trivial for a fixed mast, but lower heights are possible with

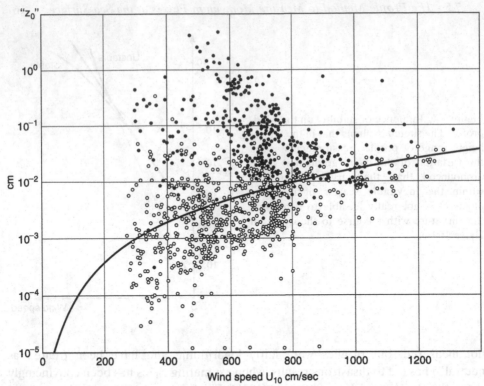

Figure 7.8. Effect of neglect of the diabatic influence on the wind profile. Taking wind profile measurement without regard to stability as neutral, "logarithmic" profiles results in an apparent "roughness length", that is considerably in error. With a preference for slightly unstable conditions in the data set, the Charnock relation, shown as a solid line, appears to be supported, unfortunately, from an unphysical treatment of data (from Brocks and Kruegermeyer 1971).

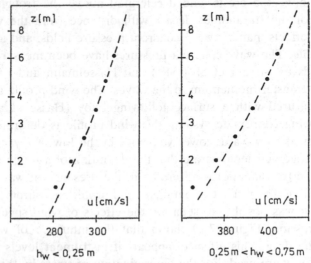

Figure 7.9. Average wind profiles observed at a surface following buoy during GATE, grouped according to the estimated wave height h_w. The broken line is the KEYPS diabatic flux profile relationship fitting the upper four levels best through selection of surface roughness, while the curvature of the profile is given from the observed stability (from Dittmer 1977).

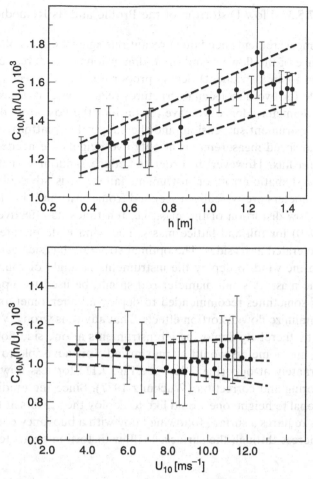

Figure 7.10. Multiple regression of the drag coefficient as a function of both windspeed and wave height. The variable $C_{10,N}$ is an expression for the neutral drag coefficient, derived solely from the wind profiles, and it is influenced by both wind speed and waveheight, as indicated by the two figures. The upper figure shows $C_{10,N}$ versus waveheight, while the lower figure shows wind speed dependency (Hasse et al. 1978b).

surface following buoys. This warning should also be kept in mind when buoy measurements of wind speeds in high sea states are used to calibrate satellite wind or stress retrieval algorithms!

The influences of waves and of non-neutral conditions on the wind profile almost always coexist. The influence of waves decays exponentially with height, scaled with wavenumber, while the influence of density increases almost linearly with height. Hence it seems permissible to treat the two influences as if they were independent. Also, the influence of density stratification is most important at low wind speeds, where waves are low, while at higher wind speeds conditions are more near neutral.

7.5.3   Flow Distortion of the Profile and its Remedies

Wind profile measurements need due precautions against effects of flow distortion. The disturbance of the flow around three-dimensional structures (e.g. taken as an approximation to an instrument) decays proportional to the second power of the diameter while for two-dimensional structures (e.g. a mast) it decays with the first power of the diameter. Hence we have to consider the flow around anemometers, masts, spars, instrument suspension, and the supporting platform, e.g. a buoy. In principle, wind speed measurements at only two heights are needed to determine the momentum flux. However, in order to reduce random variance and permit detection of systematic errors or instrument failure, it is advisable to use 5 to 7 levels. Anemometers are typically suspended from masts and should be outside of the range of flow distortion of the mast, i.e. ten times its effective diameter, see Wucknitz (1980) for full and lattice masts. The wind is decelerated in front of a mast and accelerated at its sides. The optimal structure has sidewards spars, at the end bent into the wind to deploy the instruments at angle of roughly 30 degrees forward of the mast. A small diameter rod should be used to support the anemometers. It is sometimes recommended to deploy all anemometers at one side of the mast to minimize flow distortion effects. That advice is wrong. What you do not see may still be there! By deploying all instruments at one side flow distortion by neighbouring anemometers would be maximized. It is advisable to support anemometers alternately at opposing sides (see Fig. 7.11) for less flow distortion and better monitoring and correction (Wucknitz 1977). Since the gradients are inversely proportional to height, one would like to deploy the sensors at the lowest level possible. This requires a surface following buoy with a buoyancy compartment that necessarily pierces through the surface. Its flow distortion needs to be considered too.

7.6   Instrumentation

The instruments for air–sea interaction measurements were reviewed in Dobson et al. (1980). Most stress measurements today are based on sonic anemometers. These instruments are in principle absolute instruments without need of calibration. They derive three-dimensional velocity by measurements of transmission time of sound pulses across well known gaps. Therefore they constitute an advance compared to the older mechanical turbulence instruments such as propellers and vanes. Sonics however are also susceptible to sources of inaccuracy and drift.

A number of different instruments are on the market, and it is difficult to summarize the totality of error sources for the different types of sonics. Drift in the electronics and the transducers varies from producer to producer. The sonic arrays generally come in two types: those where ruggedness and simple, reliable operation is most important, which can be left unattended most of the time, and those where flow distortion is minimized for a certain wind direction. This means that one has to turn the instrument when the wind direction changes. The former array type will often be symmetric or almost symmetric around the z-axis.

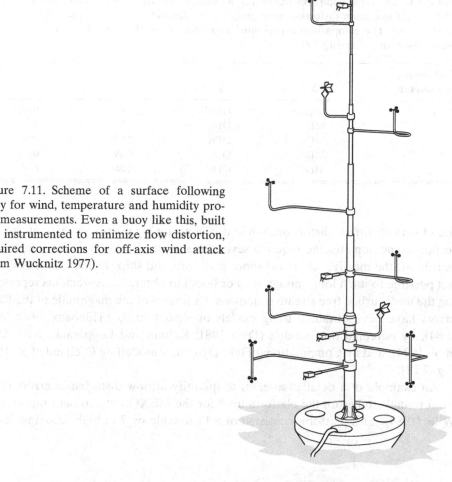

Figure 7.11. Scheme of a surface following buoy for wind, temperature and humidity profile measurements. Even a buoy like this, built and instrumented to minimize flow distortion, required corrections for off-axis wind attack (from Wucknitz 1977).

From Mortensen and Hojstrup (1995) we present in Table 7.1 a comparison between typical turbulence parameters measured by the two types of sonics and a propeller anemometer.

Table 7.1 shows typical systematic variation between the different types of instruments. They should not be considered arguments for not using the omnidirectional sonics, since operational simplicity can be very important and because different schemes for correcting the measurements have been devised, based on wind tunnel calibration studies, e.g. Mortensen and Hojstrup (1995), Mortensen et al. (1987) and Grelle and Lindroth (1994). If one follows that approach one should be aware that some producers have built corrections to the measured values in the instrumental software.

7.7 Airflow Distortion and Platform Motion

For determining the wind stress, the eddy correlation (Section 7.3), inertial dissipation (Section 7.4), and profile method (Section 7.5) each suffer from errors due to

Table 7.1. The table compares turbulence measurements from an omnidirectional sonic, ODs, and a minimum distortion array sonic, MDs. Also compared is a propeller anemometer. The comparison is presented as a linear fit to the data: $Y = aX + b$ (Mortensen and Hojstrup 1995).

Turbulence parameter	X	Y	a	b
U	propeller	ODs	0.96	0.06
u_*	MDs	ODs	0.87	0.01
σ_u	MDs	ODs	0.95	0.01
σ_v	MDs	ODs	0.89	0.02
σ_w	MDs	ODs	0.88	0.01

the effects of airflow distortion. While it is conceivably possible to mount the fast response anemometer the required several diameters from a thin instrument support mast, the massive size of off-shore platforms and ships means that it is simply not possible to use a long enough mast or boom to obtain measurements representing the undisturbed free stream conditions. Estimates of the magnitude of the likely errors have been obtained using models in wind tunnels (Thiebaux 1990; Wills 1984), by potential flow models (Dyer 1981; Kahma and Lepparanta 1981; Oost et al. 1994), and by Computational Fluid Dynamics modelling (Yelland et al. 1998; Fig. 7.12).

An example of a detailed attempt to quantify airflow disturbance errors is the wind tunnel studies on the platform used for the HEXOS experiment reported by Wills (1984). This platform consisted of a 12 m wide by 7 m high accommodation

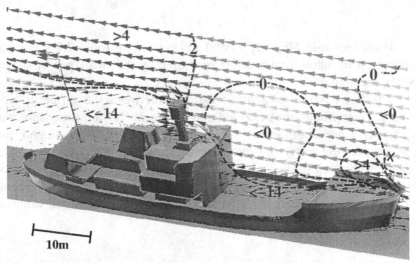

Figure 7.12. The percentage change in wind speed caused by the CSS Dawson (used for wind stress estimation by Anderson (1993), as calculated using CFD for the case where the ship is heading into the wind. The CFD results (Yelland et al. 1998) agreed well with the results of wind tunnel studies (Thiebaux 1990).

block some 10 m above the water with an instrument boom projecting horizontally to a distance of 16 m from the platform. In a plane some 16 m away from the tower the main disturbance was deflection of the mean airflow. At 10 m from the platform the turbulence structure and the mean flow were both significantly disturbed. The distortion of the stresses was significantly greater than for the variance. While the quantitative errors in the turbulence predicted in the wind tunnel were not found to apply in the field (Oost et al. 1994), the results appear qualitatively reasonable when compared, for example, to the Dyer (1981) suggestion that, whereas covariance estimates would be in error by 14% for each degree of flow misalignment, the corresponding error for variances would be only 0.4%.

Consider now the implications of the Wills (1984) results for the different methods of wind stress determination. Distortion of the mean flow occurs even at some distance from a structure and will cause unacceptable errors in stress values estimated using the profile method. The only measurements likely to be suitable for this technique would be from, for example, a very thin spar buoy. The eddy correlation method requires a covariance estimate integrated over a wide range of frequencies and hence spatial scales of turbulence; such covariance estimates will suffer badly from flow distortion induced errors. Examples of successful measurement programmes using the eddy correlation technique are Smith (1980), which used a specially constructed spar buoy installation, and the HEXOS experiment, where detailed flow distortion corrections were applied (Oost et al. 1994). In contrast, the inertial dissipation method only requires an estimate of the variance in the higher frequency region of the spectrum corresponding to small turbulence length scales; these variance estimates are much less likely to be affected by airflow distortion. Comparison of data from boom and mast mounted instruments during HEXOS suggested that the inertial dissipation method was indeed superior in this respect (Edson et al. 1991; Oost et al. 1994).

It is therefore not surprising that the inertial dissipation method has generally been adopted for estimating the wind stress using ship mounted instruments (Large and Pond 1981, 1982; Anderson 1993; Yelland and Taylor 1996). With this method, the only flow induced error in a stress estimate will be that due to any change in the height of the airflow assuming that it has occurred too rapidly for the turbulent intensity to adjust to that appropriate for the new height. Computational fluid dynamics modelling (Yelland et al. 1998) suggests that, for a foremast anemometer site and with the ship heading into the wind, any such changes are typically of order 1 m. Unfortunately, because the aim is often to relate the estimated stress to the mean wind speed, or even to combine the stress with the wind speed to calculate a drag coefficient, an accurate mean wind speed value is still required. For many operational uses it seems advantageous to measure the stress only and avoid the wind speed estimation.

Even for the inertial dissipation method, the least likely to be affected by airflow distortion, the errors in estimating the C_{D10n} to U_{10n} relationship can be significant. For example, Yelland et al. (1994) showed inertial dissipation results from four anemometers which were apparently well exposed, being situated on the foremast of the research ship *Charles Darwin*. Figure 7.13(a) shows results from the two

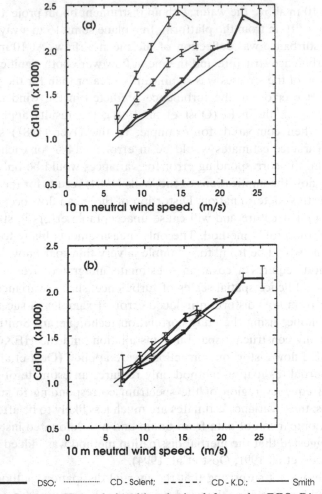

DSO;   ⋯⋯⋯ CD - Solent;   – – – – CD - K.D.;   —— Smith

Figure 7.13. The C_{D10n} to U_{10n} relationship obtained from the RRS *Discovery* in the Southern Ocean (DSO) and from two anemometers (a Solent Sonic and a Kaijo–Denki sonic) during a cruise of the research ship RRS *Charles Darwin* (CD). Also shown is the Smith (1980) relationship: (a) Results before application of CFD derived corrections; DSO from Yelland and Taylor (1996), C_D from Yelland et al. (1994).

sonic anemometers. The calculated C_{D10n} values were generally larger than those found, for example, by Yelland and Taylor (1996), and varied by 12 to 20% from one anemometer to another. Because the observed friction velocity values agreed to better than 3%, it was concluded that the C_{D10n} differences were caused by errors in estimating the wind speed. Differences in the measured mean relative wind speed ranging up to 8% were ascribed to airflow distortion in the region of the anemometer sites. Subsequently, Yelland et al. (1998) used computational fluid dynamics to determine correction factors for both the flow blockage and the elevation of the airflow at each of the anemometers. For this purpose it was assumed that the turbulence level in the inertial subrange corresponded to the height of origin of the airflow. The corrected results from the *Charles Darwin* (Fig. 7.13(b)) were now not only brought into agreement but also into agreement with the results obtained from the Discovery in the Southern Ocean. The latter were also corrected for airflow distortion but, because of the shape of the ship's superstructure, the errors were much smaller.

The Yelland et al. (1994) results are unlikely to be unique. Indeed Dobson et al. (1994, 1995) also reported differences in C_{D10n} values determined on two different ships which were ascribed to air-flow distortion effects. It is likely that all C_{D10n} values determined from ship data might be in error unless great care has been taken to minimize, and correct for, airflow distortion effects. Small changes in wind direction relative to the ship can make large changes to the correction required. Such changes could be due to the ship's officers responding to changes in the relative directions of wind and waves. In such a case a "wave induced" anomaly might be spuriously identified. If the data have been obtained using the eddy correlation method even greater caution is required since both the covariance and the mean wind estimates may contain significant errors.

As well as the flow distortion of the structure supporting the instrument, the motion of the supporting arrangement needs to be considered. Although systems have been developed and tested for on-line correction for the platform motion (Edson et al. 1998), a more normal approach to get around the combination of flow distortion and flow motion is the use of the inertial dissipation method for flux determination.

The advantage of the dissipation method becomes clear when one considers how data obtained from ship-borne measurement unavoidably must be distorted by flow perturbation by the ship itself and also by the motion of the ship. The distortion is illustrated in Fig. 7.14 showing spectra of the three velocity components during a cruise. The figure illustrates that the ship's motion often enters the spectra as additional spectral energy at 0.1–1.0 Hz.

From Fig. 7.14 it is clear that the inertial ranges of the spectra show relatively low distortion and the dissipation method can be applied here. Some uncertainty will however always exist as to the importance of more subtle flow distortions, not showing up as dramatically as in Fig. 7.14. These aspects of flow distortion are much more complicated and time consuming to correct, as described above. Therefore it becomes important to derive the fluxes from as many methods as possible and preferably from methods with different sensitivities to the flow distortion and ship motion. The drag determination from the profile method is certainly different from the dissipation method, in that it is based on mean values, while the dissipation estimates are based on high frequency turbulence measurements.

The covariance method, as given in Section 7.2, in a way is the basic definition of the flux. It can further be extended by use of the cospectrum. Typical cospectra show that within the surface layer the covariance receives its main contribution in the interval 0.01 Hz for a wide range of conditions. Since the distortion from the ship's motion often seems to have maximum between 0.1–1 Hz, the covariance method can obviously not be used directly.

The literature (Kaimal et al. 1972; McBean and Miyake 1972; Pond et al. 1971; Schmitt et al. 1979; Larsen 1986; Smith and Anderson 1984) on the cospectra

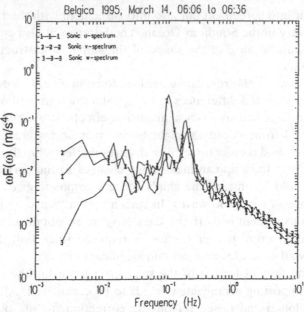

Figure 7.14. Power spectra of the three velocity components measured on a moving ship. The ship motion shows as a peak in the frequency spectra. Notice the power spectral densities are weighted by frequency. (Larsen et al. 1996.)

reveals that for unstable to neutral conditions the cospectra have a fairly flat maximum for frequencies between 0.01 and 0.1 Hz. The maximum of the spectra are found to be:

$$\omega CO_{uw}(\omega) \approx 0.22 \langle u'w' \rangle \text{ in the interval } 0.048 \; < \frac{\omega z}{U} < 0.72$$

$$\omega CO_{wT}(\omega) \approx 0.18 \langle w'T' \rangle \text{ in the interval } 0.042 \; < \frac{\omega z}{U} < 0.72$$

$$\omega CO_{wq}(\omega) \approx 0.20 \langle w'q' \rangle \text{ in the interval } 0.042 \; < \frac{\omega z}{U} < 0.72,$$

On the average we therefore can assume that all the cospectra involving w, in the range 0.01–0.1 Hz equal about 0.2 times the flux. For unstable to neutral conditions, the universal and simple form of the cospectra suggests that the amplitude around 0.01 Hz should offer reasonable conditions for determining the corresponding flux. Figure 7.15 compares the different stress estimation techniques for a few days of a cruise with unstable conditions.

For stable conditions the cospectra slide towards higher frequencies, and they simultaneously become less broad. Therefore, it is expected that the cospectral method becomes less useful during stable conditions, because the peak of the spectrum, now gets closer to the frequency interval where typically the distortion from the ship motion is dominant. Simultaneously the increased peakedness of the spectrum means that there will be relatively less energy at the lower frequencies,

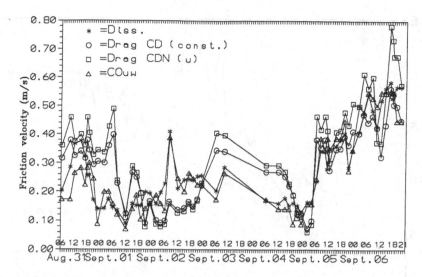

Figure 7.15. Estimates of friction velocity by four different methods: including the dissipation method; a constant drag coefficient; the measured peak of the cospectrum; and a
velocity dependent neutral drag coefficient. From the Poseidon Cruise 1995.

that could be used for flux estimation. The only point against this argument is that
the stable cospectra also tend to be better defined than the unstable to neutral ones,
with respect to both shape and frequency of the peak. Therefore it may be possible
still to use the low frequency behaviour of the cospectra for stable conditions. The
approach has so far not been tried under stable conditions.

7.8   Pressure Supported Stress

Fluctuating pressure forces that are associated with the surface waves play a role in
transporting momentum into and out of the wave profile. They fall off with distance
above the sea surface and become negligible at heights the order of the wave length
where in a horizontally homogeneous flow all the momentum flux is supported by
Reynolds stress. In Chapter 11 the contribution of large planetary boundary layer
eddies are shown to modify this simple picture.

When the waves travel faster than the wind it is to be expected that the fluctuating pressure associated with such wave components might locally transfer
momentum to the atmosphere. Such transport would be up-gradient and could
appear in the form of a positive correlation between the u and w fluctuating
velocity components. In the example of the cospectra of $u'w'$ in Fig. 7.16 the
average value of $u'w'$ implies momentum flux towards the sea surface but eddies
of frequency near 0.03 and 0.12 Hz are transporting momentum away from the
surface. Surface waves of the latter period travel at about 13 m/s and have a
wavelength of order 100 m. The cospectrum is measured at a fraction of a wavelength above the surface. This band of momentum flux away from the surface has
the effect of reducing the drag coefficient in the case of falling seas. With increasing

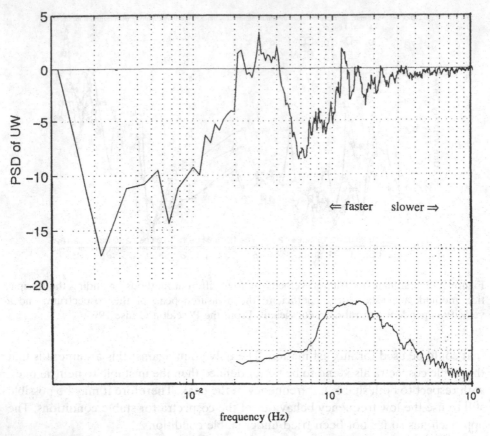

Figure 7.16. The cospectrum from a 15 minute record taken 2.25 hours after Fig. 7.3. Now the wind has momentarily slowed leaving the peak in the wave spectra travelling faster than the 10 m wind. $z = 7$ m, $\overline{u'w'} = -0.113$ m^2/s^2, SWH ~ 2.5 m. The lower curve is the wave height spectrum.

height above the waves the pressure supported stress drops off. It does show however that Reynolds stress alone is not the whole picture.

7.9   Satellite Remote Sensing

7.9.1   Introduction

In the next decade, there will be several microwave sensors in space that will provide flux modellers with global marine surface stress fields. These fields will be used extensively by climate modellers and for storms analyses. They constitute a new source of global marine data and a new perspective on global stress. The differences between these data and traditional sources are large. Methods of analysis can be quite different. For instance, the relation between a point ship or buoy stress and the satellite sensor 25 km footprint is not simple, depending on the turbulence field and synoptic tendency. For some purposes, the new scales and averaging will not be appropriate. For others, the new era will require a re-thinking of traditional methods.

7.9.2   Sensor Theory

The microwave satellite sensors respond to the active radar reflection from the surface of the ocean, or the passive emissivity from the surface. The active sensors (scatterometers, altimeters and Synthetic Aperture Radars) measure the density of short gravity or capillary waves over large footprints. The engineering parameter is called a "backscatter cross-section parameter". The passive sensors (radiometers) respond to the change in emissivity due to the same surface roughness changes. The satellite, 'SeaSat' in 1978 carried all of these sensors and provided a proof of concept experiment for their geophysical measurement capability. The phenomena of reduced reflectivity in regions of increased winds is well known to mariners. Surprisingly, the backscatter data correlate well to surface winds (e.g. Brown 1983, 1986). Multiple looks at the same region allow these instruments to detect the surface wind vector. Since the wave density is dependent on the momentum flux, which is equal to the surface stress, one might expect a surface stress algorithm for the scatterometer backscatter. This has not been done, due to the lack of surface truth with which to establish the correlation. One suggestion for such an algorithm was put forward by Weissmann et al. (1993) where the wind speed was related to the backscatter. Since 1991, there has been an operational scatterometer in orbit on the ERS-1 satellite. This is a 500 km wide swath view to one side, circling in a 90 minute polar orbit. It has a 25 km footprint with data given in 50 km gridpoints. The NASA scatterometer, NSCAT, was included on the ADEOS, launched in August 1996. This was a two-sided scatterometer with 600 km swaths separated by 250 km.

There are other microwave sensors which return a wind speed measurement from a single look, both active and passive. The SSM/I, radiometer returns surface brightness temperatures which has been successfully correlated to wind speed and could quite likely be responding to wind stress.

Finally, the Synthetic Aperture Radar, SAR, produces wave spectra from centimetres to kilometres. It has been used successfully to document the impact of the downdraft regions of 2 km wavelength rolls on the surface (see Chapter 11). The augmented short waves, and hence stress, in these regions has been used to document the effects of rolls on the surface fluxes by Gerling (1986), Thompson et al. (1983) and Alpers and Brumner (1994). Recently, ERS-1 and RADARSAT data have been used to demonstrate the ubiquity of the roll signature on the surface of the ocean (Mourad and Walker 1996; Brown 2000a). The implied periodicity of winds and stress must therefore be allowed for in modelling.

7.9.3   Large-scale Weather and Climate Modelling

Given the difficulties in obtaining a good large-scale oceanic surface stress field, recourse may be made to the new source of global surface stress from satellite

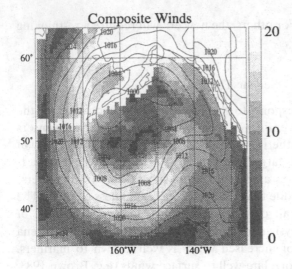

Figure 7.17. The 10 m wind speed on 17 November 1992 from a composite of ERS-1, SMM/I, ECMWF and University of Washington model output. The gray-scale range is 0–20 m/s.

microwave sensors as they have the ability to penetrate cloud cover and produce very good wind fields (stress fields) at the surface in tropical storms.

The excellent agreement between ECMWF and ERS-1 derived surface pressure fields in the Northern Hemisphere demonstrates the viability of the method of calculating the pressure field from scatterometer measurements. The departures between these two analyses are significant in: (1) Central low pressures of storms where the numerical model suffers from lack of observations and resolution problems (Brown and Zeng 1994); and (2) Southern Hemisphere storms intensity, location and occurrence (Brown 2000b). These data show that the satellite sensors provide superior storms analysis in regions of poor in situ data for model initialization. Commensurate with every wind field is a correspondingly different surface stress field.

An example of a composite synoptic-scale surface wind field including data from SSM/I, ERS-1, PBL_LIB and ECMWF analyses for a north Pacific storm system is shown in Fig. 7.17, from Dickinson and Brown (1996). The corresponding surface stress field and other flux parameters are available. As the satellite data get more sophisticated, basic planetary boundary layer variations due to variable stratification and thermal wind can be incorporated into the wind stress product (Foster et al. 2000).

PART TWO

8 The Influence of Swell on the Drag

M. A. Donelan and F. W. Dobson

8.1 Introduction

Swell is formally defined as old wind sea that has been generated elsewhere. The term "old" is meant to signify that at some past time the swell energy, propagating through a given defined point, had been directly forced by the wind elsewhere. In view of the rather specific notion of "wave age" it might be better to think of swell as "escaped" wind sea. Having come from elsewhere, bearing the imprint of a different storm, swell may propagate at any speed relative to the wind or at any angle to the wind. Indeed, the vector difference in speed of the swell and peak wind sea may provide the only unambiguous criterion for identifying and separating swell from actively growing wind sea. Frequency dispersion separates the components of swell as they propagate away from the source area, and so swell tends to have a narrower spectrum than wind sea; but this provides only a qualitative selection criterion since the bandwidth of wind sea and swell may have considerable variation. For clarity we consider only two clearly defined cases of swell: (1) a distinct peak in the spectrum having peak phase speed greater than the wind component in the direction of propagation of the peak; (2) a distinct peak in the spectrum having peak phase velocity at an angle greater than 90 degrees to the wind.

The addition of a swell component to an existing wind sea may affect the drag in two ways: (a) the direct interaction of the wind and swell could enhance the drag when the swell propagates counter to the wind (case 2 above), or could reduce the drag when the swell runs ahead of the wind (case 1 above); (b) the effect of swell on altering the wind sea spectrum will also change the aerodynamic roughness of the surface. In the following sections these two effects are discussed with samples from published work and new data.

8.2   Wind–Swell Interaction

The idea that swell running ahead of the wind can return momentum to the atmosphere and produce a wave-driven wind appears to have first been promulgated by Harris (1966). Subsequent field observations by Davidson and Frank (1973) and Donelan (reported in Holland 1981 and reproduced in Fig. 8.1) provide clear evidence for an upward transfer of momentum, and Dobson (1971) reported direct measurements of the momentum transfer from a swell group over-running the wind. It remains to parametrize this in terms of a roughness length, z_0 or drag coefficient, C_D but it is likely that the effects of over-developed swell on the drag will only be significant in very light winds when the difference in phase speed and wind speed is large and the wind sea's contribution to roughness will be small.

The case of swell running against the wind has not yet been well documented, at least in connection with the modification to the wind stress. Clearly, the largest effect would be expected when wind direction and swell direction are directly opposite. On the open ocean this circumstance is unusual and, to our knowledge,

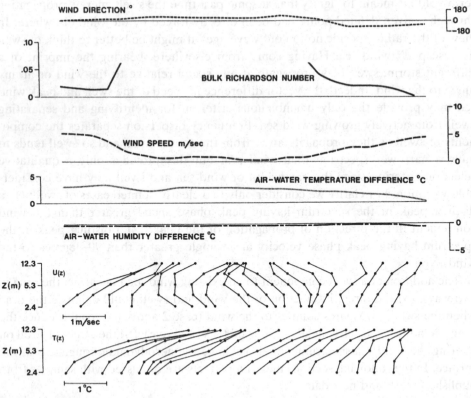

Figure 8.1. Wind and temperature profiles measured over Lake Ontario showing formation and decay of a wave-driven wind. The profiles are running averages over 30 min plotted at 10 min intervals. The humidity difference is expressed in buoyancy equivalent degrees Celsius. The time series of wind speed and air–water differences are obtained from measurements at the top level and the surface water temperature. (From Holland 1981.)

has only recently been observed with concurrent stress and wave directional measurement (Dobson et al. 1994; Donelan et al. 1997).

Lake Ontario's long axis is aligned (with the prevailing wind direction) WSW to ENE and a few times a year a low pressure centre crosses the lake from south to north bringing first east winds and then, soon after, west winds. The research platform of the Canadian National Water Research Institute (Donelan et al. 1985) is located 1.1 km from the western shore. So the east winds, blowing over 300 km of fetch, produce large waves that persist for many hours after the wind has abated or turned to the west. Under these latter conditions the east swell faces an adverse wind and produces an enhanced wind drag. Figure 8.2 shows a record of wind and wave properties under these circumstances. It is seen that the drag coefficient

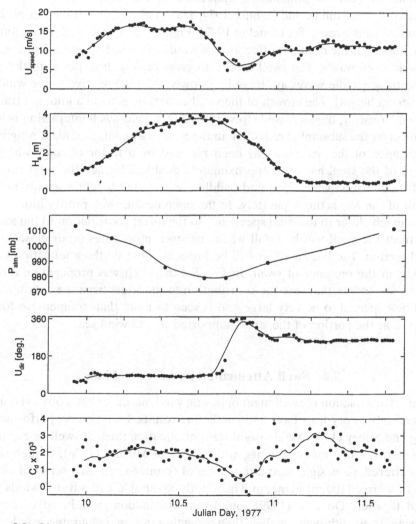

Figure 8.2. Observations of wave height, wind and the drag coefficient 1.1 km from the western shore of Lake Ontario during the passage of a low pressure centre, causing the wind to reverse direction in a few hours.

decreases as the wind speed decreases when the wind is from the east and the waves are mature, having propagated over the entire 300 km fetch of the lake. The wind abruptly turns to the north and, in blowing across the swell propagation direction, i.e. along the crests, sees very little drag. Finally, the wind turns to the west and intensifies; the rate of decay of the waves increases and the drag coefficient rises up to and above its value when the wind was from the east and 50% stronger and the waves were more than six times larger.

8.3   Swell–Wind Sea Interaction

It has often been noticed that the addition of paddle-generated swell to wind waves in a tank produces a pronounced reduction in the energy of the wind sea (Mitsuyasu 1966; Phillips and Banner 1974; Hatori et al. 1981; Bliven et al. 1986; Kusaba and Mitsuyasu 1986; Donelan 1987). Figure 8.3, reproduced from Donelan (1987), illustrates the dramatic effect on the wind sea caused by quite gentle monochromatic swell waves. The swell is seen to grow rapidly in response to the wind because these paddle waves are travelling slowly (2.2 m/s) relative to the wind (11 m/s at 26 cm height). The growth of the swell corresponds to an additional transfer or stress. However, the increased stress from direct wind–swell interaction is more than offset by the substantial reduction in the wind sea and its attendant roughness. The variance of the wind sea has been reduced by a factor of four, while the variance of the swell has been approximately doubled. In fact, the total variance of surface elevation when wind and paddle are operated together is only 67% of the sum of the two acting separately. In the open ocean, swell propagation speeds may be much closer to the wind speed and so the direct contribution to the surface stress from the swell will be small when the swell propagates in, or close to, the wind direction. The largest effect will be brought about by the attenuation of the wind sea in the presence of swell. In Fig. 8.2 the swell was propagating directly against the wind. In that case the contribution to the stress from the counter-swell would be expected to be very large and is seen to more than compensate for the reduction in the portion of the stress supported by the wind sea.

8.4   Swell Attenuation in Adverse Winds

The direct attenuation of swell in an opposing wind has never been observed in the field, and laboratory tests have had conflicting results. Experiments performed by Young and Sobey (1985) yield insignificant attenuation rates of swell as estimated from pressure-slope measurements, while those of Mizuno (1976), though somewhat scattered, show significant attenuation of counter-swell. In a series of experiments in a large (100 m) flume, in which both favourable and adverse winds were applied to swell, Donelan (1999) found that attenuation rates in adverse winds were significant although smaller than growth rates in favourable winds (Fig. 8.4). The large increase in momentum transfer described above in an opposing swell (Fig. 8.2) appears to come from drag on the swell itself since the wind sea

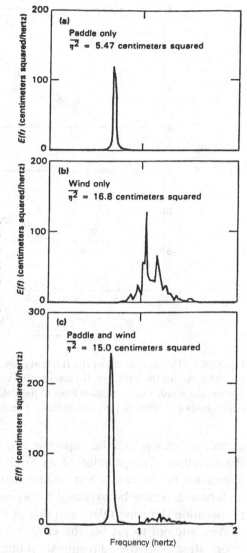

Figure 8.3. Wave spectra at 50 m fetch in a laboratory wind-wave tank. (a) The spectrum of a continuous train of 0.707 Hz paddle-generated waves of steepness $ak - 0.067$; (b) The spectrum of a pure wind sea with measured wind of 11 m/s at 26 cm height. (c) The spectrum of waves with wind and paddle excited together as in (a) and (b).

is flattened by the swell. Such a momentum transfer from the swell corresponds to its rapid attenuation.

On the other hand, observations of swell propagation, across the widest oceans and through various meteorological conditions, Snodgrass et al. (1966), seem to suggest that when the steepness of swell is reduced enough it no longer interacts directly with the wind. These very gentle long swell components may carry significant energy and on approaching the coast the energy is concentrated near the surface.

8.5   Change in the Drag due to Swell

The addition of swell to a locally wind-generated sea alters the roughness of the surface in two distinctly different ways: (1) the swell contributes directly to the

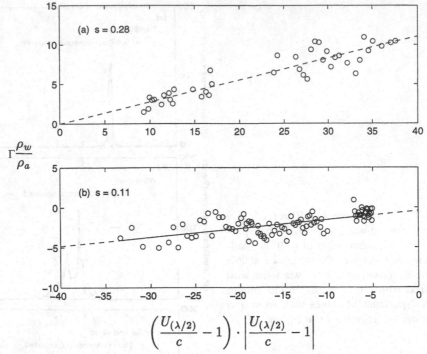

$$\Gamma\frac{\rho_w}{\rho_a}$$

$$\left(\frac{U_{(\lambda/2)}}{c}-1\right)\cdot\left|\frac{U_{(\lambda/2)}}{c}-1\right|$$

Figure 8.4. The magnitude of the fractional energy change per radian x the density ratio: (a) growth rates for the wind sea; (b) attenuation rates for the paddle-generated waves travelling against the wind. The regression lines to the data are shown and the corresponding sheltering coefficients (s = line slopes) are indicated on the figure.

surface roughness and the importance of this contribution depends sensitively on the direction of propagation of the swell relative to the local wind; (2) the swell attenuates the wind sea and, although the mechanism of attenuation is poorly understood, it may be expected to depend on the steepness of the swell and its propagation direction relative to that of the wind sea components.

Any attempt to predict the effect of swell on the drag will require detailed information on the directional properties of both wind sea and swell. Consequently, measurements at sea of the wind stress without concomitant information on the wave directional properties will exhibit considerable noise, much of which may be caused by swell. In a recent paper Yelland and Taylor (1996) have obtained a wealth of data on drag coefficients inferred from the high frequency spectrum of horizontal air velocity fluctuations via the inertial dissipation method. Their estimates of the drag coefficient, reproduced here as Fig. 8.5, show substantial scatter about some average value that is taken to depend on stability and wind speed only.

Dobson et al. (1994) made simultaneous measurements of wind stress with the inertial dissipation technique (Anderson 1993) and directional wave spectra with a pitch-roll buoy in the open ocean during the passage of several winter weather systems. Their intent was to determine a relationship between wind stress and sea state in the open sea; to do this it was necessary to partition their wave spectra into

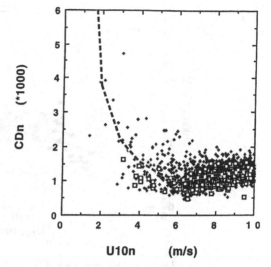

Figure 8.5. Neutral drag coefficients from two ships using the inertial dissipation method. (From Yelland and Taylor 1996.)

sea and swell, which they did using the energy and mean direction at each frequency of their buoy spectra. Wavelength and direction from the buoy spectra compared favourably with image spectra taken by ship-, air-, and space-borne radar systems.

Their findings can best be described as scattered. First, no clear wind stress versus sea state relation emerged from the wind sea parts of their wave spectra – merely a general confirmation, over a severely limited range of ages (all near $c_p/U_c = 1$, where c_p is the wave phase speed at the sea peak and U_c is the component of U_{10} in the wave direction), of the HEXOS result (Smith et al. 1992). Second, their wind stress measurements, although giving drag coefficients (Fig. 8.6a) consistent with the open-sea results of Smith (1980, 1988) and exhibiting the typical large scatter, did not stratify significantly with either the swell amplitude relative to that of the sea or with the swell direction relative to the wind (Fig. 8.6b).

They concluded that better definitions were needed of "sea" and "swell" and "wave age" in the presence of propagating and developing weather systems containing fronts. For such systems the lack of a clear understanding of the mix of physical mechanisms by which energy and momentum were transferred from the wind and from the swell into the sea severely hampered their ability to extract information from a well-calibrated, carefully made set of simultaneous wind and wave measurements.

Donelan et al. (1997) made direct observations of Reynolds stresses and wave directional properties from the SWATH ship *Frederick G. Creed* during the Surface Wave Dynamics Experiment (SWADE). Their estimates of the neutral drag coefficient are tagged with the general swell condition in Fig. 8.7. It is apparent that the presence of swell greatly increases the variability of the drag coefficient over that which obtains in a pure wind sea. Generally, when the swell is counter to the wind the drag is higher, and cross and following swell tend to produce lower drag. In Fig. 8.7 the very high drag coefficients near wind speeds of 5 m/s occurred when there was a large swell running directly against the wind. These results are in

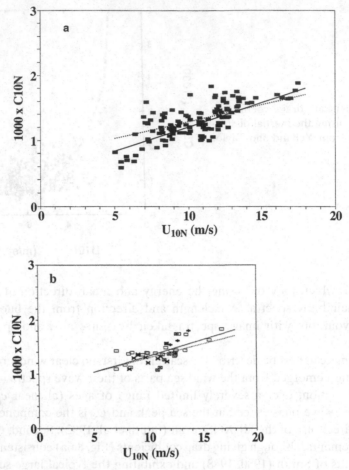

Figure 8.6. (a) Neutral 10 m drag coefficients from ship's bow anemometer runs (corrected for ship-induced distortions of the mean flow and the turbulence at the anemometer): lines are: (solid) regression to these data and (dotted) Smith (1988); (b) Neutral 10 m drag coefficients interpolated to times of wave runs: ($\times$) mature waves ($U_c/c_p < 1.1$), (open square) middle-aged waves, (+) younger waves ($U_c/c_p > 1.7$), (box with $\times$) wind sea energy > swell energy. Data points deleted for cases when there was no well-defined wind sea peak or when the wind sea direction was more than 30 degrees off the wind direction. Lines are: (solid) regression on these data and (dashed) Smith (1988).

general accord with the effects of swell on the drag discussed above. As pointed out by Donelan et al. (1997), the inertial dissipation method, which was used by Yelland and Taylor (1996) and Dobson et al. (1994), responds only to the turbulent Reynolds stress through its interaction with the wind profile. In the presence of long waves (swell) some fraction of the total stress is carried by wave-coherent (non-turbulent) motions. These tend to cause a reduction on the slope of the wind profile that is more pronounced near the surface but may reach up to heights of the order of the swell wavelength. The inertial dissipation estimates, which depend on the production of kinetic energy through the interaction of the turbulent stress with the wind profile, tend to underestimate the total stress in these conditions. Some

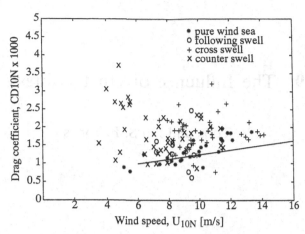

Figure 8.7. Neutral drag coefficients from the SWATH ship in SWADE from direct measurements of the Reynolds stress during various conditions of wind sea and swell. The pure wind sea relation of Smith (1980) is indicated with a solid line.

comparisons of the inertial dissipation and direct eddy correlation methods of estimating the wind stress are given in Donelan et al. (1997), but a full resolution of the matter will be realized only by the concurrent estimates of all the terms in the kinetic energy budget that are required in a rigorous determination of the stress via a balance of production with the dissipation and local divergence of kinetic energy.

8.6 Summary

When swell is relatively steep and travels with the wind, the wind waves are suppressed and the drag is lower than in swell-free cases. When swell is relatively steep and travels against the wind, the wind waves are again suppressed, but now the momentum transfer (the drag) to the swell is large enough to enhance the drag coefficient. For cross-wind swells and low-slope swells the effect on the drag coefficient appears to be small.

9 The Influence of Unsteadiness

Y. Toba and Ian S. F. Jones

9.1 Introduction

The boundary layers each side of the sea surface are forced on many time and space scales. To simplify matters, the previous chapters described situations where the external forcing was constant. However the measurements presented from the natural environment were subjected to variations in forcing conditions. Observations yielded many different values of neutral atmosphere drag coefficient. While there are some trends (in Fig. 1.4 for instance), that can be attributed to different wind speed, much of the variability remains unexplained. Could some of the remaining variability result from the unsteady nature of the boundary layers each side of the air–sea interface?

The air–sea interface is a coupled system. The momentum transfer between them occurs through the sea surface which is distorted by *wind-waves*. These are not the idealized irrotational gravity and capillary waves but are the turbulent windsea motions induced by the sheared boundary layers. The waves play a special role in the momentum transfer between the sheared air and water boundary layers, as discussed in Section 4.5.4. The time scales typical of the air and water boundary layers need to be compared with the response time of the windsea to assess the interactions. With some understanding of these time scales, the influence of unsteadiness on the momentum flux at the sea surface can be examined. The spatial variations in the boundary layers are swept past a stationary observer who may interpret the variations as temporal change. We will adopt that point of view and look at the range of time scales that occur above, at and below the surface of the sea. Such values are summarized in Table 9.1.

The planetary boundary layer is typically 1000 m thick and moves with a velocity of 7 m/s. This produces a time scale of order 100 seconds and indeed some spectra of the near surface velocities show spectral peaks in this region. Variations on this time scale are sometimes called gustiness. We will use the word gust, however, in

Table 9.1 Time scales for wind speeds of 7 m/s

	Disturbance	Speed	Time scale
Atmospheric b.l.	synoptic		7 days
	Roll cells	drift	minutes
	diurnal		24 hours
	planetary b.l.*	7 m/s	100 seconds
	turbulence at 7 m	7 m/s	1 second
Wave field	50 m wavelength	9 m/s	40 hours[†]
	10 m wavelength	4 m/s	4 hours[†]
	1 m wavelength	1 m/s	95 seconds[†]
	20 cm wavelength	0.5 m/s	7 seconds[†]
Oceanic b.l.	turbulence at 7 m	0.23 m/s	30 seconds
	Langmuir cells	drift	minutes
	mixed layer diurnal		24 hours
	internal tides		12 hours
	mesoscale eddies		1 month

* "gustiness", [†] time to reach steady state.

the more general sense of any deviation from the mean of duration more than a few seconds. Within the surface layer the turbulent eddies have length scales of the order of the distance from the surface. At a height of 10 m and a representative speed of 7 m/s such eddies induce fluctuations of period 1.3 seconds. Most of the surface layer turbulence occurs on time scales shorter than 100 times this value. The planetary boundary layer is embedded in a larger atmospheric system that responds diurnally to the sun's heating and to disturbances on a still greater scale. These longer time scales will be called unsteadiness.

The time scale for the windseas to respond fully to a sudden increase of wind can be estimated from the duration curves of Fig. 1.5. For open ocean waves these times are measured in hours, much longer than the response time of the planetary boundary layer. This mismatching of time scales between the air and the surface to which it is transferring momentum is possibly one of the causes of the difficulties the air–sea interaction community is having in explaining the fluctuations in drag coefficient shown in Fig 1.4.

In the ocean the upper ocean turbulence is of smaller magnitude than turbulence in the air and the time scales are typically 30 times longer than in the atmosphere. Superimposed on the shear induced turbulence is windsea coupled turbulence with time scales of the order of the surface wave characteristic period. The downward bursting discussed in Section 4.5.4 occurs sporadically at many characteristic wave periods (Toba and Kawamura 1996).

The approach used in this chapter is to first examine the response of the wave field to changing wind forcing or changing currents and then to use this to speculate on the change in drag coefficient. Little is known about the transient response of waves and even less about the impact of this on drag coefficient. This chapter is forced to rely on new ideas based on some general principles established in previous chapters together with a small number of observations.

9.2   Wave Field Response to Change in the External Forcing

9.2.1   Background

When there is a gust in the atmospheric surface layer or a fluctuation due to a turbulent eddy in the ocean mixed layer, the coupled system must adjust. Let us develop the concepts in terms of air velocity fluctuations. The influence of the water boundary layer is thought to be less than the atmospheric boundary layer but the truth of this will need to await further study.

9.2.2   Local Equilibrium Seas

When there is a large weather system with uniform winds over the ocean, the windsea some distance from the edge of the system will, initially, be duration limited. The wind speed, U, can be used as a scaling variable or preferably the friction velocity, u_*. For an idealized duration limited sea, the wind is zero at time $t = 0$ and there is no swell. The nondimensional wave energy is a function of the nondimensional time, tg/U, in such a situation. With time, the peak frequency of the wave spectrum becomes lower until the age, c_p/U, a measure of the peak frequency, exceeds about 1.3. The variation of c_p/U with time is shown in Fig. 1.5.

Many measurements of drag at sea are made under an approximation to duration limited seas as weather systems do not persist for the 40 hours or so, required for surface waves to become steady (see Table 9.1). Consequently the *wind-wave* field is considered spatially uniform, to the first approximation, and only a small fraction of the momentum flux to the sea surface goes to provide the growth of wave momentum with time. From the empirical duration law of Fig. 1.5 we know the fraction of momentum flux retained as wave momentum growth is a decreasing function of tg/u_*, with a maximum value of 6% according to Toba (1978).

In steady fetch-limited situations the situation is much the same with the wave field constant in time but increasing in momentum with distance from shore. As in the duration-limited case most of the wind momentum goes into the currents locally. In both cases the wave spectrum with its peak frequency ω_p takes on a form over much of its domain that depends only on a measure of the wind, such as the friction velocity, u_*. The spectrum is observed to be in *local equilibrium with the wind* and to take the form

$$F(\omega) \sim \alpha_s u_* \omega^{-4} \qquad \omega > \omega_p \tag{9.1}$$

over the equilibrium range of frequencies. This implies that:

$$\frac{\overline{\eta^2}g^2}{u_*^4} = \frac{B^2}{16}\left(\frac{2\pi g}{u_*\omega_p}\right)^3 \tag{9.2}$$

as discussed in Chapter 4.

The evidence for the friction velocity dependence of the frequency spectrum is shown in Fig. 4.20. At frequencies around the phase speed minimum and higher, the windsea spectrum depends more strongly on u_* than expressed in Eq. (9.1).

9.2.3   Unsteady Duration-Limited Situations

If there is a sudden increase or decrease in wind speed over a wide area, the windsea will initially not be in *local equilibrium with the wind*. Immediately after the change, the wave peak frequency and energy will be unchanged but u_* will have increased. The value of B in Eq. (9.2) will decrease with a sudden increase in u_*, as will the spectral constant, α_s. We will call such a windsea spectrum, where α_s is below its steady value, under saturated. In the next section we will examine the experimental evidence of how the windsea responds to being out of local equilibrium with the wind.

9.2.4   Fetch-Limited Sea

While fetch-limited experiments, conducted in wave tanks or downwind of straight coastlines such as the well known JONSWAP experiment, have been much studied, the vast reaches of the ocean are often limited by the extent of the weather system rather than the distance from a straight coastline. First an unsteady duration-limited set of observations are discussed then the response to wind change from one fetch-limited situation to another in a wind wave tank is discussed.

9.3   Field Data

The wave height, aerodynamic roughness and wind stress (friction velocity) have been measured at the Shirahama Oceanographic Tower Station of Kyoto University in Tanabe Bay and reported in Kawai et al. (1977). We considered the Shirahama data only while the direction was such that the fetch was about 60 km. At such fetches and for the wind speeds observed, the age of the wave field, c_p/U, needs to be greater than 0.8 for the windsea to be fetch-limited (see Fig. 1.5). The situation is one of duration-limited waves. The response to a change in the wind speed for a duration-limited situation can be studied with the aid of Fig. 9.1 which plots the nondimensional roughness against wave age. It also shows both the relationship $gz_0/u_*^2 \propto$ (wave age)$^{-1}$ and the relationship for gz_0/u_*^2 independent of wave age with a Charnock constant of 0.0185. The data is prepared from 3 minute averages.

The experiment started at the point 'a' and the waves grew older until point 'b' in the manner one would expect for a duration limited sea. At point 'b' the age c_p/U is near the fetch-limited value. Then the wind freshened and the wave age decreased between 'c' and 'd'. The rate of increase of the wind during the period 'c' to 'd' was greater than the increase of wave speed that one would predict from Fig. 1.5. Consequently c_p/U decreased or $\omega_p u_*/g$ increased. For very slow increases in wind speed, the wave age can continue to grow. Decreasing wind speed leads to rapid wave age increases.

In Shirahama, waves of period about 3 seconds responded to varying wind speed and direction in the manner shown in Fig 9.1. In general the wave age increased from 'a' to 'b' as a result of the wind speed falling and decreased from 'b' to 'd' as a

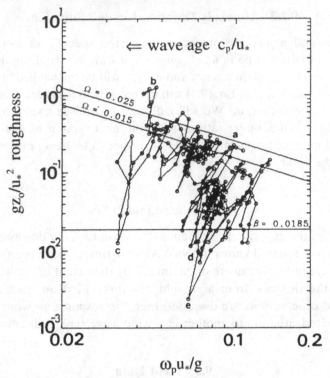

Figure 9.1. The nondimensional roughness length as a function of inverse wave age reproduced from Toba and Ebuchi (1991). Moving averages of 3 minutes were used. Record
length about 4 hours. a–b, wind decreasing, wave age increases; c–d, wind rising, wave age
decreasing. Heavy line is the Charnock relation with $gz_0/u_*^2 = 0.0185$. Thin line is the Toba
and Koga (1986) relationship with proportionality factors 0.015 and 0.020.

result of wind speed increasing. There are notable fluctuations about these most
general trends and one sees that points from 3 minute averages scatter between
$gz_0/u_*^2 = 0.18$ and $gz_0/u_*^2 = 0.025(c_p/u_*)$ and beyond.

If we take a portion of this record about point 'c' we can examine the fluctuations around the 60 minute trend. The fluctuations in the spectral constant, α_s track
the stress, u_*^2. They are in antiphase because the product $\alpha_s u_*$ in Eq. (9.1) is more
constant than its components. The windsea equilibrium range spectra changes little
during the 5 or 7 minute fluctuations which produce the variations in the stress
about the running 60 minute mean. Such results are reproduced from Toba et al.
(1988) at Fig. 9.2.

What does happen is that the spectrum becomes under or over saturated. If you
assume the spectrum for windsea in *local equilibrium with the wind* is given by Eq.
(9.1):

$$F(\omega) = \alpha_s g u_* \omega^{-4}$$

for the equilibrium range of frequencies then variations of α_s are a measure of
saturation.

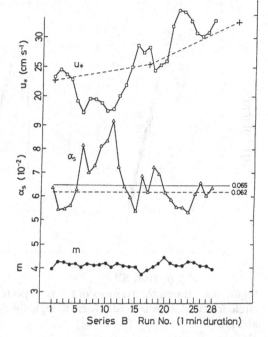

Figure 9.2. Friction velocity and spectral coefficient α_S reproduced from Toba et al. (1988). A 30 minute section of data around point c in Fig. 9.1 was used. Only the equilibrium gravity waves are considered in determining α_S. The spectral dependence on frequency $m = 5 - n$ in Eq. (4.2).

Toba et al. (1988) showed that those cases, where the wind, measured over three minutes was rising faster than the 60 minute wind trend, are ones where as we expect the wave spectrum is undersaturated. The power spectral density at the peak frequency was less than in the situations where the short term wind was near the 60 minute mean. This is shown in Fig. 9.3. During the periods where the wind was falling, the spectral peak was higher than the examples near the mean; and the peak was also sharper.

An estimate for the time for the surface boundary layer to respond to a change in aerodynamic roughness can be estimated by observing that the inner layer boundary layer grows with a slope of 1/10 to 1/100. If we take a slope of 1/50, the boundary layer perturbation in momentum can diffuse up to (or down from) 10 m in a wind speed of 10 m/s in about 50 s. Thus we averaged our profiles for 90 s and found that the measured drag coefficient was higher in the positive gusts and lower in the negative wind gusts as shown in Fig. 9.4. Under saturated windsea spectra extract more momentum from the wind than over saturated spectra.

We need to provide an explanation why the drag is higher during undersaturated conditions. It is well accepted that the drag coefficient increases with wind speed. We also showed in Chapter 4, that the short wave amplitudes and steepness are greater for large u_*. Are we seeing more than this trend in Fig. 9.4? Let us use Eq. (2.2b):

$$10^3 C_D = 0.61 + 0.063U$$
$$\approx 0.6 + 2u_*$$
$$10^3 \delta C_D = 2u_*(\delta u_*/u_*)$$

for $u_* \approx 0.5$ m/s

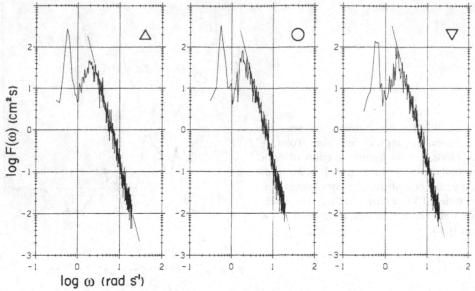

Figure 9.3.   Ensemble averages of ten raw spectra marked by triangle for increasing winds, circle for constant wind and inverse triangle for decreasing wind. Reproduced from Toba et al. (1988).

$$10^3 \delta C_D \approx \delta u_* / u_* \qquad (9.3)$$

The change observed in Fig. 9.4 is about four times that due to the value of perturbation expected from the wind speed expression itself. Thus the fluctuations are rapid enough to take the drag coefficient away from the value appropriate for flows in *local equilibrium with the wind*.

Why is the drag coefficient higher during the adjustment period? It is not likely to be the momentum needed to make the small change in retained wave momentum during this period. One possibility is that the shorter waves that respond first to the change in wind overshoot their equilibrium value and so for a positive gust, initially exceed the equilibrium steepness. Being steeper they would be more prone to microbreaking. Breaking has been shown to increase the stress by Toba and Kunishi (1970) and Banner (1990b) for example. Such speculations need to await measurements at high wavenumbers during unsteadiness to be verified.

The relative time scales of the wind speed fluctuations and the wave field response need to be considered. We will produce evidence below that wind-waves not in *local equilibrium with the wind* evolve rapidly to correct this situation. What is their adjustment time scale and how does it compare with the time for the windsea to grow to age $c_p / U = 1$?

One would imagine that the rapid adjustment time would be related to the wave energy and wave age. Young windseas are expected to adjust more rapidly than older seas with their usually larger energy. At Shirahama, the adjustment time appeared to be less than 15 minutes since Fig. 9.2 shows little evidence of increased

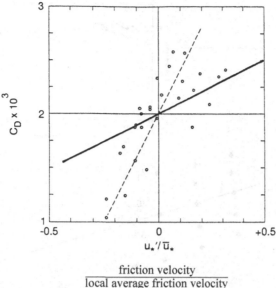

$$\frac{\text{friction velocity}}{\text{local average friction velocity}}$$

Figure 9.4. The fluctuations of drag coefficient observed in 1.5 minute (overlapping) samples of the data in Fig. 9.2. Reproduced from Toba et al. (1996). Positive wind gusts have higher drag coefficients than negative wind gusts when the time is too short for the wave field to be in local equilibrium with the wind. The solid line is Eq. (9.3).

drag, while Fig. 9.4 for 1.5 minute data, showed considerable increase in drag during periods of high wind speed. Let us use the broken line in Fig. 9.4 in the form

$$\frac{\delta C_{\mathrm{D}}}{C_{\mathrm{D}}} = \frac{2\,\delta u_*}{u_*} \tag{9.4}$$

where C_{D} and u_* are the values for a sea in *local equilibrium with the wind*.

The nondimensional roughness length $z_0 g / u_*^2$ is very sensitive to changes in z_0 in response to u_* fluctuations. Using Eq. (9.3) and the assumption that fluctuations in u_* do not change the spectral peak frequency (an issue discussed below), the variation of $z_0 g / u_*^2$ for a 5% and 10% change in u_* is shown in Fig. 9.5. The fluctuations due to surface turbulence over the ocean in neutral stability are of order 10%. Some of these fluctuations are quite fast but typical 2 minute averaged wind speeds during a storm (known as S24 in Jones and Toba (1985)) are shown in Fig. 9.5 for reference. Antonia and Chambers (1980) at the same Bass Strait site found typical downwind turbulence levels of three times the friction velocity.

One can see that fluctuations in wind stress that cause over or under saturation can lead to spread of data points that more or less cover the range shown in Fig. 2.3. It will be important in future to examine the windsea spectra in detail together with the wind fluctuations before asserting a steady-state drag coefficient has been measured.

If the measurement of wind stress is made over a period short compared with the adjustment period of the windsea, then the drag coefficient will differ from the steady-state result. As most measurements are made for 15 or 30 minute averages,

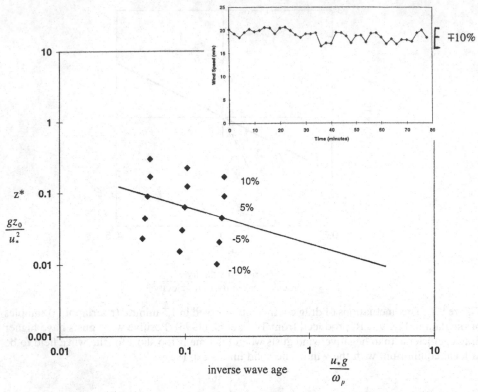

Figure 9.5. The variation in nondimensional roughness length, for fluctuations in friction velocity assuming Eq. (9.3) and a friction velocity of 0.5 m/s. The straight line (TIKEJ) is a possible expression for roughness length over waves in local equilibrium with the wind. This figure emphasizes how potentially important wind stress fluctuations may be in any particular measurement of drag coefficient. The insert shows a sample of wind speed at a height of 26 m above the sea in Bass Strait during a storm. The two minute averages have eliminated much of the marine surface layer turbulence, leaving only the planetary boundary layer gusts.

this suggests the scatter due to gusts should be largest at older windseas where the adjustment time is longer.

In steady situations there are as many positive gusts as negative gusts. If there were a symmetrical response to over and under saturation, there would be no impact on the drag coefficient averaged over a number of gust periods. Thus it is only fluctuations of duration of the averaging time that will produce a deviation from the steady-state situation.

9.4 Laboratory Data

9.4.1 Rapid Adjustment

Experiments in laboratory tanks can be performed at much lower wave ages than field data typically provide. The debate about the continuity of field and laboratory data has been addressed in Jones and Toba (1995). One difficulty with laboratory experiments where the wind speed is rapidly changed from one value to another, is that the wave energy upstream of the observation is not uniform. The general nature of the fetch limited problem can be understood by looking at the significant wave height as a function of fetch and wind speed in Fig. 9.6. The spatial distribution of wave heights are shown at the start of such an experiment and the distribution at the end of the readjustment. The waves propagate past the observation point (at a fetch of 15 m) and after the sudden increase of wind speed, they appear to be in a kind of duration limited situation. As the waves travel down the tank, they are subjected to a wind stress that makes them grow until eventually all the waves at the observation point are fetch limited at the new high wind speed.

First, there is the strong desire of the *wind-waves* to adopt the spectral form $\alpha_s g u_* \omega^{-4}$. Although the spectral value in steady state depends upon u_*, when the wind speed fluctuations are rapid, the *wind-wave* spectra cannot adjust fast enough. Thus for a rapid increase in wind speed, the energy of the windsea becomes undersaturated. In this situation it is also observed in the field that the energy decreases near the spectral peak (Fig. 9.3). The high frequency waves, say $\omega > 60$ rad/s, are the "catspaws" that quickly appear as a result of gusts, and were studied by Dormon and Mollo Christensen (1973). The higher wind speed increases the amplitude of these waves. The high frequency waves are now steeper as a result of a positive gust and can be expected to remove more momentum from the air. This can be seen from the model of Section 6.4 for wave drag, i.e.

$$\tau_w = \rho \int \omega^2 \Gamma F(\mathbf{k}) d\mathbf{k} \qquad (9.5)$$

The stress, τ^w due to short waves will increase as $F(\mathbf{k})$ increases. The short waves may also provide the increase in the equilibrium range frequencies by wind-forced strong wave interactions. If the resultant increase in friction velocity due to these changes is greater than the wind speed, the drag coefficient will rise.

When the wind gust is a reduction in wind speed, the equilibrium-range wave spectrum is initially over saturated. The energy at the peak is observed to increase without a change in peak frequency, Fig. 9.3. Thus the peak frequency waves become steeper. The high frequency waves diminish, as will be shown in Section

Figure 9.6. Wave height measurements in the Tohoku tank at two different wind velocities.

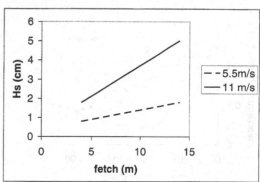

9.4.2, and if they undershoot their equilibrium value, they are expected to extract less momentum from the wind.

We speculate that the process above is the rapid readjustment of the wave field to an external perturbation. It is the process that re-establishes the *local equilibrium with the wind*. Following this we expect a period of adjustment to occur as the waves upstream of the observation point propagate past the observation point and are subjected to the new wind stress for longer and longer periods of time. In the adjustment process, shown in Fig. 9.7, we see the slower adjustment that occupies most of the time it takes the waves to reach a new fetch limited value. In the next section we will examine the initial period of adjustment.

9.4.2 Unsteady Laboratory Experiments

The recent laboratory experiments of Waseda et al. (2001) provide further understanding of the unsteady forcing. The experiment examined the adjustment processes of fetch-limited laboratory windsea, which were already in *local equilibrium with the wind*, to sudden changes of wind speed. In the experiment, a very rapid change in wind speed from 4.6 m/s to 7.1 m/s, and vice versa, were produced. The existence of two time scales in the wave response were found: the first was of the order of 4 s, and the second was of the order of 20–30 s. The wave age c_p/u_* was between 1 to 2 at fetch 5.5 m. The first time scale is that necessary for the adjust-

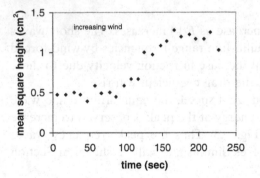

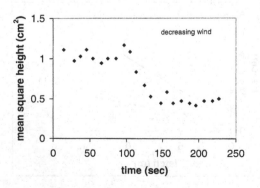

Figure 9.7. The change in wave energy $\overline{\eta^2}$ measured in the Tohoku wind wave tank at a fetch of 15 m. Wind speed changed between 5.5 m/s and 11 m/s at time equals 100 s.

ment of the windsea to the local equilibrium with the new wind to be achieved. The second time scale was related to the conventional growth and propagation in the wave field that followed the first rapid adjustment period. It was observed during the first period that the high frequency components of the spectrum, or the "physical roughness" of the windsea increased to approximately their final state. This can be seen in the spectra reproduced in Fig. 9.8.

The low frequency measures such as peak wave period or significant wave height, make only small changes towards their final state during this rapid adjustment state. The changes in bulk measures occurred mostly during the slow adjustment stage. It can be seen that, during the rapid adjustment phase, the position of the peak frequency does not change whereas the high frequency energy approaches its final level and the wave steepness increased.

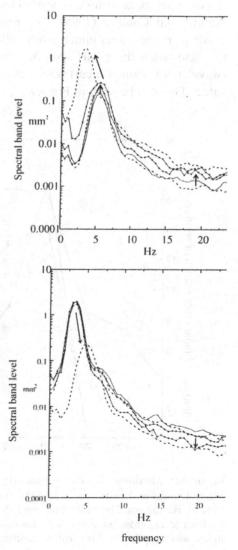

Figure 9.8. Spectral development of laboratory windsea for sudden change of wind speed from 4.6 m/s to 7.1 m/s (upper panel). The lower dotted line is the equilibrium spectrum for the lower wind speed, and the arrows indicate the change of spectrum from 0 s, 1 s and 4 s after the sudden wind change, to the upper dotted line spectrum for the equilibrium spectrum for the upper wind speed. The lower panel is the opposite case of decreasing wind. (Cited from Waseda et al. 2001.)

Waseda et al. (2001) have proposed an interpretation that the *local equilibrium with the wind* is the condition that the energy input by wind is proportional to the energy dissipation by wave breaking. Sudden change of the wind causes imbalance of these, and the time needed for relaxation of the imbalance gives the first time scale. Then the wind waves become steeper, and as a result wave dissipation becomes large enough within the first time scale, to be quasi-balanced with the new wind input or the new wind stress for stronger wind.

This first time scale is the evidence of the existence of a particular perturbation period of the windsea to the wind unsteadiness. The rapid adjustment time scale for this laboratory experiment was 4 s for the wave age, c_p/u_*, of 1 to 2, whereas it was several minutes for the field data described in Section 9.2 for the wave age of about 20. At older wave ages we suspect the initial rapid adjustment period may be in the order of an hour.

An experiment with sharp spatial increase of wind speed in a tank was made by Autard and Caullier (1996). The results showed an overshoot of the gravity–capillary range waves immediately following the sharp increase in wind speed. In the field when there was a sudden change in the wind in Bass Strait, the small (wavelength about 1.5 cm) waves over responded and overshot their equilibrium value. This can be seen in Fig 9.9.

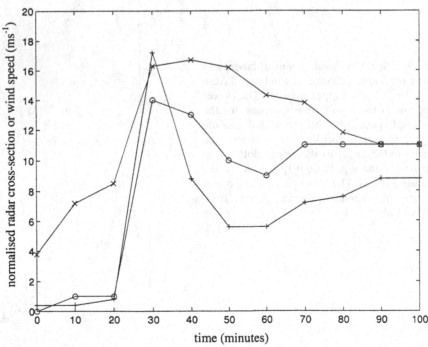

Figure 9.9. Microwave backscatter for HH and VV polarization during a wind event in Bass Strait. Details of the site are to be found in Jones and Negus (1996). Wind speed measured at 53 m above the sea. The radar angle was 45°; × is the wind speed; + is the vertically polarized backscattering cross-section; ○ is the horizontally polarized backscattering cross-section. This figure was supplied by Lisa Rufatt, University of Sydney.

9.5 The Influence on Drag

During the period of rapid readjustment we speculate that the momentum flux is much disturbed. The wave energy change is evidently small during the re-establishment of *local equilibrium with the wind* compared with the total wave energy change, which occurs only when a fetch limited state is obtained. We expect the wind input to the high frequency waves supplies extra momentum flux to these waves while nonlinear transfer redistributes the energy. The significant change in drag coefficient shown in Fig. 9.4 cannot be explained from the need for extra wave momentum. Thus we speculate during the rapid adjustment period the short waves steepen and break and the extra momentum ends up in the ocean currents.

9.6 Slow Rates of Change of Wind Speed

We now see that for time scales long compared with the rapid adjustment period of windsea, the drag coefficient is expected to be little different from steady-state values for situations in *local equilibrium with the wind*. A slow rate of change of wind speed, about 1.4×10^{-4} m/s^2, was observed on day 127 of the field experiment reported in Chapter 12. The drag coefficient fluctuated but generally took on a value that was little different from the usual value as can be seen in Fig. 12.2. This seems to suggest that at rates of change slower than 10^{-4} m/s^2, (5 m/s in 10 hr) stress at the sea surface can follow the wind speed. It is only those processes in Table 9.1 that are of order of the rapid readjustment period or faster that are important for producing deviations in the drag coefficient away from the steady-state value.

9.7 Turning Winds

When the wind shifts direction the *wind-wave* field cannot respond immediately. The response time of the *wind-wave* field can be anticipated to be a function of wave age. When the direction change is not large, a considerable part of the wave energy seems to be reorganized in the direction of new wind. Old large waves will take longer to be regenerated in the new wind direction than short waves. As with changes in wind speed, we expect that large rapid changes in wind direction will initially find the wave spectra well out of *local equilibrium with the wind*. Presumably after *local equilibrium with the wind* is established, the old sea energy still travelling in the initial direction can be treated as swell at an angle to the younger windsea. Swell was treated more generally in Chapter 8. Can a sudden shift of the wind direction be thought of as a decrease of the wave in one wind component and an increase of that component at right angles? We will assume this is a reasonable working model for changes of some magnitude.

The time scale for the change of mean wave direction can be written t. This time scale can be nondimensionalized as $t^* = tg/u_*$ and plotted against wave age. Quanduo and Komen (1993) used the observations of Holthuijsen et al. (1987)

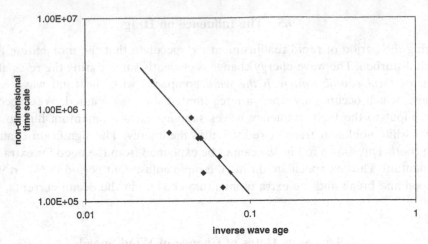

Figure 9.10. The nondimensional time scale for the wave direction change as a function of inverse wave age using the analysis of Quando and Komen (1993). These times seem much longer than the rapid adjustment period for speed changes.

to compare t^* with the inverse wave age $\omega_p u_*/g$ and these are reproduced in Fig. 9.10.

Only general results can be given for the expected change in drag coefficient with changing wind direction. If the change of direction is large (order 90°) and rapid, the directional wave spectrum in the new wind direction will be under saturated. Thus we expect a higher drag coefficient until *local equilibrium with the new wind* is restored. The previous wave spectra will be decaying and look like swell at a large angle to the new young sea. Since the representative drag coefficient is wave age dependent in the manner shown in Chapter 10, the drag coefficient can be expected to increase as the wave age in the new direction increases. As the waves become even older, the drag coefficient is expected to decrease.

The surface stress and reference wind do not have to be in the same direction. In Chapter 12 it is shown that during turning winds, quite large angles between the stress and the wind are observed. It is the slow response of the surface waves to the changing wind direction that leads to the large angles between wind and stress.

9.8   Physical Model

We can now sketch a hypothesis for the drag coefficient for unsteady winds. The earlier view, of a steady-state wind forcing, needs to be modified when we realize the boundary layers each side of the sea surface are turbulent and these boundary layers are in turn being forced by synoptic scale disturbances. Gustiness, the word we have used for the boundary layer turbulence either in the air or the water, is always causing fluctuating surface stress, but on timescales short compared with most open ocean wave timescales. We talk of steady conditions when there are no changes on timescales longer than the gustiness turbulence. In such conditions, the windsea becomes in *local equilibrium with the wind*.

When the wind suddenly changes speed or direction after a period of steady conditions, the waves are no longer in *local equilibrium with the wind*. There is a period of rapid adjustment. The high frequency, short waves, respond rapidly to take on a height appropriate for the new wind stress. The equilibrium frequencies unable to respond immediately become over or under saturated. The spectral peak energy changes rapidly, due we assume to nonlinear transfers, to bring the equilibrium frequencies to heights in *local equilibrium with the wind*. This readjustment process changes the demand for momentum from the wind leading to increased or decreased drag coefficients. Once the system is in *local equilibrium with the wind* this period of rapid adjustment gives way to the slower change as a kind of duration-limited sea evolves to a fetch-limited case.

When the wind increases, the wave age decreases but when the wind decreases there is the opportunity, especially for larger wave age seas, for the peak wave speed to far exceed the wind speed. As discussed in Chapter 7 those wave components travelling "faster than the wind" have been observed to give up momentum to the wind. In such situations one expects a lower drag coefficient.

9.9   Concluding Remarks

Fetch-limited situations. There are two time scales due to unsteady forcing of the air–sea interface. The period of rapid readjustment of the *wind-waves* to the new stress and the longer period of slow adjustment as the new wind stress provides the momentum for the lower frequencies to rise or fall in order to reach a fetch-limited state. There is little evidence to suggest that the drag coefficient should be significantly different from steady-state during the slow adjustment period.

The rapid adjustment period. Since over the ocean the boundary layers are always fluctuating we see this is an important issue. Figure 9.5 shows how sensitive the nondimensional roughness length is to fluctuations on timescales short compared with the rapid adjustment period. It appears possible that some of the measurements reported in Fig. 2.3 have been influenced by unsteadiness. While we concentrated on the fluctuations in wind forcing, there are also fluctuations in the water. These timescales are typically longer than the wind fluctuations and so are potentially likely to influence the drag coefficient. However, the small contribution of the currents to the total difference between the wind and the current make the influence of current fluctuations less important.

10 The Dependence on Wave Age

Y. Volkov

10.1 Introduction

In Chapter 2 we saw that there were many formulae which attempted to express the nondimensional sea surface roughness parameter as a function of wave age, and that no simple power laws can express observational data sets well. This difficulty was the origin of the decision of SCOR to set up a working group. Now we should try to understand physically the possible dependence of drag on wave age.

There are an array of variables that can be used to describe the momentum flux through the sea surface. First there are those that describe the surface of the sea with a *wind-wave* field without appreciable swell present. Then we have for the predominate wave

c_p phase speed
ω_p peak frequency
$\overline{\eta^2}$ mean square displacement
H_s significant wave height
T_p period

Then there are variables describing the dynamical characteristics of the neutral stability atmosphere:

τ stress
u_* friction velocity
z_0 aerodynamic roughness
U_{10} mean wind speed
C_D drag coefficient
θ angle between stress and wind

Ian S. F. Jones and Yoshiaki Toba, *Wind Stress over the Ocean*. © 2001 Cambridge University Press. All rights reserved. Printed in the United States of America.

Only some of these variables are independent. There are reasonably precise relationships as follows:

$$c_p = g/\omega_p$$
$$H_s = 4\overline{\eta^2}$$
$$\omega_p = 2\pi/T_p$$
$$\tau = \rho u_*^2$$
$$z_0 = f(C_D) \text{ for a logarithmic wind profile}$$
$$C_D \equiv (u_*/U_{10})^2$$

This reduces the number of variables needed to specify the problem to a set of four. We have chosen c_p, u_*, z_0, θ. This chapter is devoted to relating z_0 to c_p and u_* when $\theta = 0$ and *local equilibrium with the wind* provides $c_p = F(\overline{\eta^2}, u_*)$. It can be surmised that there is a strong statistical interdependence between all the above quantities.

It follows from the above that if the sea surface is an important component of the problem, u_* cannot be related to U_{10} alone. Expressions such that $C_D = F(U_{10})$ assume the state of the sea is either a universal function of U_{10} or it has no influence on the wind stress.

Over fixed rigid surfaces the aerodynamic roughness is a function of the physical roughness of the surface when these are higher than the viscous sublayer. Aerodynamic roughness is a concept that grew out of use of the log law to describe the velocity profile over a surface. Over the sea the situation is more complex and Charnock (1955) proposed from a dimensional argument that:

$$z_0 = \beta_* \frac{u_*^2}{g} \tag{10.1}$$

an expression that predicts a rise in aerodynamic roughness, z_0, for increased wind speed. Since the wave field is generally larger at higher wind speeds, the Charnock expression had an implicit dependence on the wave field. It uses only two of the three variables above. Kitaigorodskii and Volkov (1965) made the wave field explicit by using a weighted integral of the sea surface spectrum to estimate the aerodynamic roughness. However, it was Stewart (1974) who suggested that the age of the wave field was a second variable that needed to be considered, i.e.

$$z_0 = \frac{u_*^2}{g} f\left(\frac{c_p}{u_*}\right) \tag{10.2}$$

In Section 4.4 it was noted that a wind generated sea, free of swell contamination, often has a universal spectral form that results from a coupled system of wind, *wind-waves* and surface currents. In this sense the characteristic wave speed c_p is not only the phase speed of the spectral peak but characterizes the whole windsea field. Similarly the friction velocity u_* can be thought of as representing the momentum flux throughout the coupled system. Thus wave age, the ratio of c_p to u_*, is a likely candidate to describe the variations in roughness length.

In Chapter 5 the role in the transfer of momentum by viscous effects, of the long waves and the short waves, was discussed. If wave age, as a measure of the characteristic wave speed compared with that of the wind, is an important variable in expressions of the drag coefficient, it is either through the long waves, of which it is a direct measure, through the shorter waves, the character of which may be influenced by the long waves or through the combination of these two effects.

The momentum flux observations over the sea do not help resolve the issue of dependence on sea state. A recent extensive data set collected by Yelland and Taylor (1998), using the inertial dissipation approximation as discussed in Chapter 7, found that the drag coefficient was best described as a simple function of wind speed. Such representations, such as Eq. (2.6), suffer from the difficulties of having a dependence on a dimensional variable rather than a nondimensional one such as Reynolds number. The statistical interdependence of the wave age, wind speed, friction velocity and other wave spectral measures such as rms sea surface slope confound our efforts to test hypotheses about the variables that influence the magnitude of the drag coefficient.

The relative speed of the wind U to the wave crests c_p was seen as important in the early studies of waves as discussed in Chapter 1. When the waves travel much slower than the wind, considerable momentum is transferred but when the wave crest travels near the wind speed, little momentum is transferred (for the same sized wave). Once the waves run faster than the wind one would suspect some momentum is returned to the wind until the waves loose their *wind-wave* character and become swell. Let us consider the issue disaggregated into its elements.

10.2   Long Waves

For *wind-waves* in *local equilibrium with the wind*, the characteristic wave becomes longer with increasing wave age for a constant wind. The wave spectrum exhibits a large degree of universality, and the energy level of the peak frequency component increases with wave age. However, this increase in wave energy is at a slower rate than the wave length, so a diagram of characteristic wave slope, defined as significant height over significant wave length, decreases with wave age. Empirical results of wave slope were shown in Fig. 4.19 and presumably form the basis of the widely held belief that old (less steep) waves transfer less momentum than young waves (of the same magnitude).

For wavenumbers above the spectral peak, k_p and well below the phase speed minimum, k_m, a model expression used earlier is

$$F(\mathbf{k}) = \alpha^+(u_*/c_p)^m k^{-n} D(k, \theta) \qquad k_p < k < k_m \qquad (10.3)$$

This expression implies that the wave spectrum decreases with wave age for positive m. This could be a reflection of the behaviour of the peak enhancement which is not explicitly addressed in Eq. (10.3). From dimensional considerations n is 4.

The relationship between the wavenumber spectrum and the frequency spectrum is not simple because the long waves Doppler shift the short waves. In Fig. 8

of Banner et al. (1989) at a steepness of about 0.15, $m = 0$, $n = 4$ and one sees that the spectrum can be written almost equivalently to Eq. (4.42),

$$F(\omega) = \alpha_s g u_* \omega^{-4} \qquad \omega > \omega_p \qquad (10.4)$$

If the conversion to frequency had invoked the dispersion relation, a different result would have been obtained. Note also that the dependence of wave age in the wavenumber spectrum is not immediately transferable to the frequency spectrum.

Donelan et al. (1985) have suggested that the peak enhancement of the wave spectrum is a function of wave age, while Eq. (4.9) suggested for a spectrum proportional to ω^{-5} the constant of proportionality may also be a function of wave age. Assuming Eq. (10.4) is applicable, the mean square wave height, Eq. (4.31) becomes

$$\overline{\eta^2} = \frac{\alpha_s}{4} g u_* \omega_p^{-3}$$

Now wave age is defined as $c_p/u_* = g/\omega_p u_*$ and so

$$\overline{\eta^2} = \frac{\alpha_s}{4} g^2 u_*^4 \left(\frac{c_p}{u_*} \right)^{-3} \qquad (10.5)$$

This is the 3/2 law of Eq. (4.38) again. It shows that for *wind-waves in local equilibrium with the wind*, wave age, wind speed, c_p and wave height are not independent variables.

10.3 Wind-wave Drag

As discussed in Chapter 6, an exact expression, Eq. (6.6), for the wave component of the surface drag can be written in terms of the rate of change of the wave momentum flux, i.e. (where the directional term $\mathbf{k}/k$ has been omitted)

$$\tau^w = \rho_a \int_0^\infty \int_0^\infty \omega^2 F(k) \Gamma D(k, \theta) dk \qquad (10.6)$$

The wave growth, written as $\omega\Gamma$, can be estimated from measurement and is considered a function of the phase speed c of the wavenumber component under consideration. Plant and Wright (1977) made a series of measurements of the initial rate of growth of wave components. Their plot of the parameter Γ is shown as Fig. 2.3 in Komen et al. (1994). We have used a U/u_* ratio of 24 to give

$$\Gamma \propto \varepsilon^2 \left(\frac{24 u_*}{c} \cos\theta - 1 \right) \qquad c/u_* < 24 \cos\theta \qquad (10.7)$$

where ε^2 is the ratio of densities. Note this is not the wave growth of the component once dissipation and nonlinear transfer have become important. Thus using Eqs (10.6) and (10.7) involves a strong assumption.

If we assume that the energy input to the steady state is equal to the initial temporal growth rate then the empirical expressions for $\Gamma(k)$ allow the integrand in (10.6) to be evaluated. The expression is

$$\tau^{w} \propto \rho_a \alpha^+ \int_0^\infty \frac{2}{\pi} \int_0^\infty \omega^2 \left(\frac{24 u_*}{c} \cos\theta - 1 \right) (u_*/c_p)^m k^{-n} \cos^2\theta \, dk \qquad (10.8)$$

where a directional dependence $D(k, \theta) = 2/\pi \cos^2\theta$, $c = \sqrt{g/k}$ for wavenumbers below k_m, and $F(k)$ is given by Eq. (10.3). $F(k)$ is assumed zero for wavenumbers less than k_p.

The restriction of Eq. (10.7) implies there is no wind input for components with c/u_* greater than 24 in the wind direction and no input for lower wavenumber magnitudes off the wind direction. It is assumed in Eq. (10.7) that the growth of the waves are related to the friction velocity u_*. In Chapter 5 it was suggested that the surface stress consists of a wave supported component τ^w and a viscous stress component. It is not clear that the total stress is the correct parameter on which to scale the wave growth. Possibly the average viscous stress component over the wave does not contribute to the wave growth. As we have written it, the wave stress ends up dependent on the total stress.

The fraction of the stress supported under the assumptions of Eq. (10.8) is shown in Fig. 10.1. Remember the spectra we have used is not appropriate near $k = k_m$.

One can see that the power of the wave age dependence in Eq. 10.3 is critical in deciding whether more or less of the total stress is supported by wave drag. If the drag ρu_*^2 is $\tau^w + \tau^s$, then knowledge of the viscous stress τ^s and the angle between the two stresses allows the total surface stress to be determined. If we assume τ^s is independent of wave age, a quite reasonable assumption, then provided the wave age exponent, m, is less than about 0.6 the total drag increases with wave age.

The viscous drag on a flat plate decreases with wind speed and so it is postulated in Chapter 5 that the viscous drag over the sea does the same. Over the ocean in long fetch situations the total drag is found to increase with wind speed, for example Eq. (2.2b). For a wind speed change from 7 m/s to 14 m/s in Eq. (2.2b) the drag coefficient increases by 1.4, requiring from inspection of Fig. 10.1 that m take a low value if Eq. 10.8 is to explain the observation.

Kitaigorodskii and Volkov (1965) proposed a different expression for calculating the wave drag. The physical model was that the aerodynamic roughness of a moving physical roughness depended on the ratio of phase speed to wind speed.

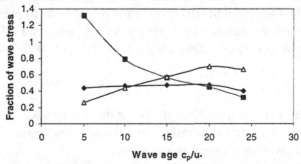

Figure 10.1. The fraction of surface stress supported by wind-wave drag for various values of m assuming $n = 4$. The calculation follows Janssen (1989). $\triangle$ $m = 0$; $\blacklozenge$ $m = 2/3$; $\blacksquare$ $m = 3/2$.

Then a particular wavenumber component of the wave field of phase speed c contributes as

$$z_0^2 = A \int F(k) \exp(-2\kappa c/u_*) \mathrm{d}k \qquad (10.9)$$

Integration over the wave spectrum gives a value of z_0, which with suitable spectral form and von Karman constant of 0.4 becomes Charnock's expression,

$$z_0 \sim u_*^2/g$$

for large wave ages as shown in Table 2.1.

10.4   Short *Wind-waves*

The spectral form of the short *wind-waves* is more complicated than for the long *wind-waves*. Radar scatterometers rely on these components having a strong dependence on wind speed or more precisely wind stress. The wave number dependence of short *wind-waves* was shown in Fig. 4.7. The crudest assumption is a form k^{-4} for $k < 1000$ rad/m and $F(k) = 0$ for larger wavenumbers. The angular distribution described by $\cos^2\theta$ is the same order of approximation. Also the growth rate given by Eq. (10.6) does not apply to very short waves. These are very poor approximations but fortunately their contribution to the integral in Eq. (10.8), as shown in Fig. 10.2, is modest.

These short waves move slowly relative to the underlying water. Thus in Chapter 5 the idea was advanced that roughness of this scale should be treated more as fixed roughness that directly supports turbulent stress by being of greater height than the viscous sublayer. Wave drag modelling like Eq. (10.8) is inappropriate if this is so.

10.5   Total Drag

We will for illustration calculate the integral of Eq. (10.8) to obtain the *wind-wave* drag. The integrand for this expression with $m = 0$, $\theta = 0$, $\mathrm{d}\mathbf{k} = k\cos\theta\mathrm{d}\theta\mathrm{d}k$ and the

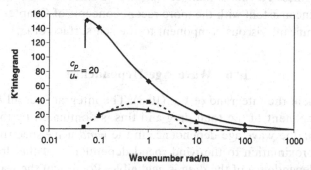

Figure 10.2. The wavenumber weighted integral in the expression for *wind-wave* drag under a number of strong assumptions. ◆ $u_* = 1$ m/s; ▲ $u_* = 0.5$ m/s; ▬ ▬ ▬ 500 times Kitaigorodskii and Volkov (1965) for z_0^2.

form of spectrum assumed to persist to the cut off wavenumber of about 1000 rad/ m, is shown in Fig. 10.2. A constant for Γ is used for capillary waves in Eq. (10.7).

To help the reader see the large contribution the lowest wavenumbers make to the integral under the above assumptions, the integrand multiplied by wavenumber is presented to allow visual integration. This compensates for the emphasis that a logarithmic plot gives to the lowest frequencies. It follows that if we wish to use the above model for wave drag, care should be taken to correctly describe the spectral peak as this calculation implies it is the largest contributor to wave drag for strong winds. Wave age becomes a significant variable as the friction velocity increases as it determines the spectral peak. As mentioned above, the poor assumptions for the short *wind-wave* spectrum may not be important in this model.

It is not the intuitive judgement of many researchers that the short *wind-waves* should carry an insignificant fraction of the drag as Fig. 10.2 implies. With all the uncertainties of the behaviour of the short *wind-waves* and the modelling of their growth as a function of wind speed, it may be best to concede that Eq. (10.8) is only appropriate for low wavenumbers.

The model of Kitaigorodskii and Volkov (1965) provides another approach. The distribution of their integrand of Eq. (10.9) is also plotted in Fig. 10.2. This model shows a strong dependence on wavenumbers near unity for $u_* = 0.5$ m/s, similar to Eq. (10.8).

The *wind-wave* induced momentum flux discussed in Section 10.2, falls off with height above the sea surface. The momentum transfer some distance above the surface averaged over many waves can be written as the sum of the *wind-wave* drag and the turbulent Reynolds stresses τ_t. The sum is assumed constant with height in a constant stress layer. Komen et al. (1994) assumed τ_t could be estimated from Charnock's expression Eq. (10.1). The justification for this assumption is not clear but is discussed in Chapter 6. Such an assumption (together with a few more) leads to a wave age dependence of surface stress discussed below.

What are the implications of assuming that the surface is aerodynamically rough? Aerodynamically rough is a concept where the drag of a fixed surface is dominated by the roughness elements that are higher than the thickness of the viscous sublayer. Wu (1969) suggested that for wind speeds greater than about 7 m/s, the roughness Reynolds number $z_0 u_*/v$ is greater than 2.3 and thus rough. Such a view is inconsistent with the more recent concepts of Chapter 5 where there is always a significant viscous component to the sea surface drag.

10.6   Wave Age Dependence

Let us first look at the integrand of Eq. (10.8). The integral over all k provides the wave drag component of the total drag and this is dominated by the long waves. For values of m, the wave age dependence in the *wind-wave* spectrum, that give a reasonable approximation to the wind speed dependence of the drag coefficient, the wave age dependence of the drag is negligible. Possibly if the wave age dependence of the peak enhancement was included it might make the model a little different.

If we abandon the formalism of Eqs (10.8) and (10.9) because of their discounting of the short waves, we can build a heuristic model that relies on Chapter 5 to place the contribution from long *wind-waves*, short *wind-waves* and skin friction all about equal. Now we assume the skin friction coefficient decreases with wind speed and is wave age independent. The long waves are assumed to have a spectral form where m is a small positive number, say $m = 1/3$. For the spectral model of $F(k) = 0$ for wave numbers below spectral peak, the effect of increasing wave age at constant wind speed (i.e. $c_p^2 = g/k_p$ increasing) is to integrate over a larger range of k. We assume that the decrease in the magnitude of the integrand implied by m just a little greater than zero does not compensate for the increased integration range and the long wave contribution to drag increases weakly with wave age. Above some wave age the curves in Fig. 10.1 decrease and the empirical evidence of Chapter 8 on swell suggests the drag becomes low. For wave ages greater than about $c_p/u_* = 24$ the long *wind-wave* drag component decreases with wave age. The short *wind-waves* are seen to be strongly wind speed dependent in Fig. 4.7. Thus we will assume their fraction of the surface stress increases with wind speed and is wave age independent. Thus the overall result is as shown in Fig. 10.3.

Let us first consider the case of constant wave age. While the skin friction fraction decreases with wind speed in this model, the combination of skin friction and short *wind-wave* drag leads to an increasing drag coefficient with wind speed. The long waves at constant wave age are a fixed fraction of the total drag and so their absolute contribution to drag coefficient increases with wind speed.

At low wind speed, the air flow may be aerodynamically smooth and the drag is dominated by skin friction. At higher steady wind speed the skin friction and short *wind-wave* contributions are constant with wave age. When the wave age is small there are no long waves. The fraction of the stress at modest wave ages carried by the long *wind-waves* increases with wave age as the spectral peak is further from the short waves which are centred on the phase speed minimum. Overall there is an increase of drag coefficient with wave age at modest wave ages. At larger wave ages

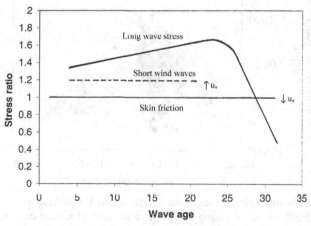

Figure 10.3. The role of short and long *wind-waves* together with the skin friction. It is assumed that u_* is large enough for the flow not to be aerodynamically smooth.

the wave age trend reverses as the long wave component of drag decreases. This occurs because the low wavenumber components are travelling faster than $24u_*$ and because the value of m is positive.

There is considerable uncertainty in the dependence of the spectral level on wave age. The direct wavenumber measurements of Banner et al. (1989) when re-analysed and supplemented, showed no evidence of wave age dependence for waves of metre length. The frequency spectra used by others to estimate α_s have a wave age dependence due to the Doppler shifting of shorter waves by the orbital motion of waves near the spectral peak. Without clarification of this issue, the approach of modelling the momentum flux by a wave component that is dependent on the surface wave spectrum cannot help resolve the debate.

10.7 Measurement of Drag as a Function of Wave Age

The data of drag over the ocean collected to date occupy a space between $z_0 g/{u_*}^2 = 0.08$ and $z_0 g/u_*^2 = 0.025(c_p/u_*)$. Some of the scatter is error, some depth effects, some unsteadiness and some related to the constant stress layer and log law assumptions. In Fig. 10.4 a recent compilation of data has been prepared using data from Suzuki et al. (1998). If the friction velocity were to vary by $\pm 10\%$ or $\pm 20\%$ while the mean velocity remained the same, the fluctuations in nondimensional roughness can easily be calculated. Let us first imagine a stress of constant magnitude fluctuating in direction with height above the surface. If it varies 36 degrees from the wind direction, then the stress component in the wind direction varies by 20%. Figure 1.14 suggests this is a very reasonable hypothesis. Thus one point of view is that diagrams such as Fig. 10.4 provide a sample that must be averaged if one wants to present an ensemble mean of observations at a particular wave age. The lower level of fluctuation is shown in Fig. 10.4 to correspond to

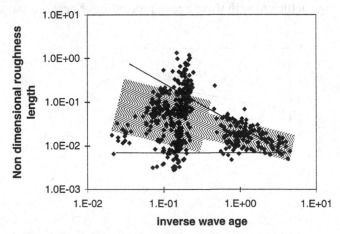

Figure 10.4. The nondimensional roughness length versus the inverse wave age. The shaded area represents $\pm 10\%$ of u_* for young wave ages and $\pm 20\%$ for field data, when ω_p and U_{10} are assumed constant. The data is from the Hiratuka tower, Japan reproduced from Suzuki et al. (1998) together with the directly measured data.

the region that shows laboratory data as we surmise that the two dimensionality is less in tanks than over the open ocean.

In Chapter 9, the fluctuation of z_0 due to unsteadiness presents a conceptual difficulty in determining a steady state value of drag coefficient as a function of wave age in a geophysical flow. The sensitivity of z_0 to measurement errors discussed in Chapter 7 is another. The third difficulty that needs to be addressed is that of using both ocean and wind tunnel results to obtain a large spread of wave ages. Can waves be thought of as a continuum between the field and the laboratory? Are there some variables, other than wave age, that change between the natural environment and the wind wave tank, that are important in determining drag? No clear candidate has emerged.

It is popular to plot the aerodynamic roughness length as the nondimensional variable $z_0 g/u_*^2$ and show it as a function of wave age c_p/u_*. Note that the friction velocity, u_*, appears on both axes. Such plots are shown in Chapter 2 where the many expressions that have been used to model this data set are shown. It is difficult to make a definitive statement from the data clouds in these figures.

10.8 Model of Aerodynamic Roughness

An important conclusion from examining the experimental data is that simple power law formulae, like Toba and Koga (1986), Toba et al. (1990), Donelan et al. (1995) or Charnock, for the dependency of $g z_0/u_*^2$ with wave age c_p/u_* are not adequate. This can be seen in Fig. 2.3. Nordeng (1991) using Eq. (10.9) as a starting point and a wave spectrum that was wave age dependent suggested an expression

$$\frac{g z_0}{u_*} = 0.11 \left(\frac{c_p}{u_*}\right)^{-3/4} \left[1 - e^{-w}\left(1 + \frac{w^2}{2} + \frac{w^3}{6}\right)\right]^{1/2} \qquad (10.10)$$

where $w = 2k c_p/u_*$. This can be seen in Fig. 2.3 and it can be claimed goes through the cloud of data points. See Fig. 10.5.

Another approach is to assume that $f(c/u_*)$ in Eq. (10.2) should be replaced by the wave slope given by a characteristic wave height and wavelength. For the characteristic height it has been suggested that wave height modified as in Eq. (10.9) by $\exp(-0.75(c_p/u_*)$ could be used. The value of $0.75 c_p$ is chosen to represent the speed of those components of the spectra most important in generating drag. Then we obtain

$$z_0^2 \propto \overline{\eta^2} \exp(-2 \times 0.75 \kappa c_p/u_*)$$

Use of Eq. (10.5) allows

$$z_0^* = 0.05 \left(\frac{c_p}{u_*}\right)^{3/2} \exp(-3\kappa c_p/4u_*) \qquad (10.11)$$

At our present state of knowledge, the best model we can offer is similar to Eq. (10.11). It is

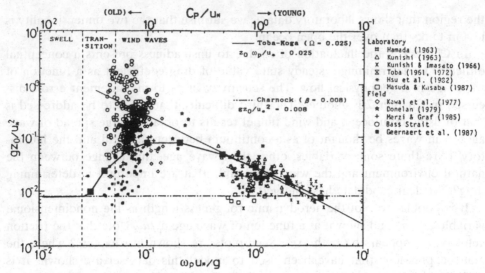

Figure 10.5. From Toba and Ebuchi (1991). The nondimensional roughness parameter gz_0/u_*^2 versus inverse wave age $\omega_p u_*/g$ diagram of Fig. 2.3. These data points are all in *local equilibrium with the wind*. The heavy solid curve is the form of Eq. (10.12). Open circles are not direct measurements.

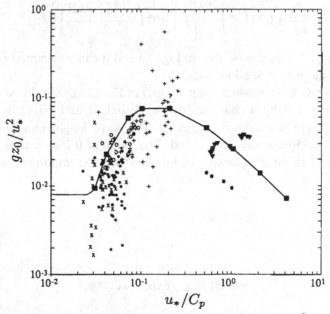

Figure 10.6. From Donelan et al. (1993). Dimensional roughness gz_0/u_*^2 versus inverse wave age. The solid curve is the form of Eq. (10.12).

$$\frac{gz_o}{u_*^2} = 0.03\left(\frac{c_p}{u_*}\right)\exp\left\{-0.14\frac{c_p}{u_*}\right\} \qquad \sim 0.35 < \frac{c_p}{u_*} < 35$$

$$= 0.008 \qquad\qquad\qquad\qquad\qquad \frac{c_p}{u_*} > 35 \tag{10.12}$$

The break in the expression at wave age of 35 that can be seen in Fig. 10.5 and Fig. 10.6 comes about because steady-state larger wave ages are only achieved by light wind over swell. Here an expression appropriate for smooth flow might be suitably limited by say Charnock's expression for transitional winds. Once the wind speed is large enough for wind waves to form, c_p is the phase speed of these waves and not the swell phase speed.

At very young wave ages gz_0/u_*^2 in the above expression follows the physical roughness diminished by the speed of propagation of the peak wave. This can be seen by noting from (10.5) that the rms wave height, when non-dimensionalized, becomes

$$\frac{g\eta}{u_*^2} \propto \left(\frac{c_p}{u_*^2}\right)^{3/2}$$

Thus for young wave ages Eq. (10.12) is of the form

$$z_0^* \propto \frac{g\eta}{u_*^2}\left(\frac{c_p}{u_*^2}\right)^{-1/2}$$

We have an expression that relates the aerodynamic roughness to the physical roughness.

10.9   Geostrophic Drag Coefficient

The above expression assumes that one knows the wind velocity at some height above the sea surface. However there are a class of problems where only the pressure gradient is known and from it the geostrophic velocity. Wind energy resource estimations and ocean circulation calculations are two examples. The height of the planetary boundary layer, H_b, where the geostrophic assumptions are reasonable, is another parameter that must be introduced to relate the roughness length to the drag coefficient.

11 The Influence of Mesoscale Atmospheric Processes

L. Hasse and R.A. Brown

11.1 Introduction

The sea surface drag has been discussed in terms of the bulk aerodynamic coefficient parametrization, C_D. There are several factors that influence stress exerted by the atmosphere on the ocean surface and these are discussed in various chapters of this book. The influence of some mesoscale phenomena is the focus here. There are, for example, the characteristic large eddies that occur in the planetary boundary layer and the sharp changes associated with storms and convective events. Finally, the sparsity of data may mean that factors contributing to ocean stress remain unmeasured, unmodelled, and even unknown.

In the context of wind stress on the sea surface we have a twofold task: first to understand processes and their influence on the stress, and second to develop parametrizations and determine the coefficients. The bulk aerodynamic method using a drag coefficient or roughness length is a local parametrization. It can be modified by local or quasi-local influences like buoyancy stratification or wave age. Atmospheric processes are typically described by numerical models with a gridscale of 50 km or 100 km. Observations at sea are considerably more sparse. Subgridscale atmospheric variations, especially of the wind, will influence the momentum transfer and thus result in apparently unexplained variations of the drag. The present chapter deals with influence of recognized structures that are not well represented in modelled surface layer variables or local measurements. We are concerned here with the influence of processes in the scale that in meteorology is often called the spectral gap, a scale that ranges from some 10 minutes to a few hours or from a few kilometres to roughly 50 km. The word gap implies that on average there is less energy in this region than at larger or smaller scales. However, there are specific situations where distinct processes occur in this spectral region. For discussion we will first look at the open ocean and second at coastal seas.

Ian S. F. Jones and Yoshiaki Toba, *Wind Stress over the Ocean*. © 2001 Cambridge University Press. All rights reserved. Printed in the United States of America.

While not discussed further the Langmuir circulations seen in the ocean may have characteristics similar to the boundary layer helical rolls discussed below.

11.2 Influences of Open Ocean Subgridscale Processes on Stress

11.2.1 Boundary Layer Helical Rolls (Organized Large Eddies)

Rolls were first inferred from observations of cloud streets in the 1920s (see Kuettner 1959; Foster 1996). They were found to have wavelengths and orientations in accordance with the instabilities of the Ekman layer (e.g. Lilly 1966). A nonlinear solution with explicit finite perturbations was found by Brown (1970). This solution was matched to the surface layer solution to obtain a planetary boundary layer model in Brown (1974). In this similarity solution, the surface drag can be related to the geostrophic flow using a single similarity coefficient, with corrections due to buoyancy stratification information, variable surface roughness and thermal wind.

Existing numerical models for the general circulation rely on approximations of the planetary boundary layer and do not yet have the capability to represent organized large eddies physics (Foster and Brown 1994). These models can be used in surface drag calculations whenever accuracy no greater than 20% is needed. For better drag values, either measured surface winds or an explicit modelling of the organized large eddies is needed. The complications to modelling of the planetary boundary layer due to the presence of large eddies is sufficient to make scientists simply ignore them as higher-order effects. Consequently, there is a long history of theory and observations of large eddies that exists outside the general drag measuring community. This brief summary is meant to apprise the reader of them, adduce their importance to measurement accuracy and completeness, and suggest ways to include them.

Organized large eddies have been reviewed by Brown (1970, 1980) and Etling and Brown (1993). These particular eddies are often referred to as rolls. They are sketched in Fig. 11.1. From the theoretical solutions and observations it is apparent that:

1. Rolls are organized counter-rotating helical eddies with axes nearly parallel to the mean flow in the planetary boundary layer. They fill the planetary boundary layer.
2. There are alternating convergence and divergence rows, with associated updraft and downdraft regions. The aspect ratio, horizontal wavelength to height, is most often in the range 2 –6, but even 20 have been observed (Etling and Brown 1993; Mourad and Brown 1990).
3. Rolls exist for windspeeds greater than a few metres/second, in unstable to slightly stable stratification.
4. A finite-perturbation solution exists for this flow. The nonlinear solution indicates that the mean flow is considerably modified.

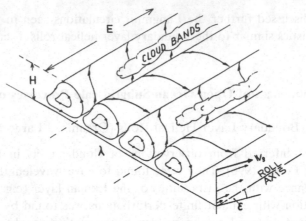

Figure 11.1.   Boundary layer helical rolls, scheme (from Etling and Brown 1993).

5. The rolls character (vigour and orientation) depends on stratification, mean wind magnitude and baroclinity. In particular, the behaviour of the rolls in barotropic conditions with stratification is: an 18° (to the left) orientation angle with the geostrophic wind in neutral stratification; greater angles in stable stratification (up to 30°); and a tendency toward alignment with the mean wind in unstable stratification.
6. Commensurate with these orientations, the rolls field exhibits a transverse drift velocity of 0.5–2 m/s in neutral to stable stratification, and no cross-wind drift at all in moderately unstable stratification (convection in the presence of shear).
7. The secondary flow is about 7–15% of the mean, and may be less. It has a crosswind and downwind variation. Cloud streets regularly appear as pearls on a string, pointing to a variation in the mean wind direction, too.

Observations by Joffre (1982, 1984) over the sea have shown that the surface wind and stress are strongly modified by baroclinity in the planetary boundary layer. These observations show typical variations of ±10 to 20° in cross-isobar angle and ±5 to 10% in surface wind speed as a function of the thermal wind orientation angle. Such variations will have a significant effect on the surface fluxes. Because of the importance of organized large eddies to the planetary boundary layer solution, Foster (1996) investigated the effect of planetary boundary layer thermal wind on the Ekman layer instabilities associated with planetary boundary layer rolls. A high-order nonlinear theory was developed that includes the effects of stratification and baroclinity. Foster found that the roll wavelength, orientation with respect to the surface isobars and perturbation magnitudes are strongly affected by thermal wind.

Since the planetary boundary layer flow solution for convective flow in the presence of wind includes rolls parallel to the wind shear with zero drift, there is a danger of a point measurement being stuck in one section of the horizontal inhomogeneity. Or, in slightly unstable stratification, the time for an entire roll wavelength to pass by the observer may be hours. In these cases, a bias to the drag is to be expected for any point measurement, even with very long averaging times. To recognize that these biases might be entering the data, one must be cognizant of the roll characteristics, and when they are present. Thus, when mean winds are greater than about 7 m/s in neutral stratification, or 4 m/s in unstable stratification, theory predicts the organized large eddies will be a feature of the planetary boundary layer flow. Cloudstreets are a certain indicator of the rolls of course. One of the best ways to sample the organized large eddies environment is with systematic aircraft flights in stacks orientated upwind/downwind and particularly crosswind. This is a difficult experimental challenge; however several flight missions have been successful (e.g. Walter and Overland 1984; Hein and Brown 1988; Kontur experiment, see Etling and Brown 1993). Typical spectra from such flights are shown in Fig. 11.2. There often is a peak in energy at the roll wavelength (about 1–2 km). While it is difficult obtaining long enough legs to identify these phenomena, it is important to note that there is the possibility of this variation. Since up- and downdrafts may be widely separated, it is arduous to identify and measure them. Also, for convectively driven rolls, the analysis shows that the updrafts are more concentrated than the downdrafts, as sketched in Fig. 11.3.

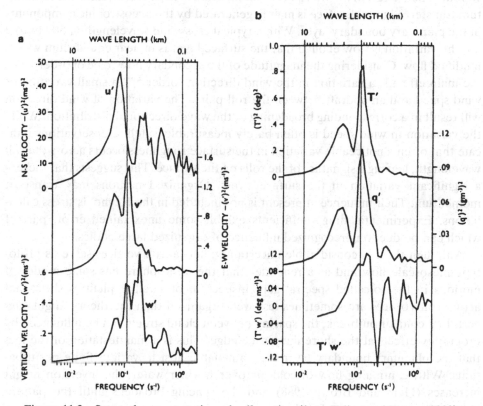

Figure 11.2   Spectral representation of roll motion (from Etling and Brown 1993).

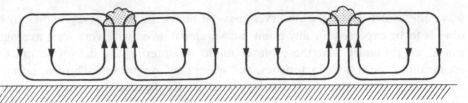

Figure 11.3. Schematic presentation of convective motion. The upward motion typically is more concentrated and more rapid than the compensating sinking motions. During GATE radar echos covered only 5 to 8% of the area, indicating strong concentrated updrafts and slower compensating sinking. Additionally, there are local cold downdrafts, produced by evaporation of rain, that reach the ground (from Hasse 1984).

Given the lack of observations over the world's oceans, recourse is often made to numerical models for surface winds, or even surface stress fields. Currently, this is a bit risky since the general circulation models do not have sufficient resolution in their planetary boundary layers to model surface winds and stress correctly (Brown and Foster 1994; Foster and Brown 1994). The alternative is to rely on measured winds. These can be obtained from dedicated ships and buoys in regional-scale experiments, or on satellite derived wind fields for global modelling. The latter is discussed in Chapter 7.

Roll vortices are effective in transporting admixtures, especially heat and water vapour. Their effect in the transport of momentum is less evident, since the momentum transferred to the surface is mainly generated by the ageostrophic components in the planetary boundary layer. With a typical crosswind wavelength of 500 m to 2 km the variation is slow enough that the surface layer is in near equilibrium with a modified flow. Considering the magnitude of the secondary flow to be of order 10%, the main effect is a variation in the wind direction (order 5°), a small variation of wind speed and an updraft between the roll pairs. The variation of wind direction will result in a corresponding broadening of the wave directional distribution. While the variation in wind speed is often barely measurable, satellite observations indicate that often a noticeable variation in the surface roughness occurs across the roll wavelength, leaving a signature of the rolls on the surface. This suggests that there is a significant variation in the surface stress. Organized motions may transport momentum. Their influence at present is not included in the Reynolds stress calculations. Experimental drag coefficients exhibit some unexplained error, part of which can be due to unrecognized influence of organized large eddies.

Although there is considerable uncertainty, helical roll vortices serve as prototype mesoscale flow and as a reminder that the atmosphere has some modes of motions in the so-called spectral gap. Inspection of satellite pictures show that actual roll vortices are sometimes less well organized than the theory might suggest. In a cold air outbreak, the spacing between cloud streets can be influenced by orography effects at the shore line or ice edge. This is a manifestation of the fact that the planetary boundary layer is an amenable host to various finite perturbations. With continuing flow of cold air over warmer water, the inversion height increases (Hein and Brown 1988) and the spacing broadens until the pattern changes to cell structures, Fig. 11.4.

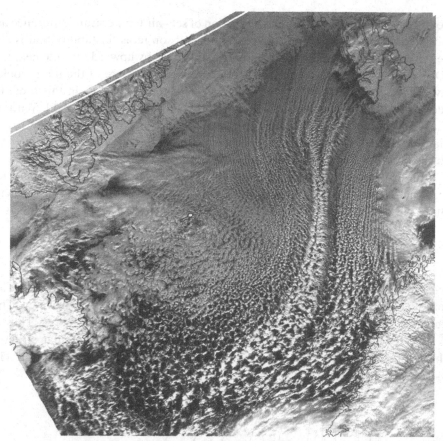

Figure 11.4. Cold air outbreak over the European Nordic Seas. This is the typical picture of the flow at the rear side of a depression. The northerly flow of cold air after crossing the ice edge or shore line moves over warmer water. While the air takes heat up from below and warms, it moves over still warmer water. The strong air–sea temperature differences therefore can prevail for several 1000 km. Near the edge of open water, boundary layer helical rolls develop, documented by the nicely organized string of clouds. At increasing distance, the clouds become larger and the structure changes to open cells. The clouds that form the walls of open cells typically reach greater heights and show stronger convective activity. (Courtesy Met. Inst. FU, Berlin.)

11.2.2 Deep Atmospheric Convection

Deep atmospheric convection is seen frequently in cold air outbreaks at the rear side of depressions. This often takes the form of open cellular convection, where the walls of cells are formed of high penetrative clouds.

The cloud induced secondary motions add to the mean velocity of the cold air outbreak. This corresponds to a structured variation of the wind field (by perhaps ±10% at higher wind speeds and ±20% at lower wind speeds) and an increased roughness of the sea. The diameter of the open cells is of order 50 km. Also, cold downdrafts carry air with higher velocity and different direction to the ground, enhancing gustiness. A picture of frequency of occurrence of open cells in cold air outbreaks is given in Fig. 11.5 as a function of sea–air temperature difference and windspeed. Deep atmospheric convection depends on moist instability that is triggered by instability from the surface, represented here by upward sensible heat flux.

Deep atmospheric convection is also the prominent feature of the inter-tropical convergence zone. With the lower wind speeds of the mean flow in this area, the cloud induced secondary motions dominate the surface layer wind field. With the passage of cloud clusters the wind rises from near calm to 8 or 9 m/s and rapid variation (scale 1 hour) by ±3 m/s and even changes of sign are imposed, correlated with the passage of precipitating deep convection (e.g. Hasse et al. 1978a). The area of the inter-tropical convergence zone shows low resultant wind speed and low directional steadiness (Isemer and Hasse 1985), making it advisable to explicitly include the effects of subgridscale wind variances into the calculation of stress.

11.2.3   Polar Lows and Explosive Cyclogenesis

Polar lows are depressions with a scale of about 100 km to 500 km diameter. They often develop when cold continental air flows over relatively warm water. They occur most often in flow from pack ice over the ocean, discussed below. They do not have fronts like midlatitude depressions, but often show a circular structure with a cloudless eye in the centre, pointing to their convective origin, Fig. 11.6.

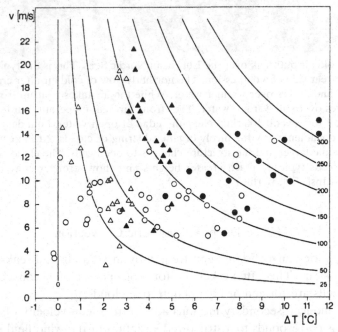

Figure 11.5. Observations of open cell activity as a function of sea–air temperature difference (ΔT positive indicates air cooler than water) from two experiments: AMTEX 1975 circles and KONTUR 1981 triangles. Full symbols with observation of open cells, open symbols without observation of open cells. The full lines give upward sensible heat flux in W/m^2 (from Hasse 1984).

Figure 11.6. Polar low between Svalbard and the Norwegian north coast. The distance between the land marks is roughly 6° latitude. The Polar low has a diameter of 200–300 km (from Rasmussen 1985).

Strong winds of gale or near gale force have been repeatedly observed. Due to their small size, these systems are not well represented in numerical weather forecast models, since isolated reports of high wind speeds or locally very low pressures are likely to be eliminated in data assembling or quality control for these models. Good examples of polar lows have been observed in the European Nordic seas (Wilhelmsen 1985), and similar lows were found in the North Pacific and at the coast of Antarctica.

In the western North Atlantic and in the Pacific the associated rapid development of such depressions has been called explosive cyclogenesis or "meteorological bombs". They deepen at a rate of 1 hPa or more per hour for 24 hours. These develop preferentially at strong sea surface temperature gradients (Kuo et al. 1991) and in the winter season. Both polar lows and explosive cyclogenesis develop to full strength within about 24 hours. Although known for some time, they have recently found more attention because of the lack of forecasting ability and resulting high damage and loss of life. Their effects on sea state and drag are manifest in their rapid development, small scale and unexpected strong winds, that produce irregular, nonequilibrium waves with locally enhanced drag.

11.3   Mesoscale Atmospheric Processes in Coastal Seas

Subgridscale variations of the wind in coastal areas are noticeable and may influence stress, important for the interpretation of wave generation experiments in coastal seas. Dobson et al. (1989) found better agreement between measured wave spectra and wave theory when considering the (hitherto neglected) variation of the wind field along the path of generating waves.

11.3.1   Land–Sea Breeze Systems

Land–sea breeze systems are often observed. Noticeable land–sea breeze variations were found even at 50 km seawards from shore, Fig. 11.7. The diurnal variation of wind speed amounted to ±3 m/sec during sunny summer days at the German Bight, 30 km from shore (Anto 1977) and ±2 m/sec, 50 km off the Oregon coast (Burt et al. 1974). In principle the land–sea breeze systems can be and have been modelled. In typical weather forecast models, however, the resolution is not yet sufficient to fully include such systems.

11.3.2   Land–Sea Roughness Transition

Generally, land surfaces are rougher than the sea surface and correspondingly winds at sea are higher than over land. The adjustment of wind speed to a roughness transition is closely related to local pressure fields. The effect is felt both with on-shore and off-shore winds. For off-shore winds the acceleration starts inland of the shoreline. On-shore winds are decelerated before reaching the shoreline. In addition to the land–sea roughness transition there is a change of roughness due to the variation of the wave field. For onshore winds, the waves steepen when running into shallower water. For offshore winds, waves are being generated. In both cases the waves are rougher than deep water waves in equilibrium with the wind field, though the exact relation is not yet known (see Chapter 15).

Observations show (Fig. 11.8) that the adjustment of surface wind speed essentially takes place within the first 10 km seawards from the coast. Because of the roughness transition winds measured at the shoreline are not very representative. In Fig. 11.8 observed winds are referenced to the local geostrophic wind. Alternatively, the observed wind at larger distances can be used as reference (Karger 1995). This example from the Baltic Sea (Bumke et al. 1998) is for illustration mainly, the actual ratios may deviate from this since the surface to geostrophic wind speed ratio depends both on windspeed and air–sea temperature difference. The transition in the wind field shown in Fig. 11.8 is part of transition of the planetary boundary layer near the shore. In principle the situation of an inhomogeneous boundary layer and associated variation of momentum balance within – including, for example, developing organized large eddies – would be

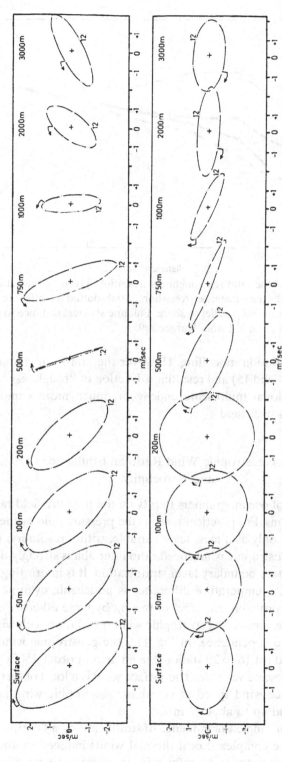

Figure 11.7. Land–sea breeze diurnal variation of wind speed over sea during month of June at the south tip of the island of Sylt, roughly 20 km distance from shore. Upper panel, fair weather; lower panel, cloudy. Even on cloudy days, the land–sea contrast induces a diurnal variation of ±2 m/s up to a height of 200 m (from Anto 1977).

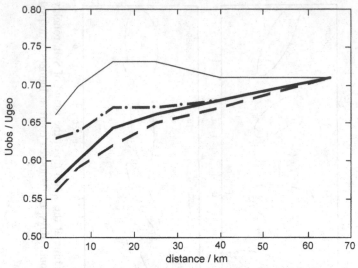

Figure 11.8. Influence of the land sea roughness transition, given as the ratio of observed wind to geostrophic wind versus distance from shore. Dash-dotted for offshore winds, dotted for onshore winds, dashed wind parallel to shore. Full line shortest distance to shore without regard of wind direction (adapted from Karger 1995).

amenable to high resolution modelling. However the non-local character of wave fields (see Chapters 14 and 15) and resulting advection of "roughness" would make this a fully 4-dimensional multi-scale endeavour, much more complicated than comparable situations over land.

11.3.3   Surface to Geostrophic Wind Ratio and Influence of Stratification and Baroclinity

It has been a custom of oceanographers to rely on the pressure field rather than on direct wind observations. For practical reasons the pressure field or the geostrophic wind can be obtained only as a mesoscale variable, with a resolution of about 100 km. The surface to geostrophic wind speed ratio inter alia is strongly dependent on wind speed and planetary boundary layer stratification. It is interesting to see that a dependence on air–sea temperature difference is noticeable even at higher wind speeds (Hasse 1974), probably an effect of mixing by large eddies.

The variation of the surface to geostrophic wind speed ratio with wind speed is a good example of scale dependence. In Fig. 11.9 the geostrophic wind is obtained from the pressure field (of 16 to 20 stations) in an area of order 150×200 km^2 and thus represents a mesoscale variable. The surface wind is a local observation. There is nonvanishing surface wind speed at vanishing geostrophic wind and relatively enhanced surface wind up to about 7 m/s.

For the ageostrophic angle the influence of stratification and thermal land–water contrasts is even more complex. Local thermal winds induced by the land–water temperature difference seem to be influential in addition to stratification effects (Luthardt and Hasse 1981). It is evident that variations influenced by land–water

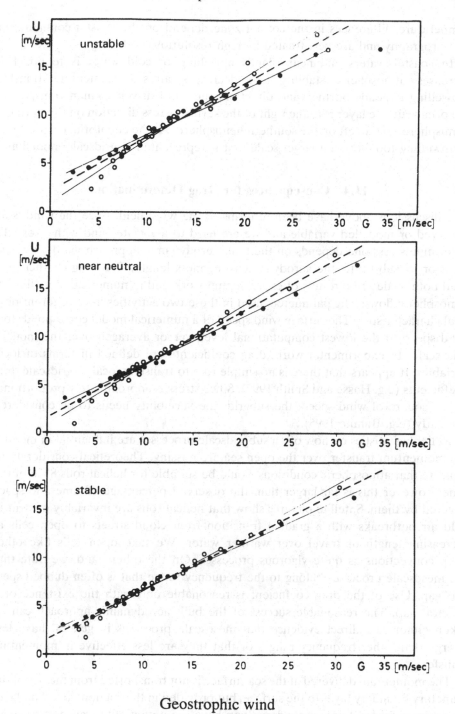

Geostrophic wind

Figure 11.9. Variation of local surface wind speed with grid scale geostrophic wind for different stabilities as given by air–sea temperature difference (from Hasse 1974).

temperature differences in the coastal zone depend on the coastal configuration and orography and are best treated by high resolution modelling.

In coastal waters and near islands upwelling of cold water is found. This increases atmospheric stability and decreases air–sea momentum transfer. Upwelling depends on the wind direction, the wind driven Ekman transport in the ocean surface layer is to the right of the surface stress direction on the northern hemisphere, to the left on the southern hemisphere. At the equatorial oceans there is upwelling too, but on a larger scale that is represented in gridscale modelling.

11.4   Consequences for Drag Determination

The discussions above have been in terms of the wind field, since the wind is an observed or modelled variable and we are used to associate wind with stress. To determine stress one depends on the bulk aerodynamic approach via drag coefficient or its substitute, the aerodynamic roughness length. The drag coefficient is used both to describe results of experimental work and in numerical modelling of atmospheric flows. The parameters used in these two activities have different physical characteristics. The surface wind speed of a numerical model corresponds to a wind speed at the lowest computational level, vector averaged over the model's grid scale. In experimental work, drag coefficients are defined in terms of local variables. It appears that there is no simple rule to translate local to gridscale drag coefficients (e.g. Hasse and Smith 1997). Since stress is approximately proportional to the square of wind speed, the subgridscale variability needs to be considered explicitly (e.g. Bumke 1995).

Reliable statistics on how often subgridscale processes are influential in modifying momentum transfer over the open sea are missing. Theoretical considerations indicate that atmospheric conditions would be suitable for helical rolls 85% of the time; however this is far larger than the observed percentage of time and space covered by them. Satellite pictures show that helical rolls are invariably present in cold air outbreaks with a gradual transition from cloud streets to open cells at increasing length of travel over warmer water. We take open cells like other deep convections as quite vigorous processes. On the other hand, we note that the mesoscale processes belong to the frequency range that is often dubbed spectral gap. Use of the drag coefficient is reasonable only with the existence of a spectral gap. The reasonable success of the bulk aerodynamic approach can be taken either as indirect evidence that mesoscale processes in general have less energy than other frequency ranges or that they are less effective in momentum transfer.

The momentum delivered at the sea surface is not transported from the top of the planetary boundary layer to the surface, but provided in the friction boundary layer by pressure forces. Large secondary flows have the potential to transport momentum towards the surface or away from the surface, but such ordered motions are impaired by continuity constraints in the immediate vicinity of the surface.

In order to improve determination of stress, explicit consideration of subgridscale processes in numerical models will be the tool of the future. At present compensations are often built into numerical models to represent unresolved processes, e.g. momentum transfer is adjusted by increasing the roughness length. The influence of subgridscale variability of wind direction is contained in the empirical description of wave directional variability in wave models. The agreement between observed variables and model variables needs to be checked before explicit descriptions of subgridscale processes are introduced in existing models.

12　Wind, Stress and Wave Directions

Carl A. Friehe, Jerry A. Smith, Karl F. Rieder, N. E. Huang,
Jean-Paul Giovanangeli and Gerald L. Geernaert

12.1　Introduction

A common assumption made in air–sea interaction is that the atmospheric surface-layer wind and stress vectors are colinear. The single drag coefficient parametrization of the stress assumes alignment of the vectors. Results of Smith (1980), Geernaert (1988) and others however have shown, on occasion, large differences of up to 40° over the ocean. The purpose of this chapter is to review previous and recent measurements, including some from the surface layer over land for comparison. Laboratory results and wind-wave results are also included.

The horizontal shear stress vector is defined in terms of the Reynolds averaged turbulent covariances:

$$\tau/\rho = -\overline{u'w'}i - \overline{v'w'}j \qquad (12.1)$$

where τ is the horizontal stress vector, ρ is air density, u', v', w' are the velocity fluctuations in the orthogonal i, j, k directions (k vertical), and the over-line represents a time average. The kinematic stress components are $\overline{u'w'}$ and $\overline{v'w'}$. For the results presented in this chapter, all stress measurements were made using the above definition (the so-called eddy correlation technique of Section 7.3), usually at ~ 10 m height, where the stress is assumed to be equal to the surface stress, τ_0. Other methods, such as the inertial dissipation technique, can only estimate the magnitude of the stress, and inherently assume colinearity.

12.2　Definitions

The horizontal wind vector is usually referred to geographic coordinates, but in most air–sea interaction experiments, it is rotated into a right-handed natural coordinate system into the mean wind direction over the averaging time interval used to calculate the stresses. The main horizontal kinematic stress component is then $\overline{u'w'}$ and v' is positive to the left of the wind direction. In the surface layer, $\overline{u'w'}$ is usually < 0, so that mean horizontal momentum is transferred downward to the ocean surface. In this coordinate system, the wind angle is zero, and the stress angle θ is defined here as:

$$\theta = \tan^{-1}\frac{-\overline{v'w'}}{-\overline{u'w'}} \tag{12.2}$$

Non-zero cross-wind stresses will produce non-zero values of θ; for $\theta > 0$, the stress vector is to the left of the mean wind vector.

Note that our definition of v' is opposite to that of Geernaert (1988) who used a left-handed coordinate system, as he defined the angle $\tan^{-1}(\overline{v'w'}/\overline{u'w'})$ and stated that the angle is positive for the stress vector pointing to the right of the wind vector. Rieder et al. (1994, 1996) appear to have followed Geernaert's convention, as well as Liu et al. (1995) who however kept the minus signs as in Eq. (12.2) to denote the quadrant of the stress vector.

12.3   Stress Directions in Boundary Layers

12.3.1   Atmospheric Boundary Layer

The variation of stress with height in the steady-state boundary layer is governed by the balance of the Coriolis, pressure gradient and stress divergence forces. The assumptions of a barotropic pressure gradient and constant eddy viscosity result in the Ekman wind spiral. At the surface, the wind and stress vectors are asymptotically colinear. The top of the surface layer is usually defined as the height where the stress is 0.9 of τ_0 (Wyngaard 1973). At this height, for a geostrophic wind speed of 10 m/s and an eddy viscosity of 1 m^2/s, Friehe et al. (2000) show θ is $-3°$. Thus, while the stress vector is predicted to be to the right of the wind at the top of the surface layer, the variation is small as expected.

12.3.2   Aerodynamic Boundary Layers

The relationship of the flow and stress vectors has been of interest in many engineering boundary-layer studies where there are three-dimensional effects.

Compton and Eaton (1995) presented extensive measurements of stress and flow angle in a three-dimensional turbulent boundary layer. Measurements were possible down to the near-wall region with the use of laser Doppler anemometry. Misalignment of stress and flow of up to 40° were found away from the wall, but became essentially equal below about 10 units of nondimensional height ($z_+ = zu_*/\nu$ where z is height, $u_* = \sqrt{\tau_0/\rho}$ and ν is molecular kinematic viscosity). While the correspondence of laboratory and atmospheric boundary layers in terms of z_+ is not clear, the laboratory results indicate alignment of wind and stress very near a solid surface. Since the flow immediately near a surface is dominated by molecular viscous stress, alignment is expected since the Newtonian viscosity constitutive equation does not allow stress to be misaligned with the rate of strain.

12.4   Laboratory *Wind-wave* Measurements

Giovanangeli et al. (1994) measured the directional aspects of wind stress and oblique swells in the IRPHE wind-wave tunnel in Marseille with hot-wire anemometers. The wind speed was 6 m/s, fetch 3 m, hot-wire measurement heights between 0.02 and 0.2 m, steepnesses from 0.08 to 0.24, and angles of incidence of the swell relative to the mean wind direction were 0, 30, 50 and 90°. Parasitical reflections of the swell on the tunnel side walls were minimized.

The wind stress direction departed from the wind direction depending on the angle of incidence and steepness of the swell and the height above the surface. The maximum value was 20° nearest the waves for the largest steepnesses. For a swell incidence angle of 50°, the angle decreased abruptly to ~ 3–5° for all other heights. For a swell incidence angle of 30°, the decrease of the angle with height was more linear, falling to $\sim 5°$. The stress vector shifted towards the direction of the swell with increasing steepness. The importance of wave-induced motions was noted, especially the correlation between pressure fluctuations and transverse wave slopes.

Branger et al. (1994) measured the misalignment of the maximum of the radar backscatter coefficient with the mean wind direction for oblique swell in a wind-wave tunnel which further supported misalignment of the stress and wind vectors. The angle of maximum backscatter coefficient was between that of the wind and swell.

As with the laboratory boundary layer over a solid surface, the exact correspondence of the results in wind-wave tunnels to the flow over the open ocean is not clear, but there is evidence of misalignment for the steepest waves.

12.5   Aircraft Results

Zemba and Friehe (1987) presented profiles of wind and stress directions obtained from a research aircraft in the marine boundary layer in a coastal up-welling area. The overall flow was characterized by strong low-level wind jets (15–22 m/s at 400–800 m) below large inversions. At the lowest flight level (~ 30 m), θ was $\sim 0°$; above 30 m the stress was rotated to the right of the wind, reaching an angle difference of 45° at 200 m. Enriquez and Friehe (1997) examined the wind-stress angle difference in the same coastal area as Zemba and Friehe and found small differences at 30 m that were however attributed to errors in the aircraft wind measurement system.

12.6   Numerical Models

Ly (1993) included the effect of cross-wind stress in a numerical simulation of air–sea interaction. The results predicted a decrease in the along-wind drag coefficient for wind speeds greater than 10 m/s. Only cross-wind stresses to the right of the wind vector were parametrized in the model.

12.7 Measurement Accuracy

Measurements of wind and turbulence are difficult over the sea. Calibration uncertainties of cup-and-vane anemometers and sonic anemometers are probably of the order of 0.1–0.2 m/s in the mean. Thus for 10 m/s mean wind components, we would expect a direction error of approximately $\pm 1°$; $\pm 2°$ for 5 m/s, etc. These are not significant, but there are other sources of mean wind direction errors such as flow distortion about the measurement platform, misalignment of the sensor, motion of the sensor, and others discussed in Chapter 7.

Accuracy of the turbulent stress components is more difficult to specify. It is also compounded by flow distortion and motion effects. An additional problem is that θ as determined from Eq. (12.2) should only be calculated when the values of $|\overline{v'w'}|$ or $|\overline{u'w'}|$ are above some estimate of a resolution or noise level. (The same restriction should also apply to the wind direction.) This is similar to the problem of the determination of phase angle between two signals. A significant level of coherence has to be defined before phase angle is meaningful.

Most measurements of the mean horizontal wind and turbulence components in air–sea interaction experiments are usually done with the same instrument – a sonic anemometer or aircraft systems – so errors in θ should not be biased due to differences in alignment between different sensors used for the mean and fluctuating velocity components. Motion corrections to sonic anemometer-measured stresses can be significant for instruments deployed on ships and buoys. Ship flow distortion effects were discussed in Chapter 7 but their influence on measurements of the wind-stress angle difference have not been thoroughly investigated. The evidence presented by Rieder et al. (1994) indicates that poor wind exposure implies stress direction errors.

In order to obtain values of wind and stress angles above resolution or noise limits, Geernaert et al. (1988) examined only data above wind speeds of 10 m/s; Geernaert et al. (1993) and Rieder et al. (1994) used a threshold of 3 m/s.

12.8 Over-land Results

For comparison to the over-ocean case, Friehe et al. (2000) examined data from a surface-layer experiment over land for which motion corrections were irrelevant and flow distortion was minimized (Oncley et al. 1996). Two nearly identical sonic anemometers at 4 m height and separated by 10 m over a surface of ploughed land were used. From examination of the data, a 10 min running averaging time and a minimum value of 0.02 m^2/s for stress components were used for the calculation of θ, which corresponds to a 10 m wind speed over the ocean of about 4.5 m/s, in between the thresholds used by Geernaert and Rieder et al. Values of θ were in general agreement between the two sonics and were +2.1 and +3.5° for 60 hours of data. Run-to-run variation was about $\pm 10°$.

In one 6-hour run, it was noticed that a 15 min transient in wind speed and stress gave θ values of 60° as calculated from the 10 min running averages. This period was isolated with block averages which showed that θ was essentially zero if the transients were excluded. Nonstationary effects have been a continual problem in turbulence, and it appears that transients due to mesoscale events can result in very large values of θ which have to be interpreted carefully.

Similarly, interpretation of covariances as stresses for short-time averages or instantaneous values (e.g., $u'w'$ with little or no averaging) may not be correct, as we are reminded (Corrsin 1957) that turbulent stress exists only by the Reynolds average definition.

12.9   Tower and Open Ocean Experiments

Smith (1980) remarked on nonzero magnitudes of cross-wind stresses of at least one-half of the along-wind stress for 27 of 147 forty-minute data segments from a stable platform in 59 m deep water. He was not able to resolve whether the nonzero cross-wind stresses were due to tower motion, measurement problems, or were real. It was noted that there were no large cross-wind stresses during a 26-day period of strong winds, but after passage of a storm with a high sea state, they were significant.

Geernaert (1988) reported early measurements of stress angle from an oceanic surface-layer experiment based on a tower in the North Sea. The first set of measurements were taken with a propeller-vane anemometer from the PISA needle at 7 m height and the second with a sonic anemometer mounted on a boom off the North Sea Platform (FPN) at 33 m. Each allowed the direct eddy correlation calculations of the wind stress components via Eq. (12.1). From 30 min averages of the wind stress components, the directional difference between wind and stress vectors was as large as $20°$. The measurements were shown not to be influenced by platform flow distortion. For both data sets, unstable stratification correlated with positive angles, and stable stratification with negative angles (note the definition of direction used by Geernaert discussed above). A mixing-length model was proposed by Geernaert where momentum at different levels has different directions, depending on thermal advection and buoyant stability, which results in nonzero stress angles (Geernaert 1993). The largest correlation of the wind-stress angle was found with the pseudo heat flux, $U_{10}(T_0 - T_{10})$, where T_0 and T_{10} are the water and 10 m air temperatures. Correlation coefficients were calculated to be 0.6. A frontal passage produced values of angle which changed from $+20°$ to $-20°$ in a period of a few hours. For one case in stable stratification conditions, the wind stress angle exhibited periodicity on the order of 3–4 hours. This periodicity was argued to be due to inertial oscillations within the boundary layer, Doppler-shifted by advection past the measurement site.

Measurements of the surface wind, wind stress, heat flux, and directional wave spectra were made from the Chesapeake Bay Light Tower by Geernaert et al. (1993). High-frequency sonic anemometers were deployed 4.5 m above mean sea level, while sub-surface pressure transducers were instrumented on the legs of the tower to measure the directional wave field. Thirteen 40 min records satisfied the requirements of zero heat flux, low wind speeds, and southerly wind directions, ensuring that there was low distortion of the flow by the tower. The surface wind stress was generally directed between the direction of the mean wind and the long-period swell (see Fig. 8 of Geernaert et al. 1993). Data were partitioned between steep and less steep swell. Wave steepness appeared to play no significant role in the ability of the swell to direct the wind stress away from the mean wind, in contrast to the laboratory results of Giovanangeli et al. (1994) referred to above. It was suggested by Geernaert et al. that the mean orbital velocity vector of the ocean surface may be an important parameter governing the wind-stress direction. Under this assumption, air-flow separations induced by wave breaking could also have a strong influence.

Measurements of the turbulent wind field, the directional wave spectra, and large-scale wave breaking were made during an open-ocean experiment from the Research Platform *FLIP* by Weller et al. (1991). Wind and wind stress vectors were calculated using 30 min averages. Wave spectral information was estimated from measurements by a four-beam surface-scanning Doppler sonar (Smith and Bullard 1995). Wave breaking speeds and directions were inferred from measurements by a free drifting sub-surface buoy, instrumented with an array of passive acoustic hydrophones (Ding and Farmer 1994). As shown in Fig. 12.1 Rieder et al. (1994) found that the direction of the wind stress was related to that of the waves across a wide range of frequencies. In general, the wind stress vector tended to lie between the directions of the mean wind and swell, in agreement with Geernaert et al. 1993 (see Fig. 6 of Rieder et al. 1994). Moreover, during high winds, a significant correlation was found between the angles of the stress and swell from the mean wind. An influence of the wave direction on the wind

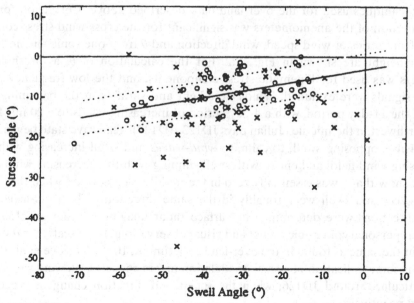

Figure 12.1. Wind-stress angles versus swell-wind angles from Rieder et al. (1994). Circles: high winds: Crosses: low winds. Solid line indicates a correlation for the high-wind data; error bounds are given by the dashed lines.

stress could be seen on a per-frequency basis as well: the directions of the periodic fluctuations in the atmospheric boundary layer which contributed to the wind stress were seen to qualitatively match those in the directional wave field (see Figs 8 and 9 of Rieder et al. 1994). These results suggest the existence of a dynamic link between the fluctuations at the sea surface and those in the overlying wind field, and hence there may be a causal relationship between the wave and stress directions. The relation of wave breaking direction to the direction of the wind stress vector was studied by Rieder et al. (1996) from the same data set. For 15 half-hour periods for which wave breaking data were available (white-cap speed, direction, duration, and dimension), the wind stress was found to be more aligned with the direction of wave breaking than with either the wind or the wave directions (see Fig. 3 of Rieder et al. 1996). As wave breaking is episodic, this observed relation to wind stress is conceptually very nonlinear, in contrast to the linear relation observed in the spectral comparison of wind stress and wave directions of Rieder et al. 1994.

Recently, Liu et al. (1995) presented results of wind, wind stress and swell and wind-wave data obtained from a ship in the Surface Wave Dynamics Experiment (SWADE). Wavelet analysis showed the modulation of the wind component in the swell direction by swell groups. Short-time turbulent stress component calculations (80 s averaging time) showed large deviations between wind and stress angles with a mean of $-41.2°$ coincident with the swell groups; the overall mean difference was $-11.6°$. Histograms of the angle difference showed large standard deviations.

Sonic anemometer data were also obtained over the open ocean in another experiment from R/P *FLIP* by Friehe et al. (2000) and were analysed with the same techniques used for the over-land data described above. Correction for the slight motion of the anemometers was significant for the cross-wind stress component. Time series of wind speed, wind direction and θ from one sonic anemometer at 9 m height are shown in Fig. 12.2. For the calculation of θ, a threshold of 0.02 m/s was used for kinematic stress components, and the low-frequency parts of the signals were excluded. Winds were light and variable at the beginning and end of the 11-day period, with a 4-day period of increasing winds to $\sim$20 m/s from the northwest in the middle (Julian days JD123–JD127). The wave state began with *wind-waves* opposing swell, to calm, to *wind-waves* and swell increasing with the increasing wind field, and ended with swell running with the decreasing wind. The results show that θ was essentially zero in the middle 4-day period when the wind, *wind-waves* and swell were roughly in the same direction. (The *wind-wave* and swell directions were determined by surface radar imaging (Frasier and Liu and Poulter, personal communications) and visual observations.) The scatter of θ values is about the same as found in the over-land experiment, 10–14$°$ (Friehe et al. 1997). There is some indication of a small modulation in θ in concert with wind variations; in particular, around JD126, when the speed and direction changed, θ became slightly negative.

The other periods in Fig. 12.2 are more transitory, where at least one variable (wind speed, direction, wave, swell) was changing. Periods of low wind (JD120–

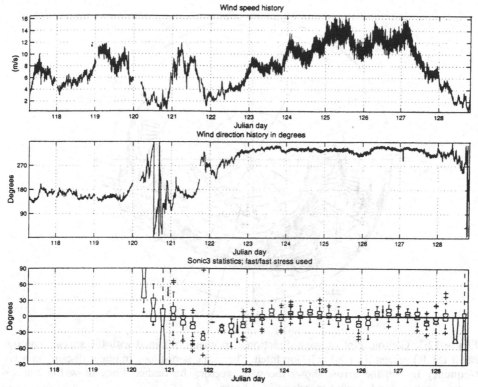

Figure 12.2. Wind speed, wind direction (meteorological convention) and boxplots of θ versus time (Julian day 1995) adapted from Friehe et al. (1997). Gaps in wind speed and direction indicate no data; periods where wind data exist but there are no values of θ are where the stress components were less than threshold values. There are no motion-corrected stress data prior to Julian Day 120.5. The boxplots for θ indicate the median (centre-line), 25% and 75% quartiles (ends of boxes). The whiskers indicate the bounds of the data and outliers are shown by +.

JD121, JD129), yet above the threshold of stress components, result in a large variability in θ, as shown by the large box-plot quartiles.

12.10 Wind and Wave Alignment

It is usually assumed, based on buoy measurements, that the wind and *wind-wave* fields are aligned. Observations from airborne radars over the ocean have challenged that assumption. The surface contour radar (Walsh et al. 1989) scans the surface of the ocean with a 10 m diameter footprint across a 400 m swath from an altitude of 400 m. Aircraft motion is removed, and directional wave spectra are calculated.

Two examples from the surface contour radar and supporting surface buoy winds are presented in Figs 12.3 and 12.4. The first shows the surface wave field in a cold-air outbreak, short fetch situation (Chou and Yeh 1987) 37 km from the coast. Spectral peaks lie along the resonant lines (dashed) according to

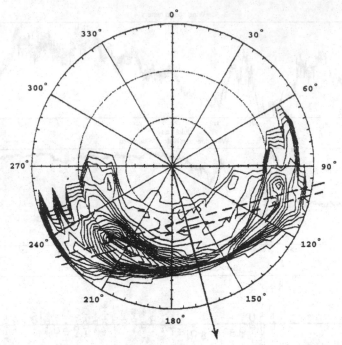

Figure 12.3. Contour plot of wave directional spectrum obtained with the surface contour radar (SCR) on Jan 20, 1983 at 37 km fetch. The radial coordinate in linear dispersion wave frequency in Hz; the arrow shows the wind vector; the dashed lines show the Phillips resonance criterion for wind speeds of 11–13 m/s.

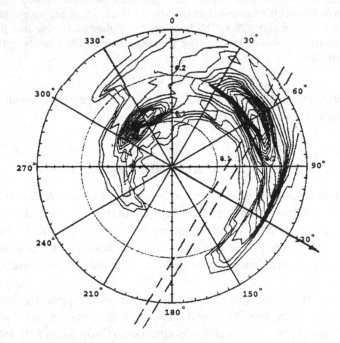

Figure 12.4   Same as Fig. 12.3 except for Jan 5, 1982 and 10 km fetch.

the Phillips–Townsend analyses (see Long et al. 1994), but are asymmetric; the peak to the right of the wind direction dominates. A second cold-air outbreak case is shown in Fig. 12.4, with a strong wind-wave resonance peak to the left of the wind and no sign of resonant waves to the right. In all surface contour radar observations, the side lobe peaks are always unequal in magnitude, the initial larger one developing faster and eventually becoming the only peak in the directional spectrum. The consequence is an inhomogeneous coastal sea state. While the wave growth and evolution of the peak frequency from the SCR measurements are similar to established empirical formulae, the nonalignment of the wind and wave field should be considered and may be related to nonalignment of the surface stress and wind.

13 The Influence of Surface Tension

H. Mitsuyasu and Erik J. Bock

13.1 Introduction

Before addressing the issues of the effects of surface tension on drag over the ocean, it is necessary to describe the physical properties and their controlling influences that make surface tension important to the air–sea boundary conditions. Surface tension is a special case of interfacial tension, that is, the pull between the molecules at the interface between two immiscible fluids. This pull exists because the molecules in the interfacial layer have fewer nearest neighbouring molecules of their own kind than do molecules in the bulk phase of either fluid. It is a physical quantity important to air–sea interaction because it affects many hydrodynamic phenomena, most notably, capillary and capillary–gravity waves. The single most important reason that these short waves (having wavelengths between about 0.1 to 30 cm) are influenced by surface tension is not really directly due to surface tension, but to surface elasticity, which is caused by the lowering of surface tension introduced by the addition of surfactants. Surfactant is a term coined by F. D. Snell that is short for surface-active agent. It describes molecular species that are more thermodynamically favoured to reside at the surface of a liquid. Typically, molecules that act as surfactants in aqueous solutions have two moieties, one hydrophilic (water-liking) and one hydrophobic (water-avoiding). In both terrestrial and aquatic environments that contain biological organisms, molecules composed of two such moieties are common. This is because they are biologically useful for performing many functions like building lipid bilayers and aiding transport across cellular membranes; they are also commonly found in the byproducts of the decomposition of biological organisms. One example of a large source of surfactants in the oceanic environment is phytoplankton, that photosynthesize surfactants from non surfactant molecules and ions (e.g. water, carbon dioxide, nitrates, and phosphates). The quantity of surfactants at the ocean surface varies as a consequence of nutrient availability, seasonal biological cycling, interfacial hydrodynamics, and probably, other mechanisms. Figure 13.1a shows a correlation between the surface activity of samples of ocean surface water and longitude for a number of samples collected by Frew and Nelson (private comm.) on several cruises off the Delaware Bay. Surface activity in this plot is determined by the polarographic technique described by Zutic et al. (1981). The increase in surface activity as the longitude nears the coast is evident. This effect is attributed to an increase in the nutrient concentration due to coastal runoff.

A similar correlation between coloured dissolved organic matter (CDOM) and longitude is shown in Fig. 13.1b. CDOM, here measured by Frew and Nelson (private comm.), is demonstrated to be a surrogate measurement for phytoplankton concentration.

Other surrogate measurements of phytoplankton concentration can be inferred from remote sensing instruments. The CZCS satellite system was used to measure phytoplankton concentration from ocean colour measurements (Feldman 1989). Averages for January and July are shown in Figs 13.2 and 13.3. The pseudo-colour scale corresponds to the concentration of phytoplankton pigment, expressed in mg/m^3. Figure 13.2 clearly shows that during the summer months, phytoplankton concentration is high on ocean-basin scales, and intercomparison between Fig. 13.2 and Fig. 13.3 shows the contrast in phytoplankton concentrations owing to differences in seasonal productivity.

13.2 Surfactants

Surfactants can loosely be classified as soluble and insoluble. When insoluble surfactants adsorb at the liquid surface, they are prevented from dissolving into the

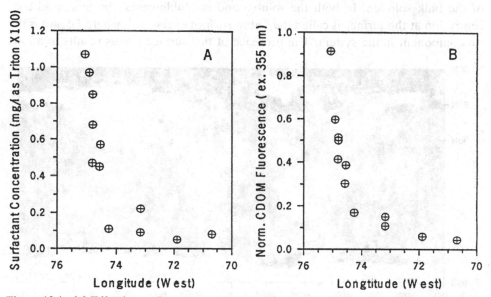

Figure 13.1. (a) Effective surfactant concentration as a function of longitude for transects off the Delaware Bay; (b) Normalized coloured dissolved organic matter (CDOM) as a function of longitude for the same transects.

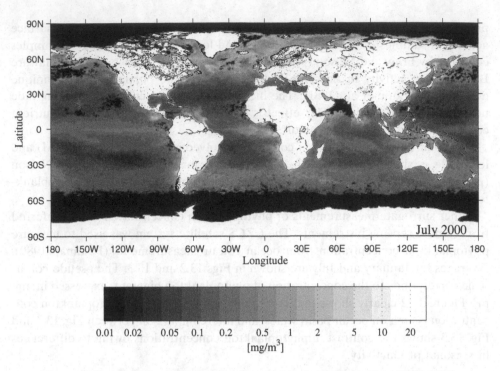

Figure 13.2.   Average chlorophyll concentration for January (data 1979–1985).

liquid and they rearrange on the surface so as to minimize intermolecular forces between the surfactant molecules. When soluble surfactants adsorb at the liquid surface they create a concentration that is relatively higher than the concentration of the bulk solution. In both the soluble and insoluble cases, the increased concentration at the surface is called the Gibbs surface excess, Γ_i^G, where i denotes the ith component in the system. The presence of this surface excess results in a low-

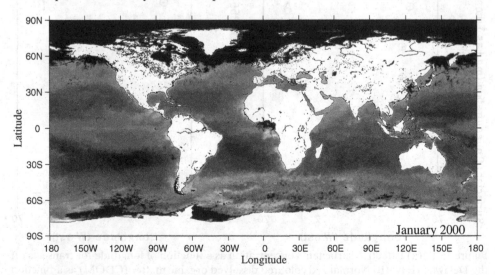

Figure 13.3.   Average chlorophyll concentration for July (data 1979–1985).

ering of surface tension that can be expressed most simply for a two component system, that is a dilute solution of a single nonionic soluble surfactant with ideal properties (component 2), in water (component 1) as:

$$-\mathrm{d}\gamma = RT \cdot \Gamma_2^G \mathrm{d}(\ln X_2)$$

where γ is the surface tension, R is the gas constant, T is temperature, and X_2 is the mole fraction of component 2. A more thorough derivation of the thermodynamics of adsorption can be found in Ross and Morrison (1988). Surface pressure, Π, can then be defined as:

$$\Pi \equiv \gamma_0 - \gamma$$

where γ_0 is the surface tension of component 1 alone and γ is the lowered surface tension resulting from the addition of component 2. For films that behave like an ideal two-dimensional gas, the equation of state for the thermodynamic relation between Π, the area between molecules, A, and temperature, T, can be given as:

$$\Pi A = k_B T$$

where k_B is Boltzmann's constant. For these films, the spreading pressure increases as the film is compressed (corresponding to an increased surface excess). Figure 13.4 shows Π versus A for this ideal gas law case. The surface elasticity, E, can be computed from the relation:

$$E \equiv \frac{\mathrm{d}\gamma}{\mathrm{d}\ln A}$$

and for the ideal gas film it is given as:

$$E = \frac{k_B T}{A} = \Pi$$

In reality, characteristics described by this simple equation of state are usually seen only at very low surface excesses, corresponding to very large area per molecule. Harkins (1952), shows classic experimental examples of phase change from ideal gas, to imperfect gas, to liquid-expanded, to liquid intermediate, to condensed phase films, each with its thermodynamically derived equation of state. Recently, equations of state for complex polymeric films have been developed, but an adequate equation of state describing the behaviour of the complex mixtures of natural oceanic surfactants has yet to be found. Despite this, much has been learned as the result of both laboratory and field observations of natural films.

The types of surfactants found in the ocean vary over a wide range of chemical composition and molecular weights. A discussion of the pertinent variables is outside the scope of this monograph, but observations from field programs can be found in the literature (Alpers et al. 1982; Bock and Frew 1993; Carlson et al. 1988; Cini et al. 1983; Fiscella et al. 1985; Herr and Williams 1986; Hühnerfuss et al. 1987). Shown in Fig. 13.5 is a typical Π versus A curve for a mixture of alkenones isolated from natural marine surfactant obtained by Frew and Nelson (private comm.), along with its mass spectrogram. This spectrogram demonstrates the complexity of its composition. Its pressure-area isotherm

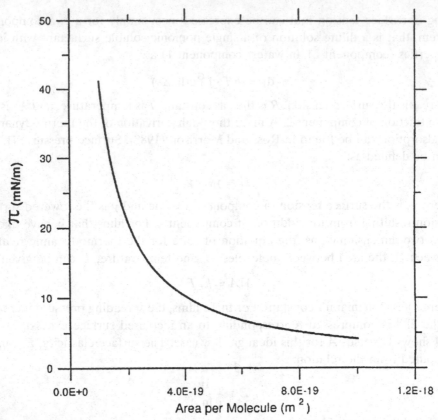

Figure 13.4. Surface pressure as a function of area per molecule for a monomolecular film exhibiting ideal gas-law behaviour.

demonstrates nonideal behaviour for all but the most expanded regions at areas above 35 Å^2 per molecule, where the film's elasticity approaches its spreading pressure.

Figure 13.6 shows two unusual Π versus A curves for surfactants obtained by Frew and Nelson (private comm.) from the waters around Woods Hole, Massachusetts. In both cases the films approach the point of collapse, although without any visual evidence of the formation of a rigid condensed phase being reached.

13.3   Damping of Waves

The damping of capillary ripples by surface films was noted early by Pliny the Elder (trans. 1634) and a physical-chemical explanation for this effect was formalized by Levich (1962), and more thoroughly by Hansen and Mann (1964). It was reviewed by Lucassen-Reynders and Lucassen (1969). The currently accepted mechanism presumes a complex-valued viscoelastic modulus that creates small changes of surface tension in response to small changes of surface area caused by the compression and dilation of the local surface as ripples propagate past. This

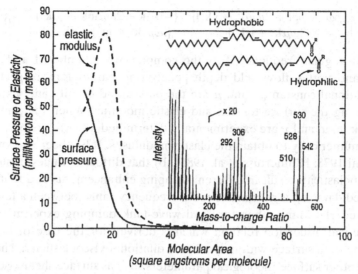

Figure 13.5. Surface pressure and static elasticity of a sample of alkenones isolated from natural marine surfactant. The inset shows the mass spectrogram to demonstrate the complex composition of the sample.

effect modifies the hydrodynamic boundary conditions and gives rise to a dispersion relation for surface waves in the presence of a surface film. By including these modified boundary conditions, the linearized Navier–Stokes equation of fluid motion can be solved to produce a dispersion relation for waves given by Bock and Mann (1989) as:

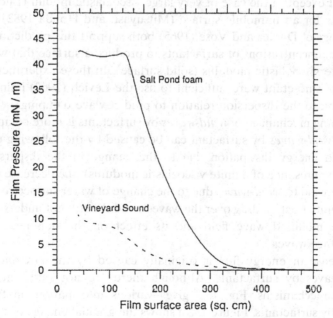

Figure 13.6. Surface pressure as a function of film area for two samples obtained in Martha's Vineyard Sound and Salt Pond. Both films exhibit what appear to be phase transitions.

$$\begin{vmatrix} \{\rho\omega^2 - \rho g\mathbf{l} + 2i\omega\mu\mathbf{l}^2 - \gamma\mathbf{k}^2\mathbf{l}\} & \{-i\rho g\mathbf{k} - 2\omega\mu\mu\mathbf{mk} - \gamma i\mathbf{k}^3\} \\ \{2\omega\mu\mathbf{kl} + i\hat{\varepsilon}\mathbf{k}^3\} & \{i\omega\mu(\mathbf{k}^2 + \mathbf{m}^2) - \hat{\varepsilon}\mathbf{k}^2\mathbf{m}\} \end{vmatrix} = 0$$

where ω is angular frequency; $\mathbf{k}$ is the complex wavenumber; $\mathbf{l}$ and $\mathbf{m}$ are the potential and vortex flow field depth penetration parameters, respectively; g is the gravitational constant; ρ and μ are the bulk liquid density and viscosity; with γ and $\hat{\varepsilon}$ being the surface tension and elastic modulus. When the physical parameters ω, $\mathbf{k}$, ρ, μ, and γ are experimentally determined, the dispersion relation can be solved numerically to obtain the elastic modulus, $\hat{\varepsilon}$. It has been shown (Cini and Lombardini 1978; Hühnerfuss et al. 1985a, b) that different values of surface dilational viscoelasticity result in different damping enhancements and that the effect of enhanced damping is maximized in the frequency range between a few Hz and a few hundred Hz. Many laboratory and wave-tank damping experiments support this theoretical dispersion relation which is derived for the case of linear waves propagating on a surface with only surface dilational viscoelasticity. The formulation omits other surface rheological parameters such as surface shear viscoelasticity and surface plasticity. Not enough is known about in situ ocean surface films to determine whether this model is appropriate to describe all oceanic films, but it is generally believed that freshly formed films can be described by this dispersion relation.

On the basis of a series of laboratory studies, the effects of a surfactant on certain air–sea interaction phenomena are discussed within the construct of the above dispersion relation. Comparisons with phenomena observed in the clean water cases (i.e. surfactant free) will be considered as well. In the case of mechanically driven long waves, the dispersion relation predicts no appreciable damping enhancement except in the case of very large viscoelastic moduli (Lucassen 1982). Model results for an immobile surface (Mitsuyasu and Honda 1982) and experimental findings of Davies and Vose (1965) both support this prediction by assuming sufficient concentrations of surfactants to produce a surface that would yield a nearly infinite viscoelastic modulus (solid surface). In these experiments, the concentrations of surfactant were sufficient to use the Levich (1962) infinite elasticity approximation to the dispersion relation to predict wave damping.

The damping mechanism of *wind-waves* by surfactants is more complicated. The damping of *wind-waves* by surfactant can be caused by the following mechanisms; the enhanced energy dissipation due to the change in the dispersion relation (caused by the presence of a finite viscoelastic modulus), the decrease of an energy flux from the wind to *wind-waves* due to the change of water surface roughness (the most significant effect on drag over the waves), and the still not understood relation between the modified wave field and its effect on the nonlinear interactions between surface waves.

The decrease in energy flux is originally caused by the attenuation of high frequency waves by surfactants. Although there are not sufficient conclusions about these mechanisms, Fig. 13.7 gives various information on the decay of *wind-waves* by surfactants. Figure 13.7a shows the gradual change of the frequency spectra of wind-waves in a laboratory experiment (wind speed $U = 7.5$ m/s, fetch

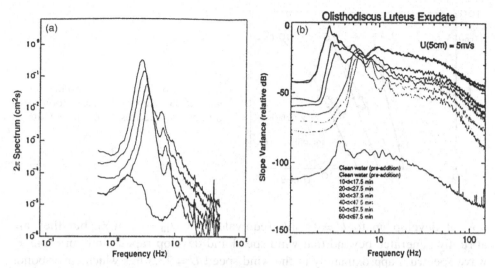

Figure 13.7. (a) Observation of wind-wave spectra as a function of successive surfactant concentration (Mitsuyasu and Honda 1986). (b) Observation of wind-wave spectra for periods of time after the addition of cultured phytoplankton exudates (Bock private communication).

$x = 6$ m), when the surfactant concentration was changed successively (Mitsuyasu and Honda 1986). The uppermost spectrum was measured on clean water surface and the lowest spectrum was measured on the water containing the maximum concentration of the surfactant (2.6×10^{-2}) used in the experiment. Figure 13.7b shows the gradual decay of the frequency spectrum for waves of unlimited fetch produced in an annular tank. In these measurements, clean water spectra were recorded for two averaged periods. After this, a volume of phytoplankton exudate (*Olisthodiscus luteus*) previously cultured under the conditions described in Frew et al. (1990) was introduced into the bulk of the tank. During the next hour, the natural surfactants spontaneously adsorbed at the air–water interface, significantly damping the waves. Wind speed was held constant during the entire run. It is noted that the spectral energy of *wind-waves* attenuates not only in a high frequency region $\omega_p < \omega$, but also in a low frequency region $\omega < \omega_p$, where ω_p is the spectral peak frequency. Such an attenuation pattern of the wind-wave spectra shown in Fig. 13.7 is considerably different from the expectation that the energy dissipation of the high frequency components of the spectrum would be much larger than those of the low frequency components. A qualitative explanation of the phenomenon is as follows. The suppression of the high frequency components of wind-waves due to the surfactant decreases the roughness of the water surface as will be shown later, and the energy transfer from the wind to the spectral components near the spectral peak decreases.

When a surfactant is present, waves are not easily generated until some critical wind speed is exceeded. The work of Scott (1972) suggests that the concept of a critical wind speed does not hold true for pure water surfaces, and our studies of surfactants indicates that when the concentration is high (2.6×10^{-2}), *wind-waves*

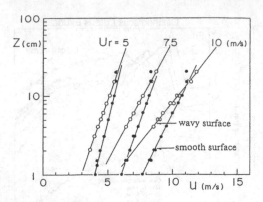

Figure 13.8. Wind speed profiles under various surface conditions. Smooth conditions are due to surface surfactants.

do not develop up to $U = 11$ m/s (equivalent to $U_{10} = 17$ m/s), but they are abruptly generated beyond that wind speed and develop rapidly into an ordinary wave spectrum approximately at the wind speed $U = 12.5$ m/s which corresponds roughly to $U_{10} = 23$ m/s. That is, the surfactant of high concentration suppresses the generation of wind-waves but its effect disappears at some high wind speeds.

As shown in Fig. 13.8 drastic changes of wind profiles are caused by the surfactant of relatively high concentration (2.6×10^{-2}) in the water (Mitsuyasu and Honda 1982). This is due to the change of water surface roughness which is caused by an almost complete suppression of the wind-waves by the surfactant. As shown in Fig. 13.9, the relationship between u_* and U_{10} changes greatly at higher wind speed. It is interesting, however, that for low wind speed ($U_{10} < 7$ m/s) this relationship is almost the same for both cases. It is only for speeds greater than 7 m/s that the surfactant reduces the drag coefficient. This observation means that small wind-waves generated by the low wind speed do not contribute to the water surface roughness.

In Fig. 13.10, for water with added surfactants, wind profiles over mechanically-generated waves are compared with wind profiles without waves (Mitsuyasu et al. 1988). It can be seen that mechanically generated waves in the presence of surfactants have little effect on the wind profile.

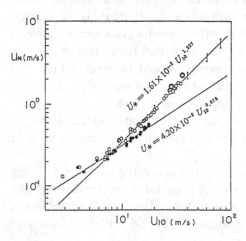

Figure 13.9. Relationship between u_* and U_{10}. Filled points are for the case of surface surfactants.

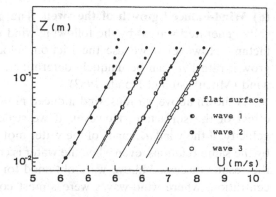

Figure 13.10. Wind profiles over mechanically-generated waves are compared with wind profiles with waves suppressed by surface surfactant.

13.4   The Impact of Surfactants

The drastic change of the surface roughness due to the effect of the surfactant causes various changes in the wind-induced water motions.

(a) Wind setup. Due to the change of the surface roughness (wind shear stress) caused by the surfactant, the water surface slope changes greatly. However, as shown in Fig. 13.11, if we use the friction velocity, the water surface slope can be described uniquely as a function of the friction velocity and the water depth (Mitsuyasu and Honda 1986).

(b) Surface drift current. The change of the wind shear stress also causes the change of the surface drift current. Here again, if we use the friction velocity of the wind u_*, the ratio of the drift velocity to the wind friction velocity U_s/u_* shows similar values, 0.57 for clean water surface and 0.67 for the water containing the surfactant (Mitsuyasu and Kusaba 1988).

(c) Wind-induced turbulence in the water. Since the turbulence in the water is closely related to the surface drift velocity, its energy spectrum can be scaled by using the friction velocity of the wind (Mitsuyasu and Kusaba 1985).

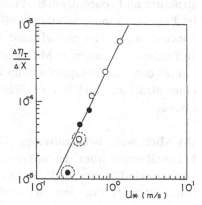

Figure 13.11. Water surface slope as a function of the friction velocity and the water depth. Filled symbols correspond to surfactant covered surface.

(d) Wind-induced growth of the swell. The growth rate of the swell (mechanically-generated waves) by the following wind is changed by the addition of a surfactant. However if we use the friction velocity of the wind the dimensionless growth rate Γ/ω can be uniquely determined by the inverse wave age u_*c of the wind (Mitsuyasu and Honda 1982).

As shown above, various wind-induced motions in the water are changed by the effect of the surfactant. However, if we scale the motions by using the friction velocity of the wind u_*, many of the water motions can be described approximately by the same relations, even when the water is contaminated by surfactants. Many of the results discussed above were obtained for a surfactant of relatively high concentration, where wind-waves were almost completely suppressed by the surfactant. Further studies will be needed to clarify the phenomena when the wind-waves are partially suppressed by a weaker effect of the surfactant.

13.5   Field Studies

Since the pioneering studies of Cox and Munk [1954], many field studies have been done of various phenomena relating to a slick, which is a thin film of oil or surfactant on the sea surface. Although the results are still qualitative due to the complexity of the phenomena, they are summarized as follows.

(a) Suppression of short waves. There are many field observations which clearly show the suppression of short waves in artificial or natural slicks (Cox and Munk 1954; Mallinger and Mickelson 1973; Hühnerfuss et al. 1981, 1983; Yermakov et al. 1985; Wei and Wu 1992). Wu (1989) summarized the results until 1989 and derived the following conclusions: (1) The suppression is most effective at relatively low wind $U \leqslant 7$ m/s and for wave components with their lengths between 2 and 40 cm; (2) The maximum suppression up to about 95% occurs at wave lengths 4–5 cm. Alpers and Hühnerfuss (1989) attributed the latter result to the Marangoni (1872) effect.

However, general features of the attenuation rate or the critical wind speed beyond which the damping effect disappears, are not clear even now. This is probably due to the fact that the wave damping effects depend on the chemical structure and concentration of the surfactant which constitutes the slick. As shown in Fig. 13.7, the attenuation effect of the surfactant strongly depends on its concentration. The critical wind speed also depends on the concentration of the surfactant. For example, Mitsuyasu and Honda (1986) showed that wind-waves are almost completely suppressed up to the wind speed $U = 17$ m/s for extremely high concentration, which is very different from the conclusion of Wu (1989) discussed above.

(b) Microwave backscattering. The effect of the surface film on the microwave backscattering from sea surface is very important for remote sensing of the sea surface wind, waves, and contamination. Although the reduction of the backscattered intensity from the slick surface is clear as compared to that from the clean sea surface (Hühnerfuss et al. 1981; Singh et al. 1986; Onstott and Rufenach 1992), its quantitative estimation is very difficult. This is due to the fact that the backscattered intensity depends not only on the structure of high frequency waves, but also on the scattering mechanism at the wavy sea surface.

(c) Sea surface wind stress. Attenuation of high frequency waves by the surface film affects the speed of the sea surface wind and the wind shear stress. Van Dorn (1953) showed clearly in his experiment in an artificial pond that the suppression of the high frequency waves by the surface film greatly reduces the wind shear stress.

According to some limited studies on the effects of slicks upon sea surface wind (Barger et al. 1970; Wei and Wu 1992), the slick reduces the sea surface roughness and slightly increases the wind speed at the sea surface. These results can well be expected from the laboratory observations that the surfactant of high concentration suppresses wind-waves and changes vertical wind profiles greatly as shown in Figs 13.7 and 13.8 (Mitsuyasu and Honda 1986). Even for the natural surface films, high frequency components of wind-waves are suppressed. However, it is difficult to make general conclusions on the effects of surface films upon sea surface wind, due to the dependence on chemical structure and concentration. More systematic studies are needed. Furthermore, in order to get the results for practical applications, say for the use of the air–sea coupling model, we may need some statistical information, e.g. probabilities of distributions of slicks in a global ocean, in addition to the information of their chemical and physical properties.

14   The Influence of Spatial Inhomogeneity: Fronts and Current Boundaries

G. L. Geernaert and N. E. Huang

14.1 Introduction

When an air mass blows over a sharp gradient in surface roughness and/or temperature, the air mass adjusts its turbulence structure to the new set of surface boundary conditions. Flow from a low roughness region to a downstream region of higher roughness produces a local increase in turbulence intensities and fluxes, exhibited by a sharp yet relatively small step function. The same pattern is found also for flow from a cold surface to a warmer surface, as occurs in the region of sea surface temperature fronts. Conversely, flow from a warm water surface to a colder one and/or from high roughness to low roughness, produces a drop in turbulence and fluxes. The drop in turbulence and flux levels is much larger than the increase in the first example. These changes in surface fluxes are more local in nature, i.e. generally covering scales of 10 km or less, and are caused by ocean currents, surface waves, and thermodynamic properties of the ocean associated with changes in depth, eddies, and overlying wind fields.

On the larger scale, the horizontal variability of surface fluxes and flux profiles must also be affected by variabilities due to processes within the atmosphere, e.g. transport and redistribution of flux with eddy scales, shear-induced gradients caused by moving storms, advection and circulation associated with sea breezes, and flow over sloping discontinuities associated with atmospheric fronts. Furthermore, the great boundary currents, e.g. the Gulf Stream and Kuroshio, exhibit such strong gradients in sea surface temperature, currents, surfactants, and atmospheric water vapour that mesoscale instabilities and storm genesis mechanisms are influenced. These processes when combined with the surface feature variabilities can provide a complex set of ocean wind patterns which influence high resolution model performance and the quality of microwave retrievals of surface quantities using space-borne sensors.

This chapter focuses on drag variability associated with fronts and current boundaries. Coastal aspects will not be emphasized, in so far as coastal processes represent a high level of complexity (e.g., alongshore Kelvin waves, fetch-limited stress, role of coastal geometry, orography, sea breeze dynamics, shoaling, etc.) that deserve special attention elsewhere. The issues concerned with eddy scales is discussed in Chapter 11. In the next section, we review the existing theory which is specifically relevant to this chapter, and follow with examples of wind drag and remote sensing signature variability over sea surface temperature fronts. We follow this discussion with a brief review of mesoscale storm variabilities which provide lower frequency signals in stress variability; and wrap up with concluding remarks.

14.2   Theory

Surface exchange processes for momentum, which are described in detail in Chapter 5, are dominated by the interaction between wave steepness and the pressure forces and local vorticity gradients acting on individual wavelets. Breaking waves often provide spikes in the momentum exchange (e.g., Melville 1982). Since waves behave as a function of water depth and fetch, the momentum flux is predicted to also depend on these parameters. Given that the heat flux also influences the eddies which affect the surface flux magnitude and distribution, variability of surface wave state, heat flux, and windspeed collectively are assumed to be responsible for much of the surface momentum flux patterns.

Heat flux is composed of two components, sensible and latent. Sensible heat flux is controlled by temperature fluctuation correlations with velocity, and the near surface is in part governed by the thermal diffusion across the thin laminar layer. The thickness of the laminar layer is inversely dependent on u_* over a long average, but locally it depends on the local velocity shear along the phase of propagating surface waves. When waves break, the laminar layer is destroyed, thus accelerating the heat exchange. Entrainment of air and creation of bubbles amplify the flux locally during the breaking process. The heat exchange caused by the difference of temperature between sea spray and the ambient air provides an additional small contribution to the heat exchange, in particular during high windspeed.

The latent heat exchange is governed by the rate of surface evaporation, which in turn is dependent upon the humidity gradient. Similar to the sensible heat flux, there is substantial heat exchange due to evaporation of sea spray during high wind events.

In the presence of surface films, there is suppression of wave energy and breaking, thereby decreasing the average fluxes of momentum and heat. Regions which contain locally dense surfactant material have correspondingly smaller fluxes, a topic dealt with in Chapter 13 in more detail.

The momentum flux, for all practical purposes, is directed downwards to the ocean surface. (See exceptions to this generalization in Chapter 7.) This flux is the consequence of both the drag of the wind by surface roughness elements (associated with the wave field) and the integral over all momentum carrying atmospheric eddies characterized by their respective *uw* and *vw* covariance contributions. There are generally no internal surface layer sources or sinks of momentum, and with relatively homogeneous conditions, the constant flux layer assumption may be invoked at some distance above the surface. It is noteworthy to point out that the constant flux layer assumption is confined to a shallower layer at short fetches downwind of a sea surface anomaly (e.g. sea surface temperature front).

This assumption of the constant flux layer also requires that the horizontally averaged downward flux of momentum by the full spectrum of atmospheric eddies is in balance with the average drag exerted by the surface waves on the wind. Since the largest momentum-carrying atmospheric eddies have time scales on the order of 10 minutes, and these eddies may also produce horizontal gradients of both momentum and momentum flux, it follows that the constant flux layer is not necessarily valid on time scales substantially less than 10 minutes. It also implies that if one wants to obtain a statistically significant estimate of the flux, one needs to sample many large eddies, thus implying that 30 minutes may be a reasonable averaging requirement to adequately obtain a flux based on the Reynolds decomposition theory. The extension of flux estimates to higher resolution than 30 minutes and/or smaller than 25 km spatial scale requires that one treats the problem as involving a height-dependent flux divergence over short averaging, and examine the fluxes analytically in terms of surface wave evolution. The estimates of high resolution fluxes has not yet been formulated in a generalized theory acceptable for application.

The wind stress vector, τ, or momentum flux, is related to the turbulent covariances terms according to:

$$\tau = -\rho \langle u'w' \rangle_i - \rho \langle v'w' \rangle_j \qquad (14.1)$$

where u', v', and w', are the fluctuating downstream, cross-stream, and vertical wind velocities, respectively; and i and j are unit vectors in the downstream and cross-stream (positive to the right) average wind direction. By definition the quantity $|\tau/\rho|$ is equivalent to the square of the friction velocity, i.e., u_*^2. In general, the cross-stream term in 14.1 is relatively small and most practitioners assume that the wind and wind stress vectors are nearly aligned. This assumption, as shown in Chapter 12, is often violated in the coastal regions. We will focus herein on the variations in magnitude of the wind stress, and assume that the angle is small.

Use of (14.1) is often cumbersome, and one generally employs similarity theory as a practical alternative to parametrize the wind stress vector. This theory relies on the assumption that the horizontally homogeneous atmospheric surface layer is a layer of constant vertical flux. Following the use of the eddy diffusivity concept, the momentum flux may be written as $\langle -u'w' \rangle = K \partial U / \partial z$, where $K = u_* \kappa z$, and one easily obtains:

$$\partial U / \partial z = (u_* / \kappa z)\, \phi_M \qquad (14.2)$$

where ϕ_M is a stratification function, and κ is the von Karman constant ($= 0.4$). Integrating (14.2) and rearranging, one obtains the bulk aerodynamic drag law, i.e.,

$$\tau = \rho C_D |U_z| U_z \qquad (14.3)$$

where τ is the wind stress vector, ρ is air density, U_z is the wind vector at height z; and C_D is the drag coefficient, defined as

$$C_D = [\kappa/(\ln[z/z_0] - \psi_M)]^2 \qquad (14.4)$$

Note that we will assume herein that the stress and wind vectors have the same direction.

The roughness length, z_0, is introduced into the integration process to preserve finite shear and is related to wave state (see below), and ψ_M is a second stratification function related to ϕ_M in the manner discussed in Eq. (3.42). The quantity ψ_M has a value of zero for neutral stratifications; and is positive/negative for unstable/stable conditions.

In order to compare data sets and to make reference calculations for step-changes in stratification, the neutral drag coefficient has been defined, i.e. representing neutral stratifications as:

$$C_{DN} = [\kappa/(\ln z/z_0)]^2 \qquad (14.5)$$

Empirical formulations for surface roughness based on wave state often reduce to the simple Charnock relation, i.e. roughness length being proportional to wind stress (Geernaert et al. 1986). Unfortunately, there is growing evidence that both the simple Charnock relation and empirical roughness length formulations perform poorly in the presence of swell. On the contrary, in the absence of swell, there have been extensive efforts placed on finding an appropriate dependence of the roughness and/or drag coefficient on wave age, given that the variations of the neutral drag coefficient have shown strong correlations with wave age. The reader is referred to Chapters 2, 8 and 10 for more details. Since surface tension can alter the wave energy density, it is likely that surfactants also act to decrease the drag coefficient (Geernaert et al. 1987); the reader is referred to Chapter 13 which discusses the role of surfactants on drag, and this topic will not be discussed further here.

14.3 Wind Stress Dynamics in the Vicinity of SST Fronts

The patterns of the wind stress changes over sea surface temperature fronts can easily be estimated by making the following simplifying assumptions: surface layer processes dominate the covariance; acceleration terms do not contribute enough to significantly violate the constant flux layer assumption; and the windspeed, air temperature and humidity in the air mass remain constant during advection over and downwind of the surface front. With these assumptions, the change in surface stress over the two sides of the front can be written simply as:

$$\Delta\tau = \rho U^2 \Delta C_D \qquad (14.6)$$

The magnitudes of C_D on the two sides of the sea surface temperature front will depend on the magnitude of the neutral drag coefficient on the two sides, which in turn depend on the two different aerodynamic roughness lengths, and Monin–Obukhov lengths. Using Eqs (14.1) to (14.6), and specifying a priori a value for the neutral drag coefficient as a constant, one easily obtains an equation for the drag coefficient based on the local air–sea temperature difference. This can be illustrated for the simple case where $z_0 = 10^{-4}$ m, and $\kappa = 0.4$, where one can arrive at:

$$C_D = (\kappa/[11.55 - \{B\Delta T/U^2 C_D^{3/2}\}])^2 \tag{14.7}$$

where B is a weak function of windspeed (see Askari et al. 1993), and ΔT is the local air–sea temperature difference. Combining (14.6) and (14.7), the change in wind stress is obtained for flow over the sea surface temperature front, where we assume for simplicity that the air temperature is a constant. Note that the use of (14.6) and (14.7) implies that if one knows the local ΔT on both sides of the front, one can compute the only remaining variable, i.e. the overlying windspeed. This allows one to use remote sensing techniques which respond to wind stress changes to make independent estimates of the overlying winds. Such estimates can be used to provide better scatterometer and SAR model functions, local estimates of surface mixing, and input to dynamical models. However, it is recognized that the simplicity of (14.7) is based on the assumption that horizontal gradients of wave processes influence the flux and the flux variability.

While (14.7) can be regarded as a simple estimate, given the assumptions imposed, more accurate estimates will require that one examine the change in the neutral drag coefficient and/or roughness length from the upwind to the downwind side of the front, which in turn will depend on the change in wave age. Referring to the summary of roughness length formulations described in terms of wave age presented in Table 2.1, one can write the roughness length in a more generalized form appropriate to many of the formulations, i.e.

$$gz_0/u_*^2 = f(c_p/u_*) = f(c_p/C_D^{1/2} U) \tag{14.8}$$

Here we introduce one more input parameter, i.e. the phase speed of the dominant surface wind wave, c_p, to estimate the change in stress. Rearranging and combining (14.4) and (14.5) with (14.8), one now has:

$$C_D^{-1/2} = C_{DN}^{-1/2} - \psi/k + k^{-1}\ln\{(C_D/C_{DN}) + k^{-1}\ln\left(f(c_p/C_{DN}^{1/2}U)/f(c_p/C_D^{1/2}U)\right)\} \tag{14.9}$$

which is a more general case than the result presented in Geernaert and Katsaros (1986), where the roughness length was defined simply in terms of the Charnock relation and the third term on the right hand side was simply written as $\ln(C_{DN}/C_D)$.

It is noteworthy to point out that there is no consensus on the specific limitations and uses of the the roughness length relationships based on wave age presented in Chapter 2, and their relevance for operational use. However, we can illustrate the application of (14.9) as an improved method to estimate the importance of improved roughness length formulations by examining the role of wave age on the drag coefficient. Since many of the wave age formulations take the form,

$$gz_0/u_*^2 = n(c_p/u_*)^p \tag{14.10}$$

where n and p are constants, one can write the normalized reference drag coefficient to be:

$$\kappa C_{DNW}^{-1/2} = \ln\left\{ gz/nC_{DNW}^{(1-p/2)} U^{2-p} c_p^p \right\} \tag{14.11}$$

The reader is reminded that (14.11) is based on the assumed functional form defined in (14.10). In (14.11), the numerical value of C_{DNW} will depend on the selected values of n and p. To make progress, we must introduce a drag coefficient which is based on both neutral conditions and idealized reference wave state. For the sake of illustration we here define long fetch to be when the ratio, c_p/u_*, is a constant K with a value of 25. Combining (14.11) with (14.4) and (14.10), one obtains:

$$C_D = C_{DNW}^{-1/2} - \psi/k + k^{-1}\ln(C_D/C_{DNW}) + k^{-1}\ln\{K/(c_p/C_D^{-1/2}U)^p\} \tag{14.12}$$

Comparisons between the drag coefficient computations with and without the third term on the right hand side (14.12) show that the third term acts to produce larger differences between C_D and C_{DNW} for every roughness length parametrization in Table 2.1 except for that of Toba and Jones, where the opposite effect occurs. It is noteworthy to point out that (14.12) will exhibit dramatic reductions of wind stress for flow from a warm to a cold water surface, while for flow in the reverse direction a smaller increase in drag will be noticed. The reader should note that these changes in stress based on wind directions across the front are based on the assumption that the upwind stratification is near neutral.

14.3.1   Radar Cross-section Changes Over SST Fronts

The detection of the ambient windspeed over sea surface temperature fronts using remote sensing has been explored by investigators since the days beginning with the SeaSat satellite sensor model function definitions (see, for example, Pierson 1990), and more recently in dedicated experiments, for example the Frontal Air–Sea Interaction Experiment (see, for example, Friehe et al. 1991; Weissman et al. 1994; and Askari et al. 1993). These studies were dedicated to exploring the subsurface dynamics governing mixing and regional heat and momentum budgets, especially near shelf breaks and semi-stationary oceanic eddies (for example, Charnock and Pollard 1983). In most of these studies, remote sensing was a primary tool and the step-change in radar cross-section observed at microwave frequencies and moderate incidence angles (in the Bragg scattering range) over sea surface temperature fronts has been attributed to the change in drag coefficient caused by a change in atmospheric stratification (for example, Li et al. 1989). Such changes have also been documented by Nilsson and Tildesley (1995) in a series of case studies. Refer to Fig. 14.1 which provides an illustration from the FASINEX project (Friehe et al. 1991) where a step-change in cross-section overlays the SST front. In Fig. 14.2, data from Li et al. (1989) are presented. In Fig. 14.3 is a scene over the East Australian current.

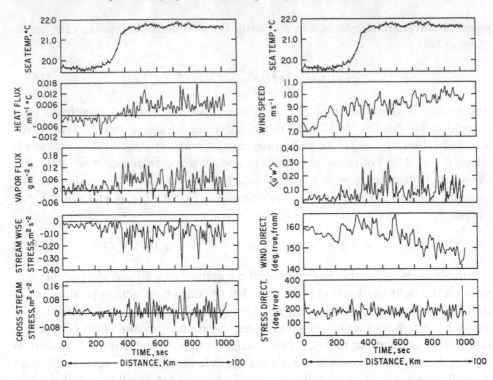

Figure 14.1. Spatial variations of (left) SST, heat flux, vapour flux, streamwise stress, and cross-stream stress; (right) SST, wind speed, kinematic stress, wind direction, and stress direction (after Friehe et al. 1991: FASINEX Feb 18, 1986, flight leg 5–6 at 30 m).

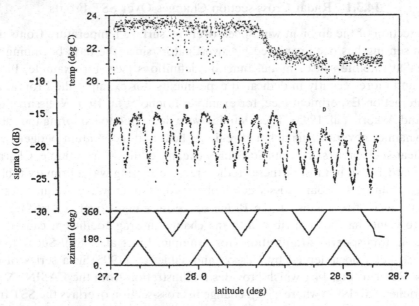

Figure 14.2. Spatial variations of the radar backscatter cross-section across the temperature front. From top: SST, radar cross-section, sigma 0, and the radar azimuthal angle.

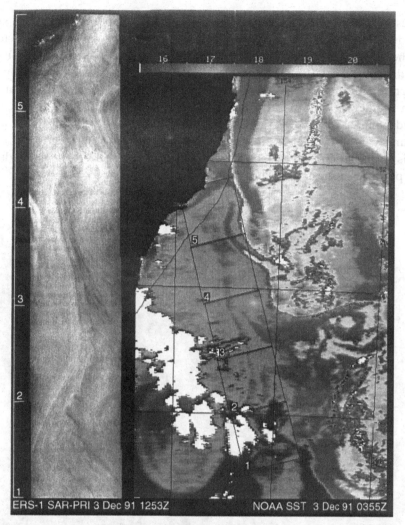

Figure 14.3. Plate 1 from Nilsson and Tildesley (1995) showing the roughness of the sea (left hand panel) and the sea surface temperature off the east coast of Australia. The SAR image location is marked on the SST image.

A general assumption invoked in most field studies is that both the mean and spatial variability of the Bragg scattering waves are correlated with the mean and spatial variability of the wind stress. Such a correlation also implies that the surface layer processes governing the stress are correlated with outer layer boundary layer processes which contribute lower frequency eddy momentum to the net drag coefficient. This suggests that there is an a priori requirement that remote sensing retrievals must have large spatial footprints to provide good estimates of the wind stress, since the correlations between outer boundary layer processes and surface layer processes require substantial temporal or spatial averaging.

On the other hand, there is substantial evidence that remotely sensed microwave signatures of the ocean surface exhibit step-changes at the same places as step-changes in surface stress. However, attributing these simply to changes in stratification is not sufficient. Horizontal gradients of current shear and surface tension alter the wave slope of Bragg scattering waves and can also alter the roughness which drags the wind. Especially in low to moderate windspeeds one can encounter a rather complicated picture of nonaligned surface stress fronts with radar cross-section gradients, which is attributable to currents, surfactants, and sea breeze induced boundary layer divergence (Askari et al. 1993). An example of surface signatures caused by changes in bathymetry and the resulting horizontal current shear is given in Fig. 14.4; as shown there is also a strong SST front in the domain, which produces a substantial step-change in cross-section.

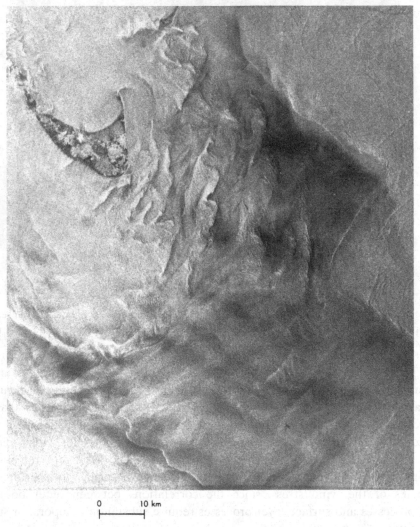

0   10 km

Figure 14.4. A synthetic aperture radar image from Nantucket Shoals. This image provides the change of surface backscatter cross-section from frontal structures as well as from bathymetric features.

14.3.2 The Role of Surface Current Gradients

Wave-current interaction remains as one of the least tested theories in the study of wave dynamics, and the role of large horizontal gradients of the interaction can be assumed to affect the stress through a complicated set of coupled processes. The difficulties in dealing with these issues have rested with an inadequate observational data base to develop a set of advanced hypotheses. Waves are generally nonstationary, nonlinear, and multidirectional, and their spectral interaction and coupling with the overlying momentum flux cospectrum are beyond simple hydrodynamics. The problem of wave–current interaction has been approached traditionally by applying the WKB approximation of the conservation of action equation. However, a singularity occurs in this equation which is associated with a phenomena called blocking. The physics governing the wave–current interaction near frontal boundaries will be quantitatively discussed here, while the extension to wave–current–stress interaction will be only qualitatively discussed.

The blocking phenomenon is the condition when the surface wave is stopped at the point where an opposing local current vector projection is larger in magnitude than the group velocity of the wave. If the simple application of the WKB approximation neglects reflection, one can calculate a dramatic increase in wave energy extracted from the currents, and the length of the waves will decrease to maintain constant frequency. The increased wave slope produces highly nonlinear waves, with possibility for increased wave breaking, and increased surface roughness. Due to the nonlinearity of the phenomena, the complete problem requires reflection to eventually be included. So far, the problem is limited to linear and weakly nonlinear approaches, and the solutions have focused in particular on the trapping and reflection of capillary waves (see, e.g. Badulin et al. 1983; Pokazeyev and Rozenberg 1983; and Shyu and Phillips 1990), as addressed in Section 4.4.

The dynamics of blocking of the full wave spectrum may be tackled by expressing the wave–current interaction in terms of its Hamiltonian, and write the conservation equations in the form:

$$(C_G + U_C)A = \text{constant} \tag{14.13}$$

where C_G and U_C are the group velocity vector and current vectors, respectively; A is the wave action defined as E/σ, where E is the wave energy density; and σ is the intrinsic wave frequency, which satisfies the dispersion relation, i.e.

$$\omega = (gk + \gamma k^3)^{1/2} \tag{14.14}$$

with γ as the kinematic surface tension, and k as wavenumber. The waves must obey the following relationship, i.e.

$$\sigma = \omega + kU_C = \text{constant} \tag{14.15}$$

where σ is the apparent frequency. Since σ is a measurable quantity, we plot σ as a function of k in Fig. 14.5, in which U with an increment of 5 cm/s, varies from zero to -35 cm/s. On the σ–k plane, the slope of any curve is the apparent group velocity, or the Doppler shifted group velocity. Thus, for positive values of $\partial\sigma/\partial k$, one has forward propagation of wave energy, and negative values indicate backward propagation of wave energy. Blocking is associated with $\partial\sigma/\partial k = 0$. The

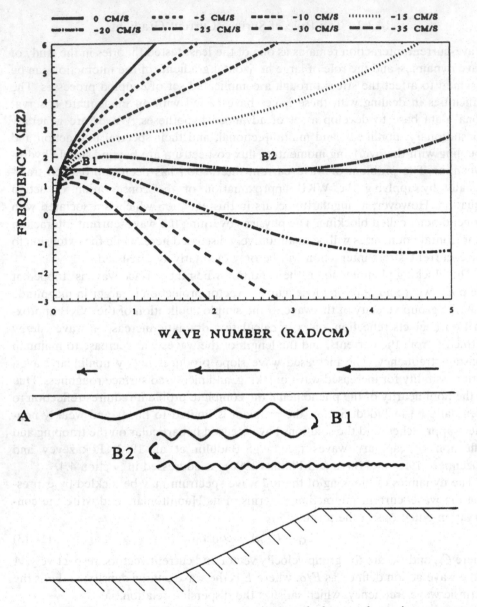

Figure 14.5. Dispersion relationship for waves in the σ-k plane, under various current conditions. Ambient adverse currents are given in 5 cm/sec increments with the solid curve indicative of zero current. A schematic diagram of wave propagation in the physical plane is given at the bottom: A indicates the initial position, B1 gives the first reflection point, and B2 gives the second reflection point. The region between B1 and B2 is the trapping zone.

existence of a horizontal tangent indicates a blocking condition. Therefore, any wave having a wavenumber larger than $(g/\gamma)^{1/2}$, could never be blocked or reflected. By the same argument, it can also be shown that for any wave having a frequency satisfying the following condition, i.e.

$$\omega > \{(7 - 4\sqrt{3})/3\}^{1/2} g^{3/4} \gamma^{-1/4} \tag{14.16}$$

the wave will never be blocked or reflected. It can also be shown that the minimum adverse current to cause blocking is given by:

$$U_{C(min)} = [(3 - \sqrt{3})g\gamma/(8\sqrt{3} - 12)]^{1/4} \tag{14.17}$$

For waves at the air–sea interface, we compute the minimum velocity is 17.58 cm/s. An example of waves propagating in an adverse current field has been studied by Long et al. (1993), where they showed the trapping by double reflection.

The consequence of the reflection caused by non-breaking blocking could have important applications in estimating the local stress and its horizontal gradients, as well as providing applications to remote sensing data interpretations. The classic wave–current interaction theory stipulates that at blocking the wavenumber can only be compressed by a factor of four. In the new reflection theory, however, the wavenumber can change from the original value all the way to capillary, the compression ratio is almost open ended. Furthermore, in the classical theory, the wavenumber compression occurs only near the immediate neighbourhood of the critical point; in the new reflection theory, the wave number compression could occur over a large region that spans over the initial blocking point and the secondary blocking point, and beyond. For example, a 1 m/s current can cause a 10 m long wave to reflect and produce waves of 25 cm in length in the trapping zone. According to the classic theory, though, the minimum wavelength at the blocking will be a 4 to 1 reduction of length, which is still too long for most microwave radars to sense. But the new theory will allow waves from the original wavelength extending down to capillary waves to exist over the current gradient region, and thus offer a much better chance for microwave radar detection. This is probably the real mechanism for the image formation for synthetic aperture radar (SAR) over many geographic locations, as reported by Fu and Holt (1982), and depicted in Fig. 14.4.

In addition to applications for SAR, the surface roughness generating mechanism can also influence the performance of the scatterometer. The scatterometer relies on the roughness satisfying the Bragg resonant condition to infer ocean surface wind stress and windspeed. Over the global ocean, the currents in the strong boundary current systems can certainly change the wave field as discussed in the introduction. This change can cause errors in the presently used scatterometer model functions (see Long et al. 1993). To exclude the currents in the scatterometer retrievals could introduce substantial error, and we highlight herein methods for reducing the errors.

Since the major current systems often coincide with the explosive growth of near-coastal storms, monitoring the wind stress over current systems becomes an important task. In order to construct the inversions over current systems, the understanding of the processes of wave–current–stress interaction is necessary, and use of such understanding must be translated into useful algorithms. To illustrate the variability, two SAR scenes are shown in Figs 14.6 and 14.7. In Fig. 14.6, the SAR image covers part of an ocean eddy, where the current in the eddy causes the long waves to change, and indirectly influences the short waves that are responsible for the surface roughness and roughness length variability. In Fig. 14.7, the

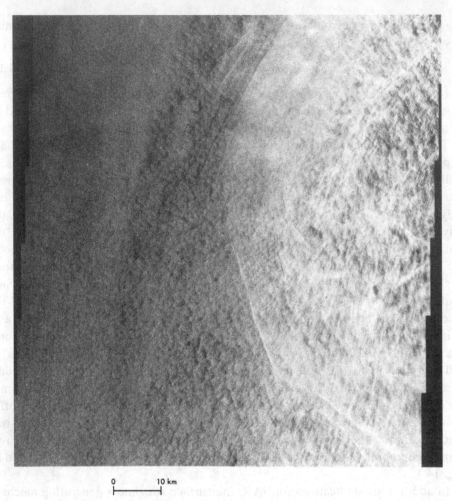

Figure 14.6.   A SAR image of an eddy in the North Atlantic Ocean.

SAR image covers the Gulf Stream front, where the shear current has caused radar backscatter power changes, again an indication of the surface roughness change. Therefore, current fronts lead to refraction of wave trains and influence the magnitude of the drag coefficient via changes in the roughness length.

14.4   Atmospheric Fronts

On scales orders of magnitude larger than sea surface temperature fronts, there are larger scale atmospheric processes which exhibit nonlinearities able to produce variabilities in the atmospheric surface layer. Storm fronts are the most obvious feature which produce such variabilities. Cold and warm fronts are associated with the advection of air masses, of different thermal, liquid and vapour water, and momentum properties than the adjacent air mass. A unique feature of the fronts is the curvature in the wind field, and the advection exhibits often a rather sharp change in near-surface air temperature, wind speed, wind direction, and degree of

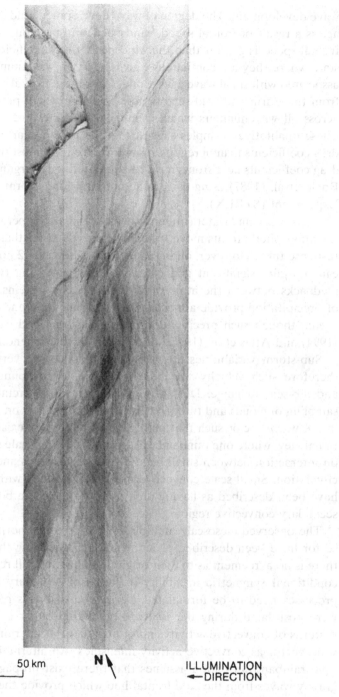

Figure 14.7   A SAR image of the Gulf Stream front near Cape Hatteras.

wave development. The degree of wave development and characterization of wave age is a result of frontal speed, windspeed, and previous wave spectra. When the frontal speed is greater than the windspeed, the wave field is in a nonequilibrium state where they are continuously growing at all wavenumbers, and the surface is associated with small wave ages. Conversely, if the frontal speed is very slow, waves from the warm and cold sectors may converge which produce a quite rough sea across all wavenumbers which cannot be characterized by simple wave spectra. These qualitative examples suggest that the region near a front may have higher drag coefficients than in regions away from fronts at most times. The magnitudes of drag coefficients near fronts have corroborated these arguments, as is illustrated in Boyle et al. (1987) using data gathered during the Storm Transfer and Response Experiment (STREX).

The surface and water column may have a weak temperature response locally, as the atmospheric fronts move typically at speeds faster than the oceanic dynamical response time. However, on scales extending beyond 12 hours, the ocean response can be quite significant and extensive, thus providing two-way interaction and feedbacks between the atmospheric and oceanic systems. Furthermore, patterns of precipitation provide a mechanism for momentum and heat exchange with the ocean, though such precipitation events are local and can be severe. See Atlas (1994) and Atlas et al. (1995) for a review of such phenomena and processes.

Sub-storm variabilities are likely to influence the vertical velocity field, and therefore such velocity variabilities will influence boundary layer convergence and air–sea exchange. During the 1960s and 1970s, remote sensing and in situ sampling of clouds and turbulence yielded information on the intricate complexity in the warm sector such that multiple rain bands can exist (contrary to the traditional view where one rainband exists) and the larger scale storm structure depends on interaction between small scale dynamical features and the larger scale storm circulation. Small scale convective motions imbedded within the larger scale flow have been described as having their genesis within the boundary layer through a secondary convective region.

The observed mesoscale multiple rainbands which normally occur in the warm sector have been described phenomenologically during the past two decades but there is no agreement as to their origin. Explanations all require that some form of conditional symmetric instability in terms of boundary layer and/or mesoscale processes need to be formulated. Formation of bands parallel to the cold front were postulated during the 1980s to be caused either by pre-existing mesoscale patterns of convective activity and/or alignment of the rainbands with the thermal wind vector as convective activity intensifies. An alternative speculation for multiple rainband formation assumes that there exist propagating ducted mesoscale gravity waves from the cold frontal line which provide the instability required for multiple rainbands and cell formation. Over land and in the marine coastal zone, orography provides an additional source and complexity for rainband genesis and storm maintenance and dissipation.

The major fronts and secondary rainbands are expected to provide substantial variability in fluxes, on scales ranging from 10 s to 100 s of kilometres in the warm sector. Evidence for such variability rests mainly with aircraft observations and indirectly using satellite imagery. In the cold sector, variability is driven by convective motions within the boundary layer, where the dominant scales of variability are on the order of 2–10 km (Geernaert and Plant 1990).

14.5 Conclusions

The spatial inhomogeneity of the ocean and atmosphere is the result of linear and nonlinear dynamical processes which act collectively to redistribute heat and momentum. These processes produce surface signatures of wave roughness and temperature, and at times also surfactant, which produce substantial variability in wind drag. In many cases, these surface features are relatively stationary, in particular when the larger scale processes are oceanic in origin. The primary illustration given in this chapter is for flow over sea surface temperature fronts, where changes in wind stress can be caused by air mass modification to a changing surface temperature and/or roughness changes due to currents and current shear. It is shown herein that use of wave age dependent roughness lengths introduces the wave phase speed, c_p, as an additional parameter to estimate the drag coefficient. Therefore, the normalized drag coefficient for neutral stratifications needs to be extended to include wave state, i.e. replacing the use of C_{DN} with C_{DNW} which considers stability and wave state together. The role of swell has not been treated here, as it was considered in Chapter 8. We state strongly that the role of a changing wave age and drag coefficient, and consequently stress, near sea surface temperature fronts is not so well understood when the flow from a high stress region encounters a low stress region. In this case the wave age drops dramatically to magnitudes where the dominant wave behaves as swell. The role of swell on a new evolving wave spectrum and overlying stress is the greatest challenge to both theoreticians and experimentalists.

While not discussed in detail, larger scale variabilities associated with storm fronts and secondary rainbands can also have substantial impact on producing wind drag variability. We provided only a brief sketch of these processes. We highlight that wave roughness is typically large in the vicinity of fronts, attributable to converging wave trains, curvature of the flow, or wave fields which are in a continually nonequilibrium state.

The theory and illustrations in this chapter highlight the potential for remote sensing of roughness over the sea, and inferring from similarity theories the magnitude of the wind drag and/or wind vector using Bragg scattering resonance at microwave frequencies. The use of scatterometers or SARs for this purpose needs a more thoughtful and complete description of remote sensing limitations and opportunities.

15 Basin Boundaries

K. B. Katsaros

15.1 Effects of Boundaries

The previous discussion has focused on the manner in which momentum is exchanged across the sea surface assuming either that there are no boundaries nearby or that the wind is blowing from a long straight coastline. The ocean has been thought of as an infinite half space in the second case with a boundary perpendicular to the wind direction. What is the influence of other shore lines? What if the wind is parallel to a long straight coast? Let us restrict the problem in this chapter to where the wind is spatially uniform.

Since *wind-waves* have a broad spread of propagation directions, the sea surface roughness at a point is influenced by the fetch from each upstream portion of land. In this situation the wave development is not only a function of fetch in the wind direction but depends also on the cross-wind dimension of the water body. In the case of a lake or bay this will be the width. The other boundary that influences waves is the sea floor. It is through the changes in the surface roughness that we can speculate about the influence of boundaries on the drag coefficient. We will not consider effects such as wave shoaling, or the spatial changes in wind speed that result from variations in surface roughness between land and sea, as this is discussed in Chapter 11.

15.2 Basin Width

The spectra of long *wind-waves* given by Eq. (4.8), i.e.

$$F(k, \theta) \propto \left(\frac{U}{c_p}\right)^{0.5} k^{-4} D(k, \theta)$$

contain a directional factor that takes the form of a hyperbolic or Gaussian (e.g. Donelan et al. 1985; Apel 1994). The function $D(k, \theta)$ is wave number dependent, with the high frequency region behaving quite differently from the waves near the peak of the spectrum. From this formula one can see that in a narrow basin (such as a bay or a lake), wave energy will be lost on the shores while no wave energy will be advected in from the sides, leading to reduced wave amplitudes near the peak of the spectrum compared to wave spectra in wider water bodies.

Evidence from lake studies by Donelan et al. (1985), and Atakturk and Katsaros (1999) indicate that for a fetch to basin width-ratio of about 2, this effect results in wave amplitudes about half of what one would observe for a wide water body. Heimbach (personal communication) calculated the wave field for Lake Washington employing the WAM model for two cases: absorbing beaches and fully reflecting beaches (corresponding to infinite width). He obtained quantitatively similar differences in wave amplitudes between these two cases, for wind speeds from 5 to 10 m/s, as the observed differences between the Lake Washington (Atakturk and Katsaros 1999) with a large fetch to width ratio and the Lake Ontario measurements (Donelan et al. 1985) for a smaller fetch to width ratio.

15.3   Shallow Water

The comments below refer to a water body of constant depth d of the order of a few metres to a few 10 s of metres. We do not discuss waves impinging on a sloping beach, but rather waves being generated over shallow water or travelling into shallow regions on the continental shelf, but not changing rapidly enough to cause shoaling, that is strong breaking.

When wind-waves develop in deep water (deep with respect to the longer waves in the spectrum, i.e. $d > \lambda$) a self-similar spectrum develops as discussed in Chapter 4. What changes in the wave spectrum and in the air–sea interaction processes are expected for a shallow water body?

Several modifications can be postulated, a priori, but further feed-back effects occurring due to nonlinear interactions require observational evidence. For long irrotational waves of wavenumber k, the phase speed is given by

$$c = \sqrt{\frac{g \tanh(kd)}{k}}$$

where the shallower the water, the slower the wave. The wave growth equation provides some guidance; for a certain value of the wind speed (or wind stress) the expected waves at the peak of the spectrum (for not too short fetch) begin to interact with the bottom. This can be characterized by the nondimensional depth $\delta = gd/U_{10}^2$. We know that these waves steepen and shorten, while maintaining their frequency. The power law dependence of the spectrum changes with water depth, e.g. see Thornton (1977). Results from Young and Verhagen (1996) are shown in Fig. 15.1.

Several consequences can be formulated:

- The wave field loses some energy to friction with the bottom of the basin.

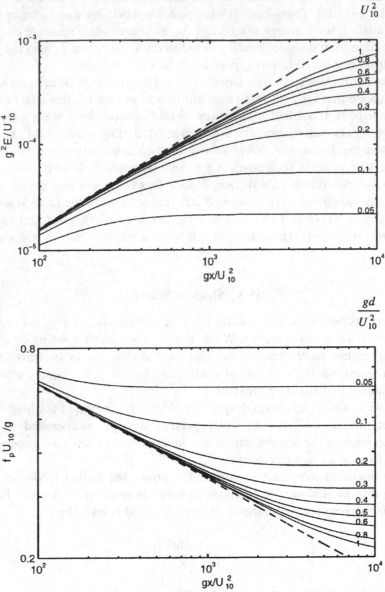

Figure 15.1. Nondimensional energy and frequency of the *wind-wave* spectrum from Young and Verhagen (1996) showing the influence of depth, *d* and fetch, *x*.

- The increased turbulence and current shear that result may modify the surface waves and also cause them to decay.
- The form drag would increase as the result of waves shorter and steeper with increased water turbulence resulting in larger net atmospheric momentum transfer to the sea surface and increased atmospheric turbulence, thus compensating for the reduction of wave energy suggested by the bottom friction.

- The increased stress, brings the atmospheric stratification parameter z/l where l is the Monin–Obukhov parameter discussed in Chapter 3, closer to 0, and thus neutral stratification, which may decrease or increase the stress transfer further depending on the sign of l.
- The increased stress and relatively younger age of the waves (wave age, c_{pd}/U_{10} is smaller since the phase speed c_{pd} is smaller than c_p) could result in more frequent occurrence of wave breaking and thus further enhanced turbulent momentum transfer from the atmosphere to the sea.
- The wave breaking is expected to result in a loss of momentum from the waves by transfer to a mean surface drift current, which is, therefore, expected to increase when wind blows over shallow water.

The net effect of increased momentum transfer to the water and the frictional loss by the waves and the drift current at the bottom cannot be predicted a priori, but would depend on the parameters of the particular situation: wind stress, atmospheric and water stratification, wave development, water depth and bottom characteristics such as roughness and sediment type.

The statement above holds for the conditions when the waves encounter shallower water, such as is found for a northerly fetch in the German Bight. Two experiments have been conducted in this region, in which some wave parameters and direct stress measurements were obtained, the Marine Remote Sensing experiment (MARSEN) and the Humidity Exchange over the Sea (HEXOS) Main Experiment (HEXMAX) after DeCosmo et al. 1996). Figure 15.2 shows that as the wind speed increases, and therefore the waves at the peak of the spectrum increase in length, the measured stress in these two experiments is significantly larger than what has been found in deep water. Further analysis by Oost (1998)

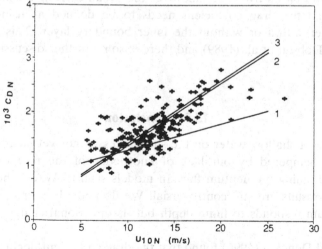

Figure 15.2. After DeCosmo et al. (1996). C_D versus wind speed from HEXOS, clearly showing higher C_D from directly measured stresses in the shallow water for high wind speeds than one finds in deep water. Line 1 Smith (1988) deep water. Line 2 Smith et al. (1992). Line 3 Reynolds stress measurements alone.

suggest that for wave fields of length greater than 80 m at the HEXOS site (water depth 18 m) there is significant enhanced stress. This is $d/\lambda = 0.2$.

As we have noted in other chapters more than one relevant measure of the wave field changes at the same time. The data in Fig. 15.2 has also been used in Chapter 10 to illustrate that for large wave ages the drag coefficient decreases with wave age. Thus both nondimensional depth and wave age are changing together in their data set.

When the whole development of the wave field has taken place over shallow water, the waves never reach the greater lengths possible for a steady wind stress in deep water, since they feel the bottom during their growth. In that case, the waves are built up to a size at which they break continuously for sufficient and constant forcing, but they are perhaps less steep than for the case of the longer waves approaching a shallow sea that steepen rapidly due to a change in depth.

The surface drift current due to momentum transfer from the air is often parametrized as being proportional to u_* (e.g. Wu 1983). One can anticipate that the drift current is still proportional to u_* in shallow water (perhaps with some modification of the proportionality constant), since over the shallow water the increased wave breaking and higher drift current enhanced the atmospheric turbulence intensity (thus the u_*).

Hicks et al. (1974) were not able to make any clear conclusions from their measurements of drag coefficient over shallow water. Surface films may be important in many shallow water situations. Baines (1974) on the other hand suggested an enhanced drag for some values of nondimensional depth.

15.4 Land–Sea Boundary

When the wind blows from the land over the ocean there is a change in surface roughness and the drag coefficient needs to be defined as using a reference velocity either within or without the inner boundary layer. This was discussed in detail by Dobson et al. (1989) and there is some further discussion in Chapter 11.

15.5 Conclusion

The influence of shallow water on the shear stress is not yet clear. A theoretical approach is hampered by our lack of knowledge of the role the long waves play in determining momentum flux. In models such as WAM they now play a role but the results are still controversial. We do know in a gross sense that the wave spectrum responds to finite depth but its directional character is even less clear.

Anctil and Donelan (1996) found that the drag coefficient depended on the sea state which in turn depended on the water depth. It is generally reported, e.g. Geernaert et al. (1986), that shallow water waves produce higher drag than deep water waves.

Deep water wind-waves in steady winds align with the wind but shallow water shoaling waves are refracted by the bottom topography to align with the bathymetry. The wind has no such constraint so in coastal situations it is possible to get wave crests systematically at an angle to the wind. It is this situation that is expected to produce shear stresses at an angle to the wind. The evidence is uncertain and further work is needed.

References

Aagard, K. and E. C. Carmack, 1989. The role of sea ice and other fresh water in the Arctic circulation. *J. Geophys. Res.* **94**, 14485–14498.

Akylas, T. R., 1984. On the excitation of long nonlinear water waves by a moving pressure distribution. *J. Fluid Mech.* **141**, 455–466.

Albrecht, B., 1993. Maritime and continental stratocumulus. In: *Workshop Proceedings Parameterization of the cloud topped boundary layer*. 8–11 June 1993, ECMWF, Reading, 21–50.

Albrecht, B. A., D. A. Randall, and S. Nicholls, 1988. Observations of marine stratocumulus clouds during FIRE. *Bull. Amer. Meteorol. Soc.* **69**, 618–626.

Alekseenko, V. N. and K. L. Egorov, 1991. On one model of the upper boundary layer of the ocean. *Soviet J. Phys. Oceanog.* **2**, 357–363.

Alpers, W. and B. Brumner, 1994. Atmospheric boundary layer rolls observed by the synthetic aperture radar aboard the ERS-1 satellite. *J. Geophys. Res.* **99**, 12613–12621.

Alpers, W. and H. Hühnerfuss, 1989. The damping of ocean waves by surface films: A new look at an old problem. *J. Geophys. Res.* **94**, 6251–6265.

Alpers, W., H. J. Blume, W. D. Garrett, and H. Hühnerfuss, 1982. The effect of monomolecular surface films on the microwave brightness temperature of the sea surface. *Intern'l. J. Remote Sen.* **3(2)**, 3642–3648.

Alves, J. O. S., 1995. Open-ocean deep convection: understanding and parameterization. *Ph.D. Thesis*, The University of Reading, 124 pp.

Al-Zanaidi, M. A. and W. H. Hui, 1984. Turbulent airflow over water waves – a numerical study. *J. Fluid Mech.* **148**, 225–246.

Anctil, F. and M. A. Donelan, 1996. Air-water momentum flux observations over shoaling waves. *J. Phys. Oceanogr.* **26**, 1344–1353.

Anderson, R. J., 1993. A study of wind stress and heat flux over the open ocean by the inertial dissipation method. *J. Phys. Oceanogr.* **23**, 2153–2161.

Andre, J.-C., 1976. A third-order-closure model for the evolution of a convective planetary boundary layer. In: *Seminars on the treatment of the boundary layer in numerical weather prediction*, Reading, 6–10 September 1976, 205–233.

Andreas, E. L., J. B. Edson, E. C. Monahan, M. P. Rouault, and S. D. Smith, 1995. The spray contribution to net evaporation from the sea: a review of recent progress. *Boundary-Layer Meteorol.* **72**, 3–52.

Anis, A. and J. N. Moum, 1992. The superadiabatic surface layer over the ocean during convection. *J. Phys. Oceanogr.* **22**, 1221–1227.

Anis, A. and J. N. Moum, 1995. Surface wave-turbulence interactions: Scaling epsilon $\varepsilon(z)$ near the sea surface. *J. Phys. Oceanog.* **25**, 2025–2045.

Anto, A. F., 1977. Observational studies on land–sea breeze phenomena around the island of Sylt. *Meteorologische Rundschau* **30**, 118–122.

Antonia, R. A. and A. J. Chambers, 1978. Note on the temperature ramp structure in the marine surface layer. *Boundary-Layer Meteorol.* **15**, 347–355.

Antonia, R. A. and A. J. Chambers, 1980. Wind-wave induced disturbances in the marine surface layer. *J. Phys. Oceanogr.* **10**, 611–622.

Apel, J. R., 1994. An improved ocean surface wave vector spectrum. *J. Geophys. Res.* **16**, 16269–16291.

Askari, F., G. L. Geernaert, and W. C. Keller, 1993. Radar imaging of thermal fronts. *Int'l J. Remote Sensing* **14**, 275–294.

Atakturk, S. S. and K. B. Katsaros, 1999. Wind stress and surface waves observed on Lake Washington. *J. Phys. Oceanogr.* **29**, 633–650.

Atlas, D., 1994. Origin of storm SAR footprints on the sea. *Science* **266**, 1364–1367, Nov 25.

Atlas, D., T. Iguchi and H. F. Pierce, 1995. Storm induced wind patterns on the sea from spaceborne synthetic aperture radar. *Bull. Amer. Meteor. Soc.* **76**, 1585–1592.

Aubry, N., P. Holmes, J. L. Lumley, and E. Stone, 1988. The dynamics of coherent structures in the wall region of a turbulent boundary layer. *J. Fluid Mech.* **192**, 115–173.

Augstein, E., 1976. Boundary layer observations over the oceans. In: *Seminars on the treatment of the boundary layer in numerical weather prediction*, Reading, 6–10 September 1976, 138–182.

Autard, L. and G. Caullier, 1996. Effect of a wind gust on a fully developed wave field. In: *The air–sea interface*, M. A. Donelan, W. H. Hui and W. J. Plant, Eds. Univ. Toronto Press, 61–67.

Badgley, F. I., C. A. Paulson and M. Miyake, 1972. Profiles of wind speed, temperature and humidity over the Arabian Sea. *Meteorol. Monog.* **6**, 66.

Badulin, S. T., K. V. Pokazeyev and A. D. Rozenberg, 1983. A laboratory study of the transformation of regular gravity-capillary waves on inhomogenous flows. *Izv. Atm. & Ocean Phys.* **19**, 782–787.

Bailey, R. J., I. S. F. Jones and Y. Toba, 1991. The steepness and shape of wind waves. *J. Oceanogr. Soc. Japan* **47**, 249–264.

Baines, P. G., 1974. On the drag coefficient over shallow water. *Boundary-Layer Meteorol.* **6**, 299–303.

Banner, M. L., 1990a. Equilibrium spectra of wind waves. *J. Phys. Oceanogr.* **20**, 966–984.

Banner, M. L., 1990b. The influence of wave breaking on the surface pressure distribuion in wind-wave interactions. *J. Fluid Mech.* **211**, 463–495.

Banner, M. L. and W. K. Melville, 1976. On the separation of air flow over water waves. *J. Fluid Mech.* **77**, 825–842.

Banner, M. L. and W. L. Peirson, 1996. Tangential stress beneath wind-driven air–water interfaces. *University of NSW School of Mathematics Report* AMR96/19.

Banner, M. L. and W. L. Peirson, 1998. Tangential stress beneath wind-driven air–sea interfaces. *J. Fluid Mech.* **364**, 115–145.

Banner, M. L. and O. M. Phillips, 1974. On the incipient breaking of small scale waves. *J. Fluid Mech.* **65**, 647–656.

Banner, M. L. and I. R. Young, 1994. Modelling spectral dissipation in the evolution of wind waves. Part 1. Assessment of existing model performance. *J. Phys. Oceanogr.* **24**, 1550–1571.

Banner, M. L., I. S. F. Jones, and J. C. Trinder, 1989. Wavenumber spectra of short gravity waves. *J. Fluid Mech.* **198**, 321–344.

Banner, M. L., E. J. Chen, E. J. Walsh, J. Jensen, S. Lee & C. B. Fandry, 1996. Sea state influence on the wind stress – initial results from SOWEX. In: *The air–sea interface*, M. A. Donelan, W. H. Hui and W. J. Plant, Eds. Univ. Miami Press, 413–419.

Barenblatt, G. I. and G. S. Golitsyn, 1974. Local structure of mature dust storms. *J. Atmos. Sci.* **31**, 1917–1933.

Barger, W. R., W. D. Garrett, E. L. Mollo-Christensen, and K. W. Ruggles, 1970. Effects of an artificial sea slick upon the atmosphere and ocean. *J. Appl. Meteorol.* **9**, 396–400.

Battjes, J. A., T. J. Zitman and L. H. Holtheijsen, 1987. A re-analysis of the spectra observed in JONSWAP. *J. Phys. Oceanogr.* **17**, 1288–1295.

Belcher, S. E. and J. C. R. Hunt, 1993. Turbulent shear flow over slowly moving waves. *J. Fluid Mech.* **251**, 109–148.

Belcher, S. E., J. A. Harris, and R. L. Street, 1994. Linear dynamics of wind waves in coupled turbulent air-water flow. Part 1. Theory. *J. Fluid Mech.* **271**, 119–151.

Bender, L. C. and L. M. Leslie, 1994. Evaluation of a third generation ocean wave model for the Australian region. *BMRC Research Report* 43, Australian Bureau of Meteorology.

Berkowicz, R. and L. P. Prahm, 1979. Generalization of K-theory for turbulent diffusion. Part I: Spectral turbulent diffusivity concept. *J. Appl. Meteorol.* **18**, 266–272.

Blackadar, A. K., 1962. The vertical distribution of wind and turbulent exchange in a neutral atmosphere. *J. Geophys. Res.* **67**, 3095–3102.

Blanchard, D. C. and A. H. Woodcock, 1980. The production, concentration, and vertical distribution of the sea-salt aerosol. *Proc. N.Y. Acad. Sci.* **338**, 330–347.

Blennerhassett, P. J. and F. T. Smith, 1987. Short-scale waves on wind-driven water ('cat's paws'). *Proc. Roy. Soc. Lond.* **A 410**, 1–17.

Bliven, L. F., N. E. Huang and S. R. Long, 1986. Experimental study of the influence of wind on Benjamin-Feir sideband instability. *J. Fluid Mech.* **162**, 237–260

Bock, E. J. and N. M. Frew, 1993. Static and dynamic response of natural multicomponent oceanic surface films to compression and dilation: laboratory and field observations. *J. Geophys. Res.* **98**, 14599–14617.

Bock, E. J. and J. A. Mann, Jr, 1989. On ripple dynamics. II, A corrected dispersion relation for surface waves in the presence of surface elasticity. *J. Coll. Interface Sci.* **129**, 501–505.

Borisenkov, E. P., 1974. Some mechanisms of atmosphere-ocean interaction under stormy weather conditions. *Problems Arctic Antarct.* 43–44, 73–83.

Bortkovskii, R. S., 1987. *Air–sea exchange of heat and moisture during storms.* D. Reidel, Dordrecht, 194 pp.

Boussinesq, J., 1877. Essai sur la theorie des courantes. *Mem. pres. par. div. Savant a l'acad. Sci.* **23**, 46.

Bouws, E., 1988. *Guide to wave analysis and forecasting.* World Meteorological Organization, 23–25.

Boyle, P. S., K. L. Davidson, and D. E. Spiel, 1987. Characteristics of over-water surface stress during STREX. *Dyn. Atmos. Oceans* **10**, 343.

Branger, H., J. P. Giovanangeli, N. Reul and F. Remy, 1994. A laboratory study of the effect of an oblique swell on air–sea momentum fluxes and scatterometry, *Second Intl. Conf. on Meteorology and Oceanography of the Coastal Zone,* Amer. Met. Soc., Lisbon.

Bretherton, C., 1993. Marine boundary layer dynamics during ASTEX. In: *Workshop Proc. Parameterization of the cloud topped boundary layer.* 8–11 June 1993, ECMWF, Reading, 51–76.

Bretherton, F. P. and C. J. F. Garrett, 1968. Wavetrains in homogeneous moving media. *Proc. R. Soc. Lond.* **A 302**, 529–554.

Brocks, K., 1959. Ein neues Geraet fuer stoerungsfreie meteorologische Messungen auf dem Meer. *Arch. Meteor. Geophys. Bioklimat.* **A11**, 227–239.

Brocks, K. and L. Kruegermeyer, 1971. The hydrodynamic roughness of the sea surface, In: *Studies in physical oceanography – a tribute to Georg Wüst,* Vol. 1. A. L. Gordon, Ed. Gordon and Breach, New York, 75–92.

Brooke Benjamin, T., 1959. Shearing flow over a wavy boundary. *J. Fluid Mech.* **6**, 161–205.

Brown, G. L. and A. Roshko, 1974. On density effects and large structure in turbulent mixing layers. *J. Fluid Mech.* **64**, 775–816.

Brown, R. A., 1970. A secondary flow model for the planetary boundary layer. *J. Atmos. Sci.* **27**, 742–757.

Brown, R. A., 1974. *Analytic methods in planetary boundary layer modelling.* Adam Hilger Ltd., London, and Halstead Press, John Wiley and Sons, New York. 150 pp.

Brown, R. A., 1980. Longitudinal instabilities and secondary flow in the planetary boundary layer: A review. *Rev. of Geophysics and Space Physics* **18(3)**, 683–697.

Brown, R. A., 1983. On a scatterometer as an anemometer. *J. Geophys. Res.* **88**, 1663–1673.

Brown, R. A., 1986. Satellite scatterometer capabilities in air–sea interaction: Review of the status and the possibilities, *J. Geophys. Res., Oceans,* **91**, No. C2, pages 2221–2232, 1986.

Brown, R. A., 2000a. On using satellite scatterometer, SAR and other sensor data and serendipity, *Johns Hopkins APL Technical Digest.* Vol. 21, 21–26.

Brown, R. A., 2000b. On satellite scatterometer model functions, *J. Geophys. Res., Atmospheres,* **D105**, 29195–29205.

Brown, R. A. and R. Foster, 1994. On large-scale PBL modelling: surface wind and latent heat flux comparisons. *The Global Atmos.-Ocean System* **2**, 199–219.

Brown, R. A. and L. Zeng, 1994. Estimating central pressures of oceanic midlatitude cyclones. *J. Appl. Met.* **33**, 1088–1095.

Budyko, M. I. and M. I. Yudin, 1946. Conditions for thermal equilibrium in the atmosphere. *Dokl. Akad. Nauk SSSR* **53**, 611–614.

Bumke, K., 1995. Spatial scales of surface wind observations and analysed wind fields over the North Atlantic Ocean. *J. Mar. Systems* **6**, 67–75.

Bumke, K., U. Karger and L. Hasse, 1998. Evaporation over the Baltic Sea as an example of a semi enclosed sea. *Contr. Atmos. Phys.* **21**, 249–261.

Burt, W., H. Crew, N. Putchak and J. Dumon, 1974. Diurnal variations of winds over an upwelling region off Oregon. *Boundary-Layer Meteorol.* **6**, 35–45.

Busch, N.E., 1972. On the mechanics of atmospheric turbulence. *Workshop on Micrometeorology,* D. A. Haugen, Ed., American Meteorological Society, 1–65.

Businger, J. and W. Seguin, 1977. Transport across the air-sea interface. *Report of U.S. GATE Central Program Workshop,* NCAR, Boulder, USA.

Businger, J. A., J. C. Wyngaard, I. Izumi, and E. F. Bradley, 1971. Flux profile relationships in the atmospheric surface layer. *J. Atmos. Sci.* **28**, 181–189.

Bye, J. A. T., 1965. Wind-driven circulation in unstratified lakes. *Limnol. Oceanogr.* **10**, 451–458.

Bye, J. A. T., 1988. The coupling of wave drift and wind velocity profiles. *J. Mar. Res.* **46**, 457–472.

Bye, J. A. T., 1995. Inertial coupling of fluids with large density contrast. *Physics Letters A,* **202**, 222–224.

Bye, J. A. T. and J-O. Wolff, 1999. Atmosphere-ocean momentum exchange in general circulation models. *J. Phys. Oceanogr.* **29**, 671–692.

Byrne, H. M., 1982. The variation of the drag coefficient in the marine surface layer due to temporal and spatial variations in the wind and sea state. *Ph.D. Dissertation,* U. of Washington, Seattle.

Byzova, N. L., V. N. Ivanov and E. K. Garger, 1989. *Turbulence in the boundary layer of the atmosphere.* Gidrometeoizdat, Leningrad, 263 pp. C3, 1663–73.

Caldwell, D. R. and W. P. Elliott, 1971. Surface stresses produced by rainfall. *J. Phys. Oceanogr.* **1**, 145–148.

Caponi, E. A., M. Z. Caponi, P. G. Saffman and H. C. Yuen, 1992. A simple model for the effect of water shear on the generation of waves by wind. *Proc. R. Soc. Lond.* **A438**, 95–101.

Carl, M. D., T. C. Tarbell and H. A. Panofsky, 1973 Profiles of wind and temperature from towers over homogeneous terrain. *J. Atmos. Sci.* **30**, 788–794.

Carlson, D. J., J. L. Cantey, and J. J. Cullen, 1988. Description and results from a new surface microlayer sampling device. *Deep Sea Res.* **35**, 1205–1213.

Caudal, G., 1993. Self-consistency between wind stress, wave spectrum, and wind-induced wave growth for fully rough air-sea interface. *J. Geophys. Res.* **98**, 22 743–22 752.

Chalikov, D. V. and M. Y. Belevich, 1993. One-dimensional theory of the wave boundary layer. *Boundary-Layer Meteorol.* **63**, 65–96.

Chalikov, D. V. and V. K. Makin, 1991. Models of the wave boundary layer. *Boundary-Layer Meteorol.* **56**, 83–99.

Chambers, A. J. and R. A. Antonia, 1981. Wave-induced effect on the Reynolds shear stress and heat flux in the marine surface layer. *J. Phys. Oceanogr.* **11**, 116–121.

Chang, M. S., 1969. Mass transport in deep-water long-crested random gravity waves. *J. Geophys. Res.* **74**, 1515–1536.

Charnock, H., 1955. Wind stress on a water surface. *Quart. J. Roy. Meteorol. Soc.* **81**, 639–640.

Charnock, H. and R. T. Pollard, 1983. Results of the Royal Society Joint Air-Sea Interaction Project (JASIN). *Proc. Royal Soc. Discussion Mtg 1982.*

Charnock, H., J. R. D. Francis and P. A. Sheppard, 1956. An investigation of wind structure in the Trades: Anegarde 1953. *Phil. Trans. Roy. Soc.* **249**, 179–234.

Chou, S. H. and E. N. Yeh, 1987. Airborne measurements of surface layer turbulence over the ocean during cold air outbreaks. *J. Atmos. Sci.* **44**, 3721–3733.

Chu, P. C., 1991. Geophysics of deep convection and deep water formation in oceans. In: *Deep convection and deep water formation in the oceans*, P. C. Chu and J. C. Gascard, Eds. Proceedings of the International Monterey colloquium on deep convection and deep water formation in the oceans, Elsevier, 3–16.

Cini. R. and P. P. Lombardini, 1978. Damping effect of monolayers on surface wave motion in a liquid. *J. Coll. Interface Sci.* **65**, 387–389.

Cini, R., P. P. Lombardini, and H. Hühnerfuss, 1983. Remote sensing of marine slicks utilizing their influence on wave spectra. *Intern'l J. Remote Sens.* **4**, 101–110.

Collineau, S. and Y. Brunet, 1993. Detection of turbulent coherent motions in a forest canopy. Part II: Time-scales and conditional averages. *Boundary-Layer Meteorol.* **66**, 49–73.

Compton, D. A. and J. K Eaton, 1995. Near-wall measurements of a three-dimensional turbulent boundary layer, *Report MD*-72, Thermosciences Division, Department of Mechanical Engineering, Stanford University, 200 pp.

Corino, E. R. and R. S. Brodkey, 1969. A visual investigation of the wall region in turbulent flows. *J. Fluid Mech.* **37**, 1–30.

Corrsin, S., 1957. Some current problems in turbulent shear flows. In: *Symposium on Naval Hydrodynamics.* F. S. Sherman, Ed. National Academy of Sciences, Publication 515, Washington, D.C.

Cox, C., 1958. Measurement of slopes of high-frequency wind waves. *J. Marine Res.* **16**, 199–225.

Cox, C. S. and W. H. Munk, 1954. Statistics of the sea surface derived from sun glitter. *J. Mar. Res.* **13**, 198–227.

Craig, P. D. and M. L. Banner, 1994. Modeling wave-enhanced turbulence in the ocean surface layer. *J. Phys. Oceanogr.* **24**, 2546–2559.

Csanady, G. T., 1974. The "roughness" of the sea surface in light winds. *J. Geophys. Res.* **79**, 2747–2751.

Csanady, G. T., 1984. The free surface turbulent shear layer. *J. Phys. Oceanogr.* **14**, 402–411.

Csanady, G. T., 1985. Air–sea momentum transfer by means of short-crested wavelets. *J. Phys. Oceanogr.* **15**, 1486–1501.

Csanady, G. T., 1990. Momentum flux in breaking wavelets. *J. Geophys. Res.* **95**, 13289–13299.

Csanady, G. T., 1991. Wavelets and air–sea transfer. In: *Air–water mass transfer*, S. C. Wilhelms and J. S. Gulliver, Eds. ASCE, New York, pp. 563–581.

Davidson, K. L., 1974. Observational results on the influence of stability and wind-wave coupling on the momentum transfer and turbulent fluctuations over ocean waves. *Boundary-Layer Meteorol.* **6**, 305–331.

Davidson, K. L. and A. K. Frank, 1973. Wave-related fluctuations in the airflow above natural waves. *J. Phys. Oceanogr.* **3**, 102–119.

Davies, J. T. and R. W. Vose, 1965. On the damping of capillary waves by surface film. *Proc. Roy. Soc. of London* **A 286**, 18–234.

Davis, R. E., 1969. On the high Reynolds number flow over a wavy boundary. *J. Fluid Mech.* **36**, 337–346.

Davis, R. E., 1970. On the turbulent flow over a wavy boundary. *J. Fluid Mech.* **42**, 721–731.

Davis, R. E., 1972. On prediction of the turbulent flow over a wavy boundary. *J. Fluid Mech.* **52**, 287–306.

Davis, R. E., 1974. Perturbed turbulent flow, eddy viscosity and generation of stresses. *J. Fluid Mech.* **63**, 673–693.

Deacon, E. L., 1962. Aerodynamic roughness of the sea. *J. Geophys. Res.* **67**, 3167–3172.

Deacon, E. L., 1988. The streamwise Kolmogoroff constant. *Boundary-Layer Meteorol.* **42**, 9–17.

Deacon, E. L. and E. K. Webb, 1962. Small scale interactions. In: *The sea*, Vol. 1, M. N. Hill, Ed. Interscience, New York, 43–87.

Deardorff, J. W. 1972. Theoretical expression for the counter-gradient vertical heat flux. *J. Geophys. Res.* **77**, 5900–5904.

Deardorff, J. W., 1973. The use of subgrid transport equations in a three-dimensional model of atmospheric turbulence. *J. Fluids Engrg.* **95**, 429–438.

Deardorff, J. W., 1966. The counter-gradient heat flux in the lower atmosphere and in the laboratory. *J. Atmos. Sci.* **23**, 503–506.

DeCosmo, J., K. B. Katsaros, S. D. Smith, R. J. Anderson, W. A. Oost, K. Bumke, and H. Chadwick, 1996. Air–sea exchange of water vapor and sensible heat: The Humidity Exchange over the Sea (HEXOS) results. *J. Geophys. Res.* **101**, 12001–12016.

de Leeuw, G., 1986. Vertical profiles of giant particles close above the sea surface. *Tellus* **38B**, 51–61.

Denman, K. L. and M. Miyake, 1973. Behavior of the mean wind, the drag coefficient, and the wave field in the open ocean. *J. Geophys. Res.* **78**, 1917–1931.

Dickinson, S. and R. A. Brown, 1996. A study of near-surface winds in marine cyclones using multiple satellite sensors. *J. Appl. Met.* **35**, 769–781.

Ding, L. and D. M. Farmer, 1994. On the dipole acoustic source level of breaking waves. *J. Acous. Soc. of Amer.* **96**, 3036–3044.

Dittmer, K., 1977. The hydrodynamic roughness of the sea surface at low wind speeds. *"Meteor"-Forschungsergebnisse* **B12**, 10–15.

Dobson, F. W., 1971. The damping of a group of sea waves. *Boundary-Layer Meteorol.* **1**, 399–410.

Dobson, F., L. Hasse, and R. Davis (Eds), 1980. *Air–sea interaction – Instruments and methods.* Plenum Press, New York, 801 pp.

Dobson, F. W., W. Perrie and B. Toulany, 1989. On the deep water fetch laws for wind generated surface gravity waves. *Atmosphere-Ocean* **27**, 210–236.

Dobson, F. W., S. D. Smith and R. J. Anderson, 1994. Measuring the relationship between wind stress and sea state in the open ocean in the presence of swell. *Atmos.-Ocean* **32**, 237–256.

Donelan, M. A., 1979. On the fraction of wind momentum retained by waves. In: *Marine forecasting*. J. C. J. Nihoul, Ed. Elsevier, 141–159.

Donelan, M. A., 1982. The dependence of the aerodynamic drag coefficient on wave parameters. *Proc. First Int'l Conf. on Meteorology and Air-Sea Interaction of the Coastal Zone*, Amer. Meteorol. Soc., Boston, 381–387.

Donelan, M. A., 1987. The effect of swell on the growth of wind waves. *Johns Hopkins APL Tech. Digest* **8**, 18–23.

Donelan, M. A., 1990. Air-sea interaction. In: B. le Méhauté and D. M. Hanes (Eds), *The Sea: Ocean Engineering Science* **9B**, 239–292. Wiley-Interscience, New York.

Donelan, M. A., 1997. Experiments on the generation and attenuation of water waves by wind. In: *Proc. of the IMA Conference on Wind-Over-Wave Couplings*, University of Salford, 8–10 April, 1997. Oxford University Press.

Donelan, M. A., 1999. Wind-induced growth and attenuation of laboratory waves. In: *Wind-over-wave couplings, perspectives and prospects*, S. G. Sajjadi, N. H. Thomas and J. C. R. Hunt, Eds. Clarendon Press, Oxford.

Donelan, M. A. and W. J. Pierson, 1987. Radar scattering and equilibrium ranges in wind-generated waves with applications in scatterometry. *J. Geophys. Res.* **92**, 4971–5029.

Donelan, M. A., J. Hamilton and W. H. Hui, 1985. Directional spectra of wind-generated waves. *Phil. Trans. Roy. Soc. London, Ser. A*, **315**, 509–562.

Donelan, M. A., F. W. Dobson, S. D. Smith and R. J. Anderson, 1993. On the dependence of sea surface roughness on wave development. *J. Phys. Oceanogr.* **23**, 2143–2149.

Donelan, M. A., F. W. Dobson, S. D. Smith and R. J. Anderson, 1995. Reply. *J. Phys. Oceanogr.* **25**, 1908–1909.

Donelan, M. A., W. M. Drennan and K. B. Katsaros, 1997. The air–sea momentum flux in conditions of wind sea and swell. *J. Phys. Oceanogr.* **27**, 2087–2099.

Dorman, C. E. and E. Mollo Christensen, 1973. Observation of the structure on moving gust patterns over a water surface ("Cat's Paws"). *J. Phys. Oceanogr.* **3**, 120–132.

Drake, T. G., R. L. Shreve, W. E. Dietrich, P. J. Whiting, and L. B. Leopold, 1988. Bedload transport of fine gravel observed by motion-picture photography. *J. Fluid Mech.* **192**, 193–217.

Duncan, J. H., 1981. An experimental investigation of breaking waves produced by a towed hydrofoil. *Proc. Roy. Soc. London* **A 377**, 331–348.

Duncan, J. H., 1983. The breaking and non-breaking resistance of a two-dimensional aerofoil. *J. Fluid Mech.* **126**, 507–520.

Dunckel, M., L. Hasse, L. Krugermeyer, D. Schriever and J. Wucknitz, 1974. Turbulent fluxes of momentum, heat, and moisture in the atmospheric surface layer at sea during ATEX: Atlantic Trade Winds Experiment. *Boundary-Layer Meteorol.* **6**, 81–106.

Dupuis, H., A. Weill, K. Katsaros and P. K. Taylor, 1995. Turbulent heat fluxes by the profile and dissipation methods: analysis of the atmospheric surface flux from shipboard measurements during the SOFIA/ASTEX and SEMAPHORE experiments. *Ann. Geophysicae* **13**, 1065–1074.

Dyer, A. J., 1967. The turbulent transport of heat and water vapour in an unstable atmosphere. *Quart J. Roy. Meteorol Soc.* **96**, 501–508.

Dyer, J. A., 1974. A review of flux-profile relationships. *Boundary-Layer Meteorol.* **7**, 363–372.

Dyer, A. J., 1981. Flow distortion by supporting structures. *Boundary-Layer Meteorol.* **20**, 243–251.

Ebuchi, N., H. Kawamura and Y. Toba, 1993. Bursting phenomena in the turbulent boundary layer beneath the laboratory wind-wave surface. In: *Natural physical sources of underwater sound*, B. R. Kerman, Ed. Kluwer Acad. Publ., pp. 263–276.

Ebuchi, N., Y. Toba, and H. Kawamura, 1992. Statistical study on the local equilibrium between wind and wind waves by using data from ocean data buoy stations. *J. Phys. Oceanogr.* **48**, 77–92.

Edson, J. B., C. W. Fairall, P. G. Mestayer, and S. E. Larsen, 1991. A study of the inertial-dissipation method for computing air–sea fluxes. *J. Geophys. Res.* **96**(C6), 10689–10711.

Edson, J. B., A. A. Hinton, K. E. Prada, J. E. Hare, and C. W. Fairall, 1998. Direct co-variance flux estimates from mobile platforms at sea. *Atmos. Ocean. Tech.* **15**, 547–562.

Ekman, V. W., 1905. On the influence of the earth's rotation on ocean-currents. *Arkiv. Math. Astron. O. Fysik.*, 11.

Ellison, T. H., 1956. Atmospheric turbulence. In: *Surveys in mechanics*, Batchelor and Davies, Eds. Cambridge University Press, 475 pp.

Enriquez, A. E. and C. A. Friehe, 1997. Bulk parameterization of momentum, heat and moisture fluxes over a coastal upwelling area. *J. Geophys. Res.* **102**, 5781–5798.

Etling, D. and R. A. Brown, 1993. Roll vortices in the planetary boundary layer, a review. *Boundary-Layer Meteorol.* **65**, 215–248.

Fairall, C. W. and J. B. Edson, 1994. Recent measurements of the dimensionless turbulent kinetic energy dissipation function over the ocean. *2nd Int. Conf. on Air–Sea Interaction and Oceanography of the Coastal Zone*, Lisbon, Portugal, 22–27 Sept. 1994. American Met. Soc. Boston, 224–225.

Fairall, C. W. and S. E. Larsen, 1986. Inertial-dissipation methods and turbulent fluxes at the air-ocean interface. *Boundary-Layer Meteorol.* **34**, 287–301.

Fairall, C. W., E. F. Bradley, D. P. Rogers, J. B. Edson, and G. S. Young, 1996. Bulk parameterization of air–sea fluxes for Tropical Ocean Global Atmosphere Coupled Ocean-Atmosphere Response Experiment. *J. Geophys. Res.* **101**, 3747–3764.

Fairall, C. W., J. B. Edson, S. E. Larsen, and P. G. Mestayer, 1990. Inertial-dissipation air-sea flux measurements: a prototype system using realtime spectral computations. *J. Atmos. & Oceanic Tech.* **7**, 425–453.

Fedorov, K. N., 1988. Layer thicknesses and effective diffusivities in the diffusive thermo-cline convection in the ocean. In: *Small-scale turbulence and mixing in the ocean*, J. C. J. Nihoul and B. M. Jamart, Eds. Elsevier, New York, 471–479.

Fiedler, B. H., 1984. An integral closure model for the vertical turbulent flux of a scalar in a mixed layer. *J. Atmos. Sci.* **41**, 674–680.

Fiscella, B., P. P. Lombardini, P. Trivero, P. Pavese, and R. Cini, 1985. Measurements of the damping effect of a spreading film on wind-excited sea ripples using a two-frequency radar. *Nuovo Cimento* **8**, 175–183.

Fleagle, R. W., J. W. Deardorff and F. I. Badgley, 1958. Vertical distribution of wind speed, temperature, and humidity above a water surface. *J. Marine Res.* **17**, 141–157.

Forristall, G. Z., 1981. Measurements of a saturated range in ocean wave spectra. *J. Geophys. Res.* **86**, 8075–8084.

Foster, R. C., 1996. An analytic model for planetary boundary roll vortices. *Ph.D. Thesis*, University of Washington.

Foster, R. C. and R. A. Brown, 1994. On large-scale PBL modelling: Surface wind and latent heat flux comparisons. *Global Atmosphere–Ocean System* **2**, 199–219.

Francis, J. R. D., 1951. The aerodynamic drag of a free water surface. *Proc. Roy. Soc. London* **A206**, 387–406.

Frenzen, P. and C. A. Vogel, 1992. The turbulent kinetic energy budget in the atmospheric surface layer: a review and an experimental reexamination in the field. *Boundary-Layer Meteorol.* **60**, 49–76.

Frenzen, P. and C. A. Vogel, 1994. On the sensitivity of the phim function to k: a corrected illustration for the turbulent kinetic energy budget in the ASL. *Boundary-Layer Meteorol.* **68**, 439–442.

Frew, N. M., J. C. Goldman, M. R. Dennett, and A. S. Johnson, 1990. Impact of phyto-plankton-generated surfactants on air–sea gas exchange. *J. Geophys. Res.* **95**, 3337–3352.

Friehe, C. A., T. Hristov, S. D. Miller, and J. B. Edson, 2000. Surface-layer wind and stress directions, *J. Geophys. Res. – Oceans* (in press).

Friehe, C. A., W. J. Shaw, D. P. Rogers, K. L. Davidson, W. G. Large, S. A. Stage, G. H. Crescenti, S. J. S. Khalsa, G. K. Greenhut, and F. Li, 1991. Air–sea fluxes and surface layer turbulence around a sea surface temperature front. *J. Geophys. Res.* **96**, 8593–8609.

Fu, L. L. and B. Holt, 1982. SEASAT views oceans and sea ice with synthetic aperture radar, *JPL Publ. 81–120*, Jet Propulsion Laboratory, Pasadena, CA.

Galushko, V. V., V. N. Ivanov, I. V. Nekrasov, V. D. Pudov, A. V. Rostkov and A. S. Shushkov, 1975. Turbulent characteristic measurements of the marine boundary layer during GATE, *ICSU/WMO GATE Rep. No. 14*, **6**, 237–262.

Galushko, V. V., V. N. Ivanov, T. F. Masagutov, V. V. Nekrasov, and A. V. Rostkov, 1977. Experimental investigations of the atmospheric boundary layer structure in the tropical latitudes. *Proc. Int'l. Sci. Conf. on the Energetics of the Tropical Atmosphere*, Tashkent, USSR.

Gargett, A. E., 1989. Ocean turbulence. *Ann. Rev. Fluid Mech.* **21**, 419–451.

Garratt, J. R., 1977. Review of drag coefficients over oceans and continents. *Mon. Weather Rev.* **105**, 915–929.

Garratt, J. R. and B. B. Hicks, 1973. Momentum, heat and water vapour transfer to and from natural and artificial surfaces. *Quart. J. Roy. Meteorol. Soc.* **99**, 680–689.

Geernaert, G. L., 1987. On the importance of the drag coefficient in air–sea interaction. *Dyn. Atmos. Oceans* **11**, 19–38.

Geernaert, G. L., 1988. Measurements of the angle between the wind vector and wind stress vector in the surface layer over the North Sea. *J. Geophys. Res.* **93**, 8215–8220.

Geernaert, G. L., 1990. Bulk parameterizations for the wind stress and heat fluxes. In: *Surface waves and fluxes*, Vol. I, G. L. Geernaert and W. J. Plant, Eds. Kluwer, Dordrecht, pp. 91–172.

Geernaert, G. L. and K. B. Katsaros, 1986. Incorporation of stratification effects in estimating the roughness length over the ocean. *J. Phys. Oceanogr.* **16**, 1580–1584.

Geernaert, G. L. and W. J. Plant, 1990. *Surface waves and fluxes: theory and remote sensing.* Vol. 1 and 2, Kluwer, Dordrecht, 807 pp.

Geernaert, G. L., K. B. Katsaros, and K. Richter, 1986. Variation of the drag coefficient and its dependence on sea state. *J. Geophys. Res.* **91**, 7667–7679.

Geernaert, G. L., S. E. Larsen, and F. Hansen, 1987. Measurements of the wind stress, heat flux and turbulence intensity during storm conditions over the North Sea. *J. Geophys. Res.* **92**, 13127–13139.

Geernaert, G. L., F. Hansen, M. Courtney, and T. Herbers, 1993. Directional attributes of the ocean surface wind stress vector. *J. Geophys. Res.* **98**, 16571–16583.

Gerling, T. W., 1986. Structure of the surface wind field from the Seasat SAR. *J. Geophys. Res.*, **91**, 2308–2320.

Gerz, T. and Schumann, U., 1996. A possible explanation of countergradient fluxes in homogeneous turbulence. *Theoret. Comput. Fluid Dynamics*, 169–181.

Giovanangeli, J. P., R. Remy, and C. Kharif, 1994. A Laboratory Study of the Effect of an Oblique Swell on Stress, *Second Intl. Conf. on Meteorology and Oceanography of the Coastal Zone*, Amer. Met. Soc., Lisbon, 1994.

Godfrey, J. S. and Beljaars, A. C. M., 1991. On the turbulent fluxes of buoyancy, heat and moisture at the air–sea interface at low wind speeds. *J. Geophys. Res.* **96**, 22043–22048.

Gong, W. M., P. A. Taylor, and A. Dornbrack, 1996. Turbulent boundary-layer flow over fixed aerodynamically rough two-dimensional sinusoidal waves. *J. Fluid Mech.* **312**, 1–37.

Grelle, A. and A. Lindroth, 1994. Flow distortion by a Solent sonic anemometer: wind tunnel calibration and its assessment for flux measurements over forest and field. *J. Atmos. Ocean. Tech.* **11**, 1529–1542.

Gulev, S. and T. M. Tonkacheev, 1996. On the parameterization of sea–air interaction over SST fronts in the North Atlantic. In: *Air–sea interface*, M. A. Donelan, W. H. Hui and W. J. Plant, Eds. The University of Toronto Press, Toronto, Canada, 535–542.

Guymer, T. H. and P. K. Taylor, 1983. The contribution of SEASAT microwave data to the Joint Air–Sea Interaction Experiment. Large-Scale Oceanographic Experiments in the WCRP. *ICSU/WMO World Climate Research Programme Pubs.* Series 1, WMO, Geneva, 245–260.

Guymer, T. H., J. A. Businger, K. B. Katsaros, W. J. Shaw, P. K. Taylor, W. G. Large, and R. E. Payne, 1983. Transfer processes at the air-sea interface. *Phil. Trans. Roy. Soc. London* **308**, 253–273.

Hamada, T., 1963. An experimental study of development of wind waves. *Report Port and Harbour Tech. Res. Inst., No.* 2, 1–41.

Hamba, F., 1995. An analysis of nonlocal scalar transport in the convective boundary layer using the Green's function. *J. Atmos. Sci.* **52**, 1084–1095.

Hanawa, K. and Y. Toba, 1987. Critical examination of estimation methods of long-term mean air–sea heat and momentum transfers. *Ocean-Air Interactions*, **1**, 79–93.

Hansen, R. S. and J. A. Mann, Jr, 1964. Propagation characteristics of capillary ripples. I, The theory of velocity dispersion and amplitude attenuation of plane capillary waves on viscoelastic films. *J. Appl. Phys.* **35**, 152–161.

Hara, T., E. J. Bock, and D. Lyzenga, 1994. In situ measurements of capillary-gravity wave spectra using a scanning laser slope gauge and microwave radars. *J. Geophys. Res.* **99**, 12593–12602.

Harkins, W. D., 1952. *Physical chemistry of surface films.* Reinhold Publishing Corporation, New York, p. 107.

Harris, D. L., 1966. The wave-driven wind. *J. Atmos. Sci.* **23**, 688–693.

Hasse, L., 1974. On the surface to geostrophic wind relationship at sea and the stability dependence of the resistance law. *Contributions Atmospheric Physics* **47**, 45–55.

Hasse, L., 1984. Cumuluskonvektion und Konvektionsrollen. *Promet, Meteorologische Fortbildung* **13(2/3)**, 38–41.

Hasse, L., 1993. Observations of air–sea fluxes. In: *Energy and water cycles in the climate system.* Raschke and Jacob, Eds. NATO ASI Series I, Springer, Berlin, Heidelberg, **5**, 263–293.

Hasse, L. and S. D. Smith, 1997. Local sea surface wind, wind stress, and sensible and latent heat fluxes. *Journal Climate* **10**, 2711–2724.

Hasse, L., J. Wucknitz, G. Kruspe, V. M. Ivanov, A. A. Shuskv et al., 1975. Preliminary report on determination of fluxes by direct and profile methods during Intercomparison IIa. *GATE Report* 14, 267–277, WMO, Geneva.

Hasse, L., M. Grünewald, J. Wucknitz, M. Dunckel, and D. Schriever, 1978a. Profile derived turbulent fluxes in the surface layer under disturbed and undisturbed conditions during GATE. *METEOR96 Forschungsergebn.* **13(B)**, 24–40.

Hasse, L., M. Grünewald, and D. E. Hasselmann, 1978b. Field observations of air flow above the waves. In: *Turbulent fluxes through the sea surface, wave dynamics and prediction*, Favre, A. and K. Hasselmann, Eds. Plenum Press, 483– 494.

Hasselmann, D. and J. Bösenberg, 1991. Field measurements of wave induced pressure over wind-sea and swell. *J. Fluid Mech.* **230**, 391–428.

Hasselmann, D. E., M. Dunckel, and J. A. Ewing, 1980. Directional wave spectra observed during JONSWAP 1973. *J. Phys. Oceanogr.* **10**, 1264–1280.

Hasselmann, K., 1963a. On the nonlinear energy transfer in a gravity wave spectrum. Part 2. *J. Fluid Mech.* **15**, 273–281.

Hasselmann, K., 1963b. On the nonlinear energy transfer in a gravity wave spectrum. Part 3. *J. Fluid Mech.* **15**, 385–398

Hasselmann, K., 1970. Wave-driven inertial oscillations. *Geophys. Fl. Dyn.* **1**, 463–502.

Hasselmann, K., 1974. On the spectral dissipation of ocean waves due to whitecapping. *Boundary-Layer Meteorol.* **6**, 107–127.

Hasselmann, K., T. P. Barnett, E. Bouws, H. Carlson, D. E. Cartwright, J. Enke, J. A. Ewing, H. Gienapp, D. E. Hasselmann, P. Kruseman, A. Meerburg, P. Muller, D. J. Olbers, K.

Richter, W. Sell, and H. Walden, 1973. Measurements of wind-wave growth and swell decay during the Joint North Sea Wave Project (JONSWAP) *Deut. Hydrogr. Z.*, Suppl. A **8**, 95 pp.

Hasselmann, S. and K. Hasselmann, 1985. Computations and parameterizations of the non-linear energy transfer in a gravity wave spectrum. Part I: A new method for efficient computations of the exact nonlinear transfer integral. *J. Phys. Oceanogr.* **15**, 1369–1377.

Hatori, M., M. Tokuda and Y. Toba, 1981. Experimental study of strong interactions between regular waves and wind waves. *J. Oceanogr. Soc. Japan* **37**, 111–119.

Hawkins, H. F. and D. T. Rubsam, 1968. Hurricane Hilda 1964, II. Structure and budgets of the hurricane on October 1, 1964. *Mon. Weather Rev.* **96**, 617–636.

Hay, J. S., 1955. Some observations of air flow over the sea. *Quart. J. Roy. Met. Soc.* **81**, 307–319.

Heathershaw, A. D., 1974. Bursting phenomena in the sea. *Nature* **248**, 394–395.

Hein, P. F. and R. A. Brown, 1988. Observations of longitudinal roll vortices during Arctic cold air outbreaks over open water. *Boundary-Layer Meteorol.* **45**, 177–199.

Herr, F. L. and J. Williams (Eds), 1986. Role of surfactant films on the interfacial properties of the sea-surface, *Report C-11–86*, Office of Naval Research, London, 283p.

Hicks, B. B., 1975. A procedure for the formulation of bulk transfer coefficients over water. *Boundary-Layer Meteorol.* **8**, 515–524.

Hicks, B. B. and A. J. Dyer, 1970. Measurements of eddy fluxes over the sea from an offshore oil rig. *Quart. J. Roy. Met. Soc.* **96**, 523–528.

Hicks, B. B. and A. J. Dyer, 1972. The spectral density technique for the determination of eddy fluxes. *Q. J. Roy. Met. Soc.* **98**, 838–844.

Hicks, B. B., R. L. Drinkrow, and G. Grauze, 1974. Drag and bulk transfer coefficients associated with shallow water surfaces. *Boundary-Layer Meteorol.*, **6**, 287–297.

Hinze, J. O., 1972. Turbulent fluid and particle interaction. *Progr. Heat and Mass Transfer* **6**, 433–452.

Holland, J. Z., 1981. Atmospheric Boundary Layer. In *IFYGL – The International Field Year for the Great Lakes*, E. J. Aubert and T. L. Richards, Eds. NOAA, Ann Arbor, MI. 410 pp.

Holt, T. and S. SethuRaman, 1988. A review and comparative evaluation of multilevel boundary layer parameterizations for first order and turbulent kinetic energy closure schemes. *Report No. NC 27695–8208*, Depart. of Marine, Earth and Atmos. Sci., North Carolina State Univ., 54 pp.

Holthuijsen, L. J., 1983. Observations of the directional distribution of ocean-wave energy in fetch-limited conditions. *J. Phys. Oceanogr.* **13**, 191–207.

Holthuijsen, L. H., A. J. Kuik and E. Mosselman, 1987. The response of wave directions to changing wind directions. *J. Phys. Oceanogr.* **17**, 845–853.

Hsu, C. T., H. Y. Wu, E. N. Hsu, and R. L. Street, 1982. Momentum and energy transfer in wind generated waves. *J. Phys. Oceanogr.* **12**, 929–951.

Huang, N. E., 1979. On surface drift current in the ocean. *J. Fluid Mech.* **91**, 191–208.

Huang, N. E., 1986. An estimate of the influence of breaking waves on the dynamics of the upper ocean. In *Wave dynamics and radar probing of the ocean surface*. O. M. Phillips and K. Hasselmann, Eds. Plenum Press. NY, 295–314.

Huang, N. E., S. R. Long, C. C. Tung, Y. Yuan, and L. F. Bliven, 1981. A unified two-parameter wave spectral model for a general sea state. *J. Fluid Mech.* **112**, 203–224.

Huang, N. E., S. R. Long and L. F. Bliven, 1981a. On the importance of significant slope in empirical wind-wave studies. *J. Phys. Oceanogr.* **11**, 569–573.

Huang, N. E., L. F. Bliven, S. R. Long, and P. S. De Deonibus, 1986. A study of the relationship amongst wind speed, sea state, and drag coefficient for a developing wave field. *J. Geophys. Res.* **91**, 7733–7742.

Huang, N. E., C.-C. Tung, and S. R. Long, 1990a. The probability structure of the ocean surface. *The Sea*, **9**, 335–366.

Huang, N. E., C.-C. Tung, and S. R. Long, 1990b. Wave spectra. In: *The sea*, Vol. 9, Part A, 197–237.

Huang, N. E., S. R. Long, and Z. Shen, 1996. Frequency downshift in nonlinear water wave evolution. *Advances in Appl. Mech.* **32**, 59–117.

Huang, N. E., Z. Shen, S. R. Long, M. C. Wu, E. H. Smith, Q. Zheng, C. C. Tung and H. H. Liu, 1998. The empirical mode decomposition method and the Hilbert spectrum for non-stationary time series analysis. *Proc. Roy. Soc. London*, **A454**, 903–995.

Hughes, B. A., H. L. Grant and R. W. Chappel, 1977. A fast response surface-wave slope meter and measured wind-wave moments. *Deep-Sea Res.* **24**, 1211–1233.

Hühnerfuss, H., W. Alpers, W. L. Jones, P. A. Lange and K. Lichter, 1981. The damping of ocean surface waves by monomolecular film measured by wave staffs and microwave radars. *J. Geophys. Res.* **86**, 429–438.

Hühnerfuss, H., W. Alpers, W. D. Garrett, P. A. Lange and S. Stolte, 1983. Attenuation of capillary and gravity waves at sea by monomolecular organic surface films. *J. Geophys. Res.* **88**(C14), 9809–9816.

Hühnerfuss, H., P. A. Lange and W. Walter, 1985a. Relaxation effects in monolayers and their contribution to water wave damping. 1, Wave-induced phase shifts. *J. Coll. Interface Sci.* **108**, 430–441.

Hühnerfuss, H., P. A. Lange and W. Walter, 1985b. Relaxation effects in monolayers and their contribution to water wave damping, 2. The Marangoni phenomena and gravity wave attenuation. *J. Coll. Interface Sci.* **108**, 442–450.

Hühnerfuss, H., W. Walter, P. Lange and W. Alpers, 1987. Attenuation of wind waves by monomolecular sea slicks by the Marangoni effect. *J. Geophys. Res.* **92**, 3961–3963.

Hwang, P. A., D. B. Trizna and J. Wu, 1993. Spatial measurements of short wind waves using a scanning laser slope sensor. *Dyn. Atm. and Oceans* **20**, 1–23.

Hwang, P. A., S. Atakturk, M. A. Sletten and D. B. Trizna, 1996. A study of the wavenumber spectra of short water waves in the ocean. *J. Phys. Oceanogr.* **26**, 1266–1285.

Isemer, H. J. and L. Hasse, 1985. *The Bunker climate atlas of the North Atlantic Ocean. 1. Observations*. Springer-Verlag, Heidelberg, 218 pp.

Jackson, F. J., W. T. Walton and P. L. Baker, 1985. Aircraft and satellite measurement of ocean wave directional spectra using scanning beam microwave radars. *J. Geophys. Res.* **90**, 987–1004.

Jähne, B., 1985. On the transfer processes at a free air–water interface. *Habilitation Thesis*, University of Heidelberg.

Jähne, B. and K. S. Riemer, 1990. Two dimensional wave number spectra of small-scale water surface waves. *J. Geophys. Res.* **95**, 11531–11546.

Jähne, B. and S. Waas, 1989. Optical measuring technique for small scale water surface waves. Advanced Optical Instrumentation for Remote Sensing of the Earth's Surface from Space, *SPIE Proceeding 1129*, International Congress on Optical Science and Engineering, Paris, 24–28 April 1989. pp. 147–152.

Jähne, B., S. Waas and J. Klinke 1992. A critical theoretical review of optical techniques for short ocean wave measurements. *Proc. Optics of the air-sea interface: Theory and measurements*, L. Estep, Ed. *SPIE Proc. 1749*. pp. 204–215.

Jähne, B., J. Klinke and S. Waas, 1994. Imaging of short ocean wind waves: a critical theoretical review. *J. Optical Soc. Amer. A* **11**, 2197–2209.

Janssen, P. A. E. M., 1989. Wave-induced stress and the drag of air flow over sea waves. *J. Phys. Oceanogr.* **19**, 745–754.

Janssen, P. A. E. M., 1999. Note on the effect of ocean waves on the kinetic energy balance and consequences for the inertial dissipation technique. *J. Phys. Oceanogr.* **29**, 530–534.

Janssen, P. A. E. M. and P. Viterbo, 1996. Ocean waves and the atmospheric climate. *J. Climate* **9**, 1209–1287.

Jeffreys, H., 1924. On the formation of waves by wind. *Proc. Roy. Soc.* **A107**, 189–206.

Jeffreys, H., 1925. On the formation of waves by wind. II. *Proc. Roy. Soc.* **A110**, 341–347.

Jenkins, A. D., 1986. A theory for steady and variable wind and wave induced currents. *J. Phys. Oceanogr.* **16**, 1370–1377.

Jenkins, A. D., 1987. Wind and wave induced currents in a rotating sea with depth-varying eddy viscosity. *J. Phys. Oceanogr.* **17**, 938–951.

Jenkins, A. D., 1989. The use of a wave prediction model for driving a near-surface current model. *Deutsches Hydrographisches Zeitschrift* **42**, 133–149.

Jenkins, A. D., 1992. A quasi-linear eddy-viscosity model for the flux of energy and momentum to wind waves, using conservation-law equations in a curvilinear coordinate system. *J. Phys. Oceanogr.* **22**, 843–858.

Jenkins, A. D., 1994. A stationary potential-flow approximation for a breaking-wave crest. *J. Fluid. Mech.* **280**, 335–347.

Joffre, S., 1982. Assessment of the separate effects of baroclinity and thermal stability in the atmospheric boundary layer over the sea. *Tellus* **34**, 567–578.

Joffre, S., 1984. Effects of local accelerations and baroclinity on the mean structure of the atmospheric boundary layer over the sea. *Boundary-Layer Meteorol.* **32**, 237–255.

Johnson, D. W., 1993. Parametrisation of the cloud topped boundary layer: aircraft measurements. In: *Workshop Proc. Parameterization of the cloud topped boundary layer*. 8–11 June 1993, ECMWF, Reading, 77–117.

Johnson, H. K., J. Hfjstrup, H. J. Vested and S. E. Larsen, 1998. On the dependence of sea surface roughness on wind waves. *J. Phys. Oceanogr.* **28**, 1702–1716.

Jonas, P. R., 1993. Radiation and cloud physical aspects of boundary-layerclouds. In: *Workshop Proc. Parameterization of the cloud topped boundary layer*. 8–11 June 1993, ECMWF, Reading, 151–168.

Jones, H. and J. Marshall, 1993. Convection with rotation in a neutral ocean: A study of open-ocean deep convection. *J. Phys. Oceanogr.* **23**, 1009–1039.

Jones, I. S. F. and H. Kawamura, 1990. Stress fluctuations over ocean waves. Unpublished manuscript. Ocean Technology Group, University of Sydney.

Jones, I. S. F. and B. C. Kenny, 1977. The scaling of velocity fluctuations in the ocean mixed layer. *J. Geophys. Res.* **82**, 1392–1396.

Jones, I. S. F. and D. Negus, 1996. The dependence of the drag coefficient on the marine planetary boundary layer. *Ocean Sciences Institute Report No. 69*, University of Sydney.

Jones, I. S. F. and Y. Toba, 1985. Wave data from three Bass Strait storms. *Marine Studies Centre Report 2/85*, University of Sydney, Sydney.

Jones, I. S. F. and Y. Toba, 1995. Comments on "The dependence of sea surface roughness on wave development". *J. Phys. Oceanogr.* **25**, 1905–1907.

Joseph, P. S., S. Kawai, and Y. Toba, 1981. Ocean wave prediction by a hybrid model – Combination of single-parameterized wind waves with spectrally treated swells. *Tohoku Geophys. J. (Sci. Rep. Tohoku Univ. Ser. 5)* **28**, 27–45.

Juszko, B.-A., R. F. Marsden and S. R. Waddell, 1995. Wind stress from wave slopes using Phillips equilibrium theory. *J. Phys. Oceanogr.* **25**(2), 185–203.

Kader, B. A., 1992. Determination of turbulent momentum and heat fluxes by spectral methods. *Boundary-Layer Meteorol.* **61**, 323–347.

Kahma, K. K., 1981. A study of the growth of the wave spectrum with fetch. *J. Phys. Oceanogr.* **11**, 1503–1515.

Kahma, K. K. and M. A. Donelan, 1988. Laboratory study of the minimum wind speed for wind wave generation. *J. Fluid Mech.* **192**, 339–364.

Kahma, K. K. and M. Lepparanta, 1981. On errors in wind speed observations on R/V Aranda. *Geophysica* **17**(1–2), 155–165.

Kaimal, J. C., J. C. Wyngaard, Y. Izumi, and O. R. Coté, 1972. Spectral Characteristics of the surface layer turbulence. *Quart. J.R. Met. Soc.* **98**, 563–589.

Karger, U., 1995. Kuesteneinfluss auf mittlere Bodenwindgeschwindigkeiten ueber der Ostsee. Diplom-thesis, Inst. f. Meereskunde, University of Kiel, 82 pp.

Katsaros, K. and K. J. Buettner, 1969. Influence of rainfall on temperature and salinity of the ocean surface. *J. Appl. Meteorol.* **8**, 15–18.

Katsaros, K. B., P. K. Taylor, J. C. Alishouse and R. G. Lipes, 1981. Quality of Seasat SMMR (Scanning Multi-Channel Microwave Radiometer) atmospheric water determinations. *Oceanography from Space, COSPAR/SCOR/IUCRM symposium*, Venice, 1980. Plenum, New York, 691–706.

Katsis, C. and T. R. Akylas, 1987. On the excitation of long nonlinear water waves by a moving pressure distribution. Part 2. Three-dimensional effects. *J. Fluid Mech.* **175**, 333–349.

Kawai, S., 1979. Generation of initial wavelets by instability of a coupled shear flow and their evolution to wind waves. *J. Fluid Mech.* **93**, 661–703.

Kawai, S., 1981. Visualization of airflow separation over wind-wave crests under moderate wind. *Boundary-Layer Meteorol.* **21**, 93–104.

Kawai, S., 1982. Structure of air flow separation over wind wave crests. *Boundary-Layer Meteorol.* **23**, 503–521.

Kawai, S., K. Okada and Y. Toba, 1977. Field data support of the three-seconds power law and $gu_* \sigma^{-4}$ spectral form for growing wind waves with field observational data. *J. Oceanogr. Soc. Jap.* **33**, 137–150.

Kawamura, H. and Y. Toba, 1988. Ordered motion in the turbulent boundary layer over wind waves. *J. Fluid Mech.* **197**, 105–138.

Kazakov A. L. and V. N. Lykossov, 1982. On parameterization of the interaction between the atmosphere and the underlying surface in the numerical modelling of atmospheric processes. *Proc. ZapSIBNII* No. 55, Gidrometeoizdat, Moscow, 3–20.

Keller, W. and B. L. Gotwols, 1983. Two-dimensional optical measurement of wave slope. *Appl. Opt.* **22**, 3476–3478.

Kendall, J. M., 1970. The turbulent boundary layer over a wall with progressive surface waves. *J. Fluid Mech.* **41**, 259.

Kenney, B. C., 1982. Beware of spurious self-correlations! *Water Resour. Res.* **18**, 1041–1048.

Kenyon K. E., 1969. Stokes drift for random gravity waves. *J. Geophys. Res.* **74**, 6991–6994.

Khalsa, S. J. S. and J. A. Businger, 1977. The drag coefficient as determined by the dissipation method and its relation to intermittent processes in the surface layer. *Boundary-Layer Meteorol.* **12**, 273–297.

Khalsa, S. J. S. and G. K. Greenhut, 1989. Atmospheric turbulence structure in the vicinity of an oceanic front. *J. Geophys. Res.* **94**, No. C4, 4913–4922.

Khundzhua, G. G., A. M. Gusev, Ye. G. Andreyev, V. V. Gurov and N. A Skorokhvatov, 1977. Structure of the cold surface film of the ocean and heat transfer between the ocean and the atmosphere. *Izv. Atmos. And Oceanic Phys.* (Engl. Transl.), **13**, 506–509.

Killworth, P. D., 1983. Deep convection in the world ocean. *Rev. Geophys. and Space Phys.* **21**, 1–26.

Kinsman, B., 1965. *Wind waves*. Prentice-Hall, Englewood Cliffs, NJ.

Kitaigorodskii, S. A., 1961. Application of the theory of similarity to the analysis of wind-generated wave motion as a stochastic process. *Bull. Acad. Nauk SSSR Geophys. Ser.*, 105–117.

Kitaigorodskii, S. A., 1968. On the calculation of the aerodynamic roughness of the sea surface. *Izv. Atmos. Ocean. Phys.* **4**, 870–878.

Kitaigorodskii, S. A., 1970. *Fizika vzaim odestviya atmosferi i okeana*, Gidromet. Izdatel'stvo, Leningrad. Translated from Russian by A. Baruch. 1973 *The physics of air–sea interaction*. Israel Program for Scientific Translations, Jerusalem, 237 pp.

Kitaigorodskii, S. A., 1983. On the theory of the equilibrium range of wind-generated gravity waves. *J. Phys. Oceanogr.* **13**, 816–827.

Kitaigorodskii, S. A. and Yu. A. Volkov, 1965. On the roughness parameter of the sea surface and the calculation of momentum flux in the near-water layer of the atmosphere. *Izv. Atmos. Ocean. Phys.* **1**, 973–988.

Kitaigorodskii, S. A., V. P. Krasitskii, and M. M. Zaslavskii, 1975. On Phillips' theory of equilibrium range in the spectra of wind-generated gravity waves. *J. Phys. Oceanogr.* **5**, 410–420.

Kline, S. J. and S. K. Robinson, 1989. Quasi-coherent structures in the turbulent boundary layer: Part 1. Status report on a community-wide summary of the data. *Proc. Zoran Zaric Memorial Int. Seminar on Near-Wall Turbulence*, Dubrovnik, Croatia, Hemisphere, 200–217.

Kline, S. J., W. C. Reynolds, F. A. Schraub and P. W. Rundstadler, 1967. The structure of turbulent boundary layers. *J. Fluid Mech.* **30**, 741–773.

Klinke, J., 1991. 2-D Wellenzahlspektren von kleinskaligen winderzeugten Wasseroberflächenwellen. *Diploma Thesis*, Univ. Heidelberg.

Klinke, J. 1996. Optical measurements of small-scale wind-generated water surface waves in the laboratory and the field. *Ph.D Thesis*, University of Heidelberg.

Klinke, J. and B. Jähne, 1995. Measurement of short ocean wind waves during the MBL-ARI west coast experiment. *Air–water gas transfer, Selected Papers*, 3rd Intern. Symp. on Air-Water Gas Transfer, B. Jähne and E. Monahan, Eds. AEON, Hanau, pp. 165–173.

Komatsu, K. and A. Masuda, 1996. A new scheme of nonlinear energy transfer among wind waves: RIAM Method – Algorithm and performance. *J. Oceanogr.* **52**, 509–537.

Komen, G. J., S. Hasselmann and K. Hasselmann, 1984. On the existence of a fully developed wind-sea spectrum. *J. Phys. Oceanogr.* **14**, 1271–1285.

Komen, G. J., L. Cavaleri, M. Donelan, K. Hasselmann, S. Hasselmann and P. A. E. M. Janssen, 1994. *Dynamics and modelling of ocean waves.* Cambridge University Press. 532 pp.

Kondo, J., 1975. Air–sea bulk transfer coefficients in diabatic conditions. *Boundary-Layer Meteorol.* **9**, 91–112.

Kondo, J., O. Kanechika and N. Yasuda. 1978. Heat and momentum transfers under strong stability in the atmospheric surface layer. *J. Atmos. Sci.* **35**, 1012–1021.

Kraus, E. B., 1977. Ocean surface drift velocities. *J. Phys. Oceanogr.* **7**, 606–609.

Kraus, E. B. and J. A. Businger, 1994. *Atmosphere–ocean interaction,* Oxford University Press, Oxford.

Kruegermeyer, L., M. Gruenewald and M. Dunckel, 1977. The influence of the significant wave height on the wind profile. *Boundary-Layer Meteorol.* **14**, 403–414.

Kuettner, J. P., 1959. Cloud bands in the atmosphere. *Tellus* **11**, 267–294.

Kunishi, H., 1963. An experimental study on the generation and growth of wind waves. *Bull. Disaster Prev. Res. Inst., Kyoto Univ.* No. 61, 1–41.

Kunishi, H. and N. Imasato, 1966. On the growth of wind waves by high-speed wind flume. *Disaster Prev. Res. Inst., Kyoto Univ., Annals* **9**, 1–10 (in Japanese with English abstract).

Kuo, Y-H., J. Reed and S. Low-Nam, 1991. Effects of surface energy fluxes during the early development and rapid intensification stages of seven explosive cyclones in the western Atlantic. *Monthly Weather Review* **119**, 457–476.

Kurbatskii, A. F., 1988. *Modelling of non-local turbulent transport of momentum and heat.* Novosibirsk, Nauka, 240 p.

Kusaba, T. and A. Masuda, 1988. The roughness height and drag law over the water surface based on the hypothesis of local equilibrium. *J. Oceanogr. Soc. Japan* **44**, 200–214.

Kusaba, T. and H. Mitsuyasu, 1986. Nonlinear instability and evolution of steep water waves under wind action. *Report. Inst. Appl. Mech. Kyushu Univ.* **33**, 33–64.

Lamb, H., 1957. *Hydrodynamics.* Cambridge University Press, Cambridge, 738 pp.

Landau, L. D. and E. M. Lifshitz, 1963. *Fluid Mechanics.* Pergamon Press, London.

Lange, P. A., B. Jähne, J. Tschiersch and J. Ilmberger, 1982. Comparison between an amplitude-measuring wire and a slope-measuring laser water wave gauge. *Rev. Sci. Instrum.* **53**, 651–655.

Large, W. G., 1979. The turbulent fluxes of momentum and sensible heat over the sea during moderate to strong winds. *Ph.D. Thesis*, University of British Columbia, 180 pp.

Large, W. G. and J. A. Businger, 1988. A system for remote measurements of the wind stress over the oceans. *J. Atmos. & Oceanic Tech.* **5**, 274–285.

Large, W. G. and S. Pond, 1981. Open ocean momentum flux measurements in moderate to strong winds. *J. Phys. Oceanogr.* **11**, 324–336.

Large, W. G. and S. Pond, 1982. Sensible and latent heat flux measurements over the ocean. *J. Phys. Oceanogr.* **12**, 464–482.

Large, W. G., J. C. McWilliams, and S. C. Doney, 1994. Oceanic vertical mixing: a review and a model with a nonlocal boundary layer parameterization. *Rev. Geophys.* **32**, 363–403.

Larsen, S. E., 1986. Hot-wire measurements of atmospheric turbulence. Risø-R-232, Risø, Roskilde, Denmark, 342 pp.

Larsen, S. E., 1993. Observing and modelling the planetary boundary layer. In: *Energy and water cycles in the climate system.* E. Raschke and D. Jacob, Eds. NATO ASI Series I, Vol. 5, 365–418.

Launiainen, J., 1995. Derivation of the relationship between the Obukhov stability parameter and the bulk Richardson number for flux-profile studies. *Boundary-Layer Meteorol.* **76**, 165–179.

Ledwell, J. R., A. J. Wilson and C. S. Low, 1993. Evidence for slow mixing across the pycnocline from an open-ocean tracer-release experiment. *Nature* **364**, 701–703.

Lee, P. H. Y., J. D. Barter, K. L. Beach, C. L. Hindman, B. M. Lake, H. Rungaldier, J. C. Schatzman, J. C. Shelton, R. N. Wagner, A. B. Williams, R. Yee, and H. C. Yuen, 1992. Recent advances in ocean surface characterization by a scanning laser slope gauge. *Proc. Optics of the air–sea interface: Theory and measurements,* L. Estep, Ed. SPIE Proc. 1749, pp. 234–244.

Leibovich, S. A., 1983. The form and dynamics of Langmuir circulations. *Annual Reviews of Fluid Mech.* **15**, 391–427.

Le Méhauté, B. and T. Khangaonkar, 1990. Dynamic interaction of intense rain with water waves. *J. Phys. Oceanogr.* **20**, 1805–1812.

Levich, V. G., 1962. *Physicochemical hydrodynamics.* Prentice-Hall, Englewood Cliffs, N.J. p. 610.

Li, F., W. Large, W. Shaw, E. J. Walsh, and K. Davidson, 1989. Ocean radar backscatter relationship with near-surface winds: a case study during FASINEX. *J. Phys. Oceanogr.* **19**, 342–353.

Lilly, D. K., 1966. On the instability of Ekman boundary flow. *J. Atmos. Sci.* **23**, 481–494.

Lin, J-T. and M. Gad-el-Hak. Turbulent current measurements in a wind-wave tank. *J. Geophys. Res.* **89**, 627–636.

Ling, S. C., 1993. Effect of breaking waves on the transport of heat and water vapour fluxes from the ocean. *J. Phys. Oceanogr.* **23**, 2360–2372.

Liu, A. K., C. Y. Peng, B. Chapron, E. Mollo-Christensen, and N. Huang, 1995. Direction and magnitude of wind stress over wave groups observed in SWADE. *The Global Atmos. and Ocean Sys.* **3**, 175–194.

Liu, P. C., 1989. On the slope of the equilibrium range in the frequency spectrum of wind waves. *J. Geophys. Res.* **94**, 5017–5023.

Liu, W. T. and W. G. Large, 1981. Determination of surface stress by Seasat-SASS: a case study with JASIN data. *J. Phys. Oceanogr.* **11**, 1603–1611.

Lock, R. C., 1954. Hydrodynamic instability of the flow in the laminar boundary layer between parallel streams. *Proc. Cambridge Phil. Soc.* **50**, 105–124.

Long, S. R., N. E. Huang, E. Mollo-Christensen and F. C. Jackson, 1994. Directional wind wave development. *J. Geophys. Res. Letters* **21**, 2503–2506.

Longuet-Higgins, M. S., 1952. On the statistical distribution of heights of sea waves. *J. Mar. Res.* **11**, 245–266.

Longuet-Higgins, M. S., 1953. Mass transport in water waves. *Philosophical Transactions of the Royal Society of London* **A245**, 535–581.

Longuet-Higgins, M. S. 1960. Mass transport in the boundary layer at a free oscillating surface. *J. Fluid Mech.* **8**, 293–306.

Longuet-Higgins, M. S., 1973. A model of flow separation at a free surface. *J. Fluid Mech.* **57**, 129–148.

Longuet-Higgins, M. S., 1983. Bubbles, breaking waves and hyperbolic jets at a free surface. *J. Fluid Mech.* **127**, 103–121.

Longuet-Higgins, M. S., 1990. Flow separation near the crests of short gravity waves. *J. Phys. Oceanogr.* **20**, 600–609.

Longuet-Higgins, M. S., 1992 Capillary rollers and bores. *J. Fluid Mech.* **240**, 659–679.

Longuet-Higgins, M. S. and R. W. Stewart, 1960. Changes in the form of short gravity waves on longwaves and tidal currents. *J. Fluid Mech.* **8**, 565–583.

Lucassen, J., 1982. Effect of surface-active material on the damping of gravity waves: A reappraisal. *J. Coll. Interface Sci.* **85**, 52–58.

Lucassen-Reynders, E. H. and J. Lucassen, 1969. Properties of capillary waves. *Advs. in Coll. Interface Sci.* **2**, 347–395.

Lumley, J. L., 1967. The structure of inhomogeneous turbulent flows. In: *Atmospheric turbulence and radio wave propagation.* A. M. Yaglom and V. I. Tatarsky, Eds. Nauka, Moscow, 166–178.

Lumley, J. L. and I. Kubo, 1984. Turbulent drag reduction by polymer additives: A survey. In: *The influence of polymer additives on velocity and temperature fields.* IUTAM, Symposium Essen 1984. B. Gampert, Ed. Springer Verlag, 3–21.

Lumley, J. L. and H. A. Panoski, 1964. *The structure of atmospheric turbulence.* Interscience, NY. 239 pp.

Luthardt, H. and L. Hasse, 1981. The relationship between surface and geostrophic wind in the region of the German Bight. *Beitr. Phys. Atm.* **54**, 222–237.

Ly, L. N., 1993. On the effect of the angle between wind stress and wind vectors on the aerodynamical drag coefficient. *J. Phys. Oceanogr.* **23**, 159–163.

Lykossov, V. N., 1992. The momentum turbulent counter-gradient transport in jet-like flows. *Adv. Atmos. Sci.* **9**, 191–200.

Lykossov, V., 1995. Turbulence closure for the boundary layer with coherent structures: an overview. *Berichte aus dem Fachbereich Physik, Rep. No. 63*, Alfred-Wegener-Institute for Polar and Marine Research, Bremerhaven, Germany, 27 pp.

Madec, G. and M. Crepon, 1991. Thermohaline-driven deep water formation in the north-western Mediterranean Sea. In: *Deep convection and deep water formation in the oceans.* Proc. of the International Monterey colloquium on deep convection and deep water formation in the oceans, P. C. Chu and J. C. Gascard, Eds. Elsevier, 241–265.

Mahrt, L. and W. Gibson, 1992. Flux decomposition into coherent structures. *Bound.-Layer Meteorol.* **60**, 143–168.

Makin, V. K., 1989. The dynamics and structure of the boundary layer above sea. *Sen. Doct. Thesis.* Inst. of Oceanology Acad. Sci. USSR, Moscow.

Makin, V. K., V. N. Kudryavstev, and C. Mastenbroek, 1995. Drag of the sea surface. *Boundary-Layer Meteorol.* **73**, 159–182.

Makova, V. I., 1975. Features of the dynamics of turbulence in the marine atmospheric surface layer at various stages in the development of waves. *Atmos. Ocean. Physics*, **11**, 177–182.

Mallinger, W. D. and T. P. Mickelson, 1973. Experiment with monomolecular films on the surface of the open sea. *J. Phys. Oceanogr.* **3**, 326–338.

Manton, M., 1973. On the attenuation of sea waves by rain. *J. Geophys. Fluid Mech.* **5**, 249–260.

Marangoni, C., 1872. Sur principio della visconsita superficiale del liquide stabilito. *Nuovo Cimento, Ser.* **2**, 5–6, 239–273.

Martinsen, R. J. and E. J. Bock, 1992. Optical measurements of ripples using a scanning laser slope gauge, part I: instrumentation and preliminary results. *Proc. Optics of the air–sea interface: Theory and measurements*, L. Estep, Ed. SPIE Proc. 1749, pp. 258–271.

Masuda, A. and T. Kusaba, 1987. On local equilibrium of winds and wind-waves in relation to surface drag. *J. Oceanogr. Soc. Japan* **43**, 28–36.

Masuko, H., K. Okamoto, M. Shimada, and S. Niwa, 1986. Measurement of microwave backscattering signatures of the ocean surface using X band and K_a band airborne scatterometers. *J. Geophys. Res.* **91**, 13065–13083.

McBean, G. A. and M. Miyake, 1972. Turbulent transfer mechanisms in the atmospheric surface layer. *Quart. J. R. Met. Soc.* **98**, 383–398.

McLean, J. W., 1983. Computation of turbulent flow over a moving wavy boundary. *Phys. Fluids* **26**, 2065–2073.

Melville, W. K., 1982. Wind stress and roughness length over breaking waves. *J. Phys. Oceanogr.* **7**, 702–710.

Merzi, N. and W. F. Graf, 1985. Evaluation of the drag coefficient considering the effects of mobility of the roughness elements. *Ann. Geophys.* **3**, 473–478.

Mikhailova, L. A. and A. Ye. Ordanovich, 1991. Coherent structures in the atmospheric boundary layer (a survey). *Izvestya, Atmospheric and Oceanic Physics* **27**, 413–428.

Miles, J. W., 1957. On the generation of surface waves by shear flows, Part 1, *J. Fluid Mech.* **3**, 185–204.

Miles, J. W., 1959a. On the generation of surface waves by shear flows, Part 2. *J. Fluid Mech.* **6**, 568–582.

Miles, J. W., 1959b. On the generation of surface waves by shear flows, Part 3. *J. Fluid Mech.* **6**, 583–598.

Miles, J. W., 1962. On the generation of surface waves by shear flows Part 4. *J. Fluid Mech.* **13**, 443–448.

Miles, J. W., 1967. On the generation of surface waves by shear flows Part 5. *J. Fluid Mech.* **30**, 163–175.

Mitsuta, Y., Ed. 1977–1979. *Collected Scientific Papers of the AMTEX*. Japanese National Committee for GARP and Disaster Prevention Research Inst., Kyoto Univ., No. 1, 181 pp.; No. 2, 270 pp.; No. 3, 181 pp.; No. 4, 249 pp.

Mitsuyasu, H. 1966. Interactions between water waves and wind (1), *Report Inst. Appl. Mech. Kyushu Univ.* **14**, 67–88.

Mitsuyasu, H., 1968. On the growth of the spectrum of wind generated waves. (1). *Rep. Res. Inst. Appl. Mech., Kyushu Univ.* **16**, 459–482.

Mitsuyasu, H., 1985. A note on the momentum transfer from wind to waves. *J. Geophys. Res.* **90**, 3343–3345.

Mitsuyasu, H. and T. Honda, 1974. The high frequency spectrum of wind-generated waves. *J. Oceanogr. Soc. Japan* **30**, 185–198.

Mitsuyasu, H. and T. Honda, 1982. Wind-induced growth of water waves, *J. Fluid Mech.* **123**, 425–442.

Mitsuyasu, H. and T. Honda, 1986. The effects of surfactant on certain air–sea interaction phenomena. In: *Wave dynamics and radio probing of the ocean*, O. M. Phillips and K. Hasselmann, Eds. Plenum Publishing Company, 95–115.

Mitsuyasu H. and T. Kusaba, 1985. Wind waves and wind-generated turbulence in the water. In: *The ocean surface*, Y. Toba and H. Mitsuyasu, Ed. 389–394.

Mitsuyasu, H. and T. Kusaba, 1988. The momentum transfer from the air to the water (2). *Bull. Res. Inst. Appl. Mech. Kyushu Univ.* (in Japanese) **66**, 21–35.

Mitsuyasu, H., R. Nakamura, and T. Komori, 1971. Observations of the wind and waves in Hakata Bay. *Rep. Res. Inst. Appl. Mech., Kyushu Univ.* **19**, 37–74.

Mitsuyasu, H., H. Tasai, F. Suhara, T. Mizuno, S. Ohkusu, T. Honda, and K. Rikiishi, 1980. Observation of power spectrum of ocean waves using a clover-leaf buoy. *J. Phys. Oceanogr.* **10**, 286–296.

Mitsuyasu, H., T. Kusaba, and Y. Yoshida, 1988. The effects of water waves on the sea surface wind (abstract). *Annual Meeting of the Oceanographical Society of Japan*, 85–86.

Mizuno, S., 1976. Pressure measurements above mechanically generated water waves (I), *Rep. Inst. Appl. Mech. Kyushu Univ.* **23**, 113–129.

Monin, A. S. and A. M. Obukhov, 1954. The basic regularities of turbulent mixing in the atmospheric surface layer. *Tr. Akad. Nauk SSSR, Geofiz. Inst.* **24**, 163–187.

Monin, A. S. and R. V. Ozmidov, 1981. *Oceanic turbulence*. Gidrometeoizdat, Leningrad, 320 pp.

Monin, A. S. and A. M. Yaglom, 1971. *Statistical fluid mechanics: mechanics of turbulence*, Part 1. The MIT Press, Cambridge.

Montgomery, R. B., 1936. Measurements of vertical gradient of wind over water. *Papers Phys. Oceanog. Met. Mass. Inst. Tech. and Woods Hole Oceanog. Inst.* **4**, no. 3.

Mortensen, N, G. and J. Højstrup, 1995. The Solent sonic – response and associated errors. *Ninth Symposium on meteorological observations and instrumentation*, Charlotte, NC, March 27–31, 501–506.

Mortensen, N. G., Larsen, S. E., Troen, I. and Mikkelsen, T., 1987. Two-years-worth of turbulence data recorded by a sonic-anemometer–based data acquisition system. *Proceedings of Sixth Symposium on meteorology and instrumentation*. Jan. 12–16, New Orleans, Louisiana (AMS, Boston, Mass, USA), 393–396.

Mourad, P. and R. A. Brown, 1990. Multiscale large eddy states in weakly stratified planetary boundary layers. *J. Atmos. Sci.* **47**, 414–438.

Mourad, P. D. and B. A. Walker, 1996. Viewing a cold air outbreak using satellite-based synthetic aperture radar and advanced very high resolution radiometer imagery. *J. Geophys. Res.* **101**, 16391–16400.

Moyer, K. A. and G. S. Young, 1993. Buoyant forcing within the stratocumulus-topped boundary layer. *J. Atmos. Sci.* **50**, 2759–2771.

Müller-Glewe, J. and H. Hinzpeter, 1975. Turbulent fluxes in the ITCZ during GATE Phase III at Station 27. *GATE Report* **14**, 1, WMO/ICSU, Geneva, 224–232.

Muraviev, S. S. and R. V. Ozmidov, 1994 Synergetic mechanisms of formation of ordered structures in the ocean: an overview. *Oceanology* (English translation) **34**, 293–303.

Narahari Rao, K. N., R. Narasimha, and M. A. Badri Narayana, 1971. The bursting phenomenon in a turbulent boundary layer. *J. Fluid Mech.* **48**, 339–352.

Narasimha, R., 1984. The turbulence problem: a survey of simple turbulent flows. *GALCIT Report FM 84-01*, Grad. Aero. Lab., California Inst. of Technology, Pasadena, USA, 68 pp.

Narasimha, R., 1988. Turbulent bursts in the atmosphere. *Project Document* DU 8808, National Aeronautical Laboratory, Bangalore, India, pp. 40.

Narasimha, R. and S. Kailas, 1987. Energy events in the atmospheric boundary layer. In: *Perspectives in turbulence studies*. H.U. Meier and P. Bradshaw, Ed. Springer-Verlag, 188–222.

Neugum, A., 1996. Systematic influences in the determination of stress at sea with the so-called dissipation technique. *Ph.D. Thesis*, Berichte IfM, Inst. fuer Meereskunde, University of Kiel.

Newell, A. C. and V. Zakharov, 1992. Rough sea foam. *Physical Review Letters* **69**, 1149–1151.

Nicholls, S., 1985. Aircraft observations of the Ekman layer during the Joint Air–Sea Interaction Experiment. *Quart J. Roy. Meteorol. Soc.* **111**, 391–426.

Nicholls, S., 1989. The structure of radiatively driven convection in stratocumulus. *Quart. J. R. Meteorol. Soc.* **115**, 487–511.

Nicholls, S. and C. Readings, 1979. Aircraft observations of the structure of the lower boundary layer over the sea. *Quart. J. Roy. Meteorol. Soc.* **105**, 785–802.

Nilsson, C. S. and P. C. Tildesley, 1995. Imaging of oceanic features by ERS1 synthetic aperture radar. *J. Geoph. Res.* **100**, 953–967.

Nordeng, T. E., 1991. On the wave age dependent drag coefficient and roughness length at sea. *J. Geophys. Res.* **96**, 7167–7174.

Okuda, K., 1982a. Internal flow structure of shot wind waves. Part I. On the internal vorticity structure. *J. Oceanogr. Soc. Japan* **38**, 28–42.

Okuda, K., 1982b. Internal flow structure of shot wind waves. Part II. The streamline pattern. *J. Oceanogr. Soc. Japan* **38**, 313–322.

Okuda, K., S. Kawai, M. Tokuda and Y. Toba, 1976. Detailed observation of the wind-exerted surface flow by use of flow visualization methods. *J. Oceanogr. Soc. Japan* **32**, 53–64.

Okuda, K., S. Kawai and Y. Toba, 1977. Measurement of skin friction distribution along the surface of wind waves. *J. Oceanogr. Soc. Japan* **33**, 190–198.

Olmez, H. S. and J. H. Milgram, 1992. An experimental study of attenuation of short water waves by turbulence. *J. Fluid Mech.* **239**, 133–156.

Oncley, S. P., C. A. Friehe, J. C. LaRue, J. A. Businger, E. C. Itsweire and S. Chang, 1996. Surface layer fluxes, profiles and turbulence measurements over uniform terrain under near-neutral conditions. *J. Atmos. Sciences*, **53**, 1029–1044.

Onstott, R. and C. Rufenach, 1992. Shipboard active and passive microwave measurement of ocean surface slicks off the Southern California coast. *J. Geophys. Res.* **97**(C4), 5315–5323.

Oost, W. A., 1998. The KNMI HEXMAX stress data – A reanalysis. *Boundary-Layer Meteorology* **86**, 447–468.

Oost, W. A., C. W. Fairall, J. B. Edson, S. D. Smith, R. J. Anderson, J. A. B. Wills, K. B. Katsaros, and J. DeCosmo, 1994. Flow distortion calculations and their application in HEXMAX. *J. Atmos. & Oceanic Tech.* **11**(2), 366–386.

Panofsky, H. A., 1963. Determination of stress from wind and temperature measurements. *Quart. J. Roy. Meteorol. Soc.* **89**, 85.

Panofsky, H. A. and Dutton, J. A. 1984 *Atmospheric turbulence. Models and methods for engineering applications*. Wiley-Interscience Publ., New York, 397 pp.

Papadimitrakis, Y. A., N. E. Huang, L. F. Blivan and S. R. Long, 1987. An estimate of wave breaking probability for deep water waves. In: *Sea surface sound*, B. R. Kerman, Ed. Kluwer Academic Publishers, Dordrecht.

Paulson, C. A. 1970. The mathematical representation of wind speed and temperature profiles in the unstable atmospheric surface layer. *J. Appl. Meteorol.* **9**, 857–861.

Pennel, W. T. and M. A. Le Mone, 1974. An experimental study of turbulent structure in the fair-weather trade wind boundary layer. *J. Atmos. Sci.* **31**, 1308–1323.

Peregrine, D. H. and I. A. Svendsen, 1978. Spilling breakers, bores and hydraulic jumps. *Proc. 16th Coastal Eng. Conf.* 540–550.

Peters, H., M. C. Gregg, J. M. Toole, 1988. On the parameterization of equatorial turbulence. *J. Geophys. Res.* **93**, 1199–1218.

Phillips, O. M., 1957. On the generation of waves by turbulent wind. *J. Fluid Mech.* **2**, 417–445.

Phillips, O. M., 1958. The equilibrium range in the spectrum of wind-generated waves. *J. Fluid Mech.* **4**, 426–434.

Phillips, O. M., 1977. *The dynamics of the upper ocean*. Cambridge University Press. 336 pp.

Phillips, O. M., 1981. The dispersion of short ocean wave components in the presence of a dominant long wave. *J. Fluid Mech.* **107**, 465–485.

Phillips, O. M., 1985. Spectral and statistical properties of the equilibrium range in the spectrum of wind-generated gravity waves. *J. Fluid Mech.* **156**, 505–531.

Phillips, O. M. and M. L. Banner, 1974. Wave breaking in the presence of wind drift and swell. *J. Fluid Mech.* **66**, 625–640.

Pielke, R. A. and T. J. Lee, 1991. Influence of sea spray and rainfall on the surface wind profile during conditions of strong winds. *Bound.-Layer Meteorol.* **55**, 305–308.

Pierson, W. J., 1990. Dependence of radar backscatter on environmental parameters. Chapter 13, In: *Surface waves and fluxes*: Vol. 2. *Remote sensing*, Kluwer Academic Press, 173–220.

Plant, W. J., 1982. A relation between wind stress and wave slope. *J. Geophys. Res.* **87**, C1961–1967.

Plant, W. J. and J. W. Wright, 1977. Growth and equilibrium of short gravity waves in a wind tank. *J. Fluid Mech.* **82**, 767–793.

Plinius Secundus, C., 1634. *The history of the World: Commonly called, The Natural Historie of C. Plinius Secundus*. Translated by Philemon Hollond, London, p. 46.

Pokazeyev, K. V. and A. D. Rozenberg, 1983. Laboratory studies of regular gravity-capillary waves in currents. *Izv. Atm. & Ocean Phys.* **23**, 429–435.

Pollard, R. T., T. H. Guymer and P. K. Taylor, 1983. Summary of the JASIN 1978 field experiment. *Phil. Trans. Roy. Soc. London* **A308**, 221–230.

Pond, S., G. T. Phelps, J. E. Paquin, G. McBean, and R. W. Stewart, 1971. Measurements of the turbulent fluxes of momentum, moisture and sensible heat over the ocean. *J. Atmos. Sci.* **28**, 901–917.

Pond, S., W. G. Large, M. Miyake, and R. W. Burling, 1979. A gill twin propeller-vane anemometer for flux measurements during moderate and strong winds. *Boundary-Layer Meteorol.* **16**, 351–364.

Poon, Y.-K., S. Tang, and J. Wu, 1992. Interactions between rain and wind waves. *J. Phys. Oceanogr.* **22**, 976–987.

Prandtl, L., 1924. Über flüssigkeitsbewegung bei sehr kleiner reibung. *Proc. Third Int'l Math. Congress*, Heidelberg, 1924, 484–491.

Prandtl, L., 1932. Meteorologische Anwendung der Strömungslehre. *Beitrge zur Physik der freien Atmos.* **19**, 188–202.

Priestley, C. H. B. and W. C. Swinbank, 1947. Vertical transport of heat by turbulence in the atmosphere. *Proc. Roy. Soc., Ser. A* **189**, 543–561.

Quanduo, G. and G. Komen, 1993. Directional response of ocean waves to changing wind direction. *J. Phys. Oceanog.* **23**, 1561–1566.

Rasmussen, E., 1985. A case study of a polar low development over the Barents Sea. *Tellus* **37A**, 407–418.

Reynolds, O., 1883. An experimental investigation of the circumstances which determine whether the motion of water shall be direct or sinuous, and of the law of resistance in parallel channels. *Phil. Trans. Roy. Soc. London* **174**, 935–982.

Reynolds, O., 1900. On the action of rain to calm the sea. *Papers on mechanical and physical subjects*, 1, Cambridge University Press, 86–88.

Richardson, L. F., 1920. The supply of energy from and to atmospheric eddies. *Proc. Roy. Soc.* **A97**, 354–373.

Rieder, K. F., J. A. Smith and R. A. Weller, 1994. Observed directional characteristics of the wind, wind stress and surface waves on the open ocean. *J. Geophys. Res.* **99**, 22596–22598.

Rieder, K. F., J. A. Smith, and R. A. Weller, 1996. Some evidence of co-linear wind stress and wave breaking. *J. Phys. Oceanogr.* **26**, 2519–2524. (See also, ibid. *J. Phys. Oceanog.*, vol. 27, no. 1, 213, 1997.)

Robinson, A. R., 1963. *Wind-driven ocean circulation*. Blaisdell Publishing Co., New York, NY.

Rogers, D. P., 1989. The marine boundary layer in the vicinity of an ocean front. *J. Atmos. Sci.* **46**, 2044–2062.

Rogers, D. P., X. Yang, P. M. Norris, D. W. Johnson, G. M. Martin, C. A. Friehe and Berger, B. W. 1995. Diurnal evolution of the cloud-topped marine boundary layer. Part I: Nocturnal stratocumulus development. *J. Atmos. Sci.* **52**, 2953–2966.

Rossby, C. G. and R. Montgomery, 1935. The layer of frictional influence in wind and ocean currents. *Papers Phys. Oceanogr. Meteorol.* No. 3, MIT and WHOI, 101 pp.

Schacher, G. E., K. L. Davidson, T. Houlihan and C. W. Fairall, 1981. Measurements of the rate of dissipation of turbulent kinetic energy over the ocean. *Boundary-Layer Meteorol.* **20**, 321–330.

Schlichting, H., 1960. *Boundary layer theory*. Transl. J. Kestin, McGraw Hill Book Co., New York, 647 pp.

Schmitt, K. F., C. A. Friehe and C. H. Gibson, 1978. Sea surface stress measurements. *Boundary-Layer Meteorol.* **15**, 215–228.

Schmitt, K. F., C. A. Friehe and C. H. Gibson, 1979. Structure of marine surface layer turbulence. *J. Atmos. Sci.* **36**, 602–618.

Schubauer, G. B. and H. K. Skramstad, 1947. Laminar boundary layer oscillations and stability of laminar flow. *J. Aero. Sci.* **14**, 69–78.

Scott, J. C., 1972. The influence of surface-active contamination on the initiation of wind waves. *J. Fluid Mech.* **56**, 591–606.

Shaw, W. J. and J. A. Businger, 1985. Intermittency and the organization of turbulence in the near-neutral marine atmospheric boundary layer. *J. Atmos. Sci.* **42**, 2563–2584.

Shemdin, O. H., 1972. Wind-generated current and phase speed of wind waves. *J. Phys. Oceanogr.* **2**, 411–419.

Shemdin, O. H., H. M. Tran, and S. C. Wu, 1988. Directional measurements of short ocean waves with stereophotography. *J. Geophys. Res.* **93**, 13891–13901.

Sheppard, P. A. and M. H. Omar, 1952. The wind stress over the ocean from observations in the trades. *Quart. J. Roy. Meteorol. Soc.* **78**, 583–589.

Shyu, J. H. and O. M. Phillips, 1990. The blockage of gravity and capillary waves by longer waves and currents. *J. Fluid Mech.* **217**, 115–141.

Singh, K. P., A. L. Gray, R. A. Hawkins, and R. A. O'Neil, 1986. The influence of surface oil on C-band and Ku-band ocean backscatter. *IEEE Trans., Geosci. and Remote Sens.* **GE-24**, 738–744.

Smedman, A.-S., M. Tjernström and U. Högström, 1994. The near-neutral marine atmospheric boundary layer with no surface shearing stress: a case study. *J. Atmos. Sci.* **51**, 3399–3411.

Smedman, A.-S., H. Bergström and U. Högström, 1995. Spectra, variances and length scales in a marine stable boundary layer dominated by a low level jet. *Boundary-Layer Meteorol.*, **76**, 211–232.

Smith, J. A. and G. T. Bullard, 1995. Directional surface-wave estimates from Doppler sonar data. *J. Atmos. Ocean. Tech.* **12**, no. 3, 617–632.

Smith, P. C. and R. E. Stewart, Eds., 1989. Canadian Atlantic Storms Program. *Atmosphere-Ocean* **27**, 1–278.

Smith, P. C., C. L. Tang, J. I. MacPherson and R. F. McKenna, 1994. Investigating the marginal ice zone on the Newfoundland shelf. *EOS Trans. Amer. Geophys. Union* **75**, 57 and 60–62.

Smith, S. D., 1980. Wind stress and heat flux over the ocean in gale force winds. *J. Phys. Oceanogr.* **10**, 709–726.

Smith, S. D., 1988. Coefficients for sea surface wind stress, heat flux and wind profiles as a function of wind speed and temperature. *J. Geophys. Res.* **93**, 15467–15474.

Smith, S. D. and R. J. Anderson, 1984. Spectra of humidity, temperature and wind over the sea at Sable Island, Nova Scotia. *J. Geophys. Res.* **89**, 2029–2040.

Smith, S. D. and E. G. Banke, 1975. Variation of the sea surface drag coefficient with wind speed. *Quart. J. Roy. Meteorol. Soc.* **101**, 665–673.

Smith, S. D., R. J. Anderson, W. A. Oost, C. Kraan, N. Maat, J. DeCosmo, K. B. Katsaros, K. L. Davidson, K. Bumke, L. Hasse and H. M. Chadwick, 1992. Sea surface wind stress and drag coefficients: the HEXOS results. *Boundary-Layer Meteorol.* **60**, 109–142.

Smith, S. D., C. W. Fairall, G. L. Geernaert and L. Hasse, 1996. Air–sea fluxes: 25 years of progress. *Boundary-Layer Meteorol.* **78**, 247–290.

Snodgrass, F. E., G. W. Groves, K. F. Hasselmann, G. R. Miller, W. H. Munk and W. H. Powers, 1966. Propagation of ocean swell across the Pacific. *Phil. Trans.* **259**, 431–497.

Snyder, R. L., F. W. Dobson, J. A. Elliot, and R. L. Long, 1981. Array measurements of atmospheric pressure fluctuations above surface gravity waves. *J. Fluid Mech.* **102**, 1–59.

Soloviev, A. V., 1990. Coherent structures at the ocean surface in convectively unstable condition. *Nature* **346**, 157–160.

Sorbjan, Z., 1989. *Structure of the atmospheric boundary layer*. Prentice-Hall, New Jersey. 317 pp.

Stevenson, T., 1852. Observations on the relation between the height of waves and their distance from the windward shore. *Edinburgh New Philosophical Journal* **53**, 358.

Stewart, R. W., 1974. The air–sea momentum exchange. *Boundary-Layer Meteorol.* **6**, 151–167.

Stilwell, D., Jr, 1969. Directional energy spectra of the sea from photographs. *J. Geophys. Res.* **74**, 1974–1986.

Stull, R. B., 1988. *An introduction to boundary layer meteorology*. Kluwer Academic Publ., Dordrecht, The Netherlands, 666 pp.

Stull, R. B., 1991. Static stability – an update. *Bull. Am. Meteorol. Soc.* **72**, 1521–1529.

Stull, R. B., 1993. Review of non-local mixing in turbulent atmospheres: transilient turbulence theory. *Bound.-Layer Meteorol.* **62**, 21–96.

Stull, R. B., 1994. A review of parameterization schemes for turbulent boundary-layer processes. In: *Parameterization of sub-grid scalephysical processes*, ECMWF, Workshop, 5–9 September 1994, Reading, England, 163–174.

Suzuki, N., N. Ebuchi, M. Akiyama, J. Suwa and Y. Sugimori, 1998. Relationship between nondimensional roughness length and wave age investigated using tower-based measurements. *J. Adv. Mar. Sci. Tech. Soc.* **4**, 217–224.

Sverdrup, H. U. and W. H. Munk, 1947. Wind, sea and swell. Theory of relations for forecasting. *U.S. Hydrogr. Office, Wash., Publ. No. 601.*

Sverdrup, H. U., M. W. Johnson and R. H. Fleming, 1942. *The oceans, their physics, chemistry and general biology*. Prentice-Hall, Englewood Cliffs, N.J., 1087 pp.

Taylor, G. I., 1915. Eddy motion in the atmosphere. *Phil. Trans. Roy. Soc.* **A215**, 1–26.

Taylor, P. A. and Dyer, K. R., 1977. Theoretical models of flow near the bed and their applications for sediment transport. In: *The sea*, 6, E. D. Goldberg, Ed. Wiley-Interscience, New York, 579–601.

Taylor, P. A. and P. R. Gent, 1978. A numerical investigation of variations in the drag coefficient for air flow above water waves. *Q. J. Roy. Meteor. Soc.*, **104**, 979–988.

Taylor, P. K., K. B. Katsaros and R. G. Lipes, 1981. Determinations by Seasat of atmospheric water and synoptic fronts. *Nature* **294**, 737–739.

Taylor, P. K., T. H. Guymer, K. B. Katsaros and R. G. Lipes, 1983. Atmospheric water distributions determined by the Seasat multi-channel microwave radiometer. *Variations in the global water budget*. A. Street-Perrott, M. Beran and R. Ratcliffe, Eds. Reidel, 93–106.

Thais, L. and J. Magnaudet, 1995. A triple decomposition of the fluctuating motion below laboratory wind water waves. *J. Geophys. Res.* **100**, 741–755.

Thiebaux, M. L., 1990. Wind tunnel experiments to determine correction functions for shipborne anemometers. *Canadian Contractor Report of Hydrography and Ocean Sciences* 36. Bedford Inst. Oceanography, Dartmouth, Nova Scotia, 57 pp.

Thompson, T. W., D. E. Weissman and F. I. Gonzales, 1983. L band radar backscatter dependence upon surface wind stress: A summary of new Seasat-1 and aircraft observations. *J. Geophys. Res.* **88**, 1727–1735.

Thornton, E. B. 1977. Rederivation of the saturation range in the frequency spectrum of wind-generated gravity waves. *J. Phys. Oceanogr.* **7**, 137–140.

Tjernstrom, M. and D. P. Rogers, 1996. Turbulence structure in decoupled marine stratocumulus: a case study from the ASTEX field experiment. *J. Atmos. Sci.* **53**, 598–619.

Toba, Y., 1965a. On the giant sea-salt particles in the atmosphere. I. General features of the distribution. *Tellus* **17,** 131–145.

Toba, Y., 1965b. On the giant sea-salt particles in the atmosphere. II. Theory of the vertical distribution in the 10-m layer over the ocean. *Tellus* **17,** 365–382.

Toba, Y., 1966. On the giant sea-salt particles in the atmosphere. III. An estimation of the production and distribution over the World ocean. *Tellus* **18,** 132–145.

Toba, Y., 1972. Local balance in the air–sea boundary processes I. On the growth process of wind waves. *J. Oceanogr. Soc. Japan* **28,** 109–121.

Toba, Y., 1973. Local balance in the air–sea boundary process. III. On the spectrum of wind waves. *J. Oceanog. Soc. Japan* **29,** 209–220.

Toba, Y, 1978. Stochastic form of the growth of wind waves in a single-parameter representation with physical implications. *J Phys. Oceanogr.* **8,** 494–507.

Toba, Y., 1985. Wind waves and turbulence. In: *Recent studies on turbulent phenomena.* T. Tatsumi, H. Maruo and H. Takami, Eds. Assoc. for Sci. Doc. Inform., Tokyo. 277–296.

Toba, Y., 1988. Similarity laws of the wind-wave, and the coupling process of the air and water turbulent boundary layers. *Fluid Dyn. Res.* **2,** 263–279.

Toba, Y., 1998. Wind forced strong wave interactions and quasi-local equilibrium between wind and windsea with the friction velocity proportionality. *Advances in fluid mechanics,* Vol 17; *Nonlinear ocean waves,* W. Perrie, Ed. Computational Mechanics Pub., UK.

Toba, Y. and N. Ebuchi, 1991. Sea-surface roughness length fluctuating in concert with wind waves. *J. Oceanogr. Soc. Jap.* **47,** 63–79.

Toba, Y. and H. Kawamura, 1996. Wind-wave coupled downward-bursting boundary layer (DBBL) beneath the sea surface. *J. Oceanogr.* **52,** 409–419.

Toba, Y. and M. Koga, 1986. A parameter describing overall conditions of wave breaking, whitecaps, sea-spray production and wind stress. In *Oceanic whitecaps,* E. C. Monahan and G. MacNiocaill, Eds. Reidel, 37–47.

Toba, Y. and H. Kunishi, 1970. Breaking of wind waves and the sea surface wind stess. *J. Oceanogr. Soc. Japan* **26,** 71–80.

Toba, Y., M. Tokuda, K. Okada and S. Kawai, 1975. Forced convection accompanying wind waves. *J. Oceanogr. Soc. Japan* **31,** 192–198.

Toba, Y., K. Okuda and I. S. F. Jones, 1988. The response of wave spectra to changing winds. Part 1: Increasing winds. *J. Phys. Oceanogr.* **18,** 1231–1240.

Toba, Y., N. Iida, H. Kawamura, N. Ebuchi, and I. S. F. Jones, 1990. The wave dependence of sea-surface wind stress. *J. Phys. Oceanogr.* **20,** 705–721.

Toba, Y., I. F. S. Jones, N. Ebuchi and H. Kawamura, 1996. The role of gustiness in determining the aerodynamic roughness of the sea surface. In *The air–sea interface,* M. A. Donelan, W. H. Hui and W. J. Plant, Eds. Univ. Miami Press. 407–412.

Tober, G., R. C. Anderson, and O. H. Shemdin, 1973. Laser instrument for detecting water ripple slopes. *Appl. Opt.* **12,** 788–794.

Tokuda, M. and Y. Toba, 1981. Statistical characteristics of individual waves in laboratory wind waves. I. Individual wave spectra and similarity structure. *J. Oceanogr. Soc. Japan* **37,** 243–258.

Toms, B. A., 1948. Some observations on the flow of linear polymer solutions trough straight tubes at large Reynolds number. *Proc. Intern. Rheolog. Congress, Holland,* p. 149.

Townsend, A. A., 1961. Equilibrium layers and wall turbulence. *J. Fluid Mech.* **11,** 97–120.

Townsend, A. A., 1972. Flow in a deep turbulent boundary layer over a surface distorted by water waves. *J. Fluid Mech.* **55,** 719–735.

Trukenmüller, A., 1988. Einflüsse von Viskosität und Oberflächenspannung auf winderzeugte Wasseroberflächenwellen. *Diploma Thesis,* University of Heidelberg.

Tsuji, T. and Y. Morikawa, 1982. LDV measurements of an air-solid two-phase flow in a horizontal pipe. *J. Fluid Mech.* **120,** 385–409.

Tulin, M. P., 1996. Breaking of ocean waves and downshifting. In: *Waves and nonlinear processes in hydrodynamics*, J. Grue, B. Gjevich and J. E. Weber, Eds. Kluwer. pp. 177–190.

Turner, J. S., 1985. Multicomponent convection (1985). *Ann. Rev. Fluid Mech.* **17**, 11–44.

Unna, D. T., 1947. Sea waves. *Nature* **159**, 239–242.

Valenzuela, G. R., 1976. The growth of gravity-capillary waves in a coupled shear flow. *J. Fluid Mech.* **76**, 229–250.

van Dorn, W. G., 1953. Wind stress on an artificial pond. *J. Mar. Res.* **12**, 249–276.

Volkov, Y. A., 1970. Turbulent flux of momentum and heat in the atmospheric surface layer over a disturbed sea surface. *Izv. Atmos. Oceanic Phys.* **6**, 1295–1302; English Edition, 770–774.

Volkov, Y. A., L. G. Elagina, B. M. Koprov and T. K. Kravchenko, 1974. Turbulent fluxes of heat and moisture and some statistical characteristics of turbulence in the surface layer of the atmosphere in the tropical zone of the Atlantic. *TROPEX*-1972. Gidrometeoizdat, Leningrad, 305–312.

Volkov, Y. A., L. G. Elagina, B. M. Koprov, B. A. Semenchenko and E. M. Feigelson, 1976. Heat and moisture exchange on the equator. *TROPEX-1974*, Vol. 1, Gidrometeoizdat, Leningrad.

Volkov, Y. A., E. Augstein and H. Hinzpeter, 1982. The structure of the atmospheric boundary layer under different convective conditions. *The GARP Atlantic Tropical Experiment (GATE) Monograph, GARP Publ. Ser.*, **25**, 345–387.

von Karman, T., 1930. Mechanische ähnlichkeit und turbulenz. *Nach. Ges. Wiss. Göttingen, Math. Phys. Klasse*, 58.

Wallace, J. M., H. Eckelmann and R. S. Brodkey, 1972. The structure of turbulent wall layer. *J. Fluid Mech.* **54**, 39–61.

Walsh, E. J., D. W. Hancock, D. E. Hines, R. N. Swift and J. F. Scott, 1989. An observation of the directional spectrum evolution from shoreline to fully developed. *J. Phys. Oceanogr.* **19**, 670–690.

Walter, B. A. and J. E. Overland, 1984. Observations of longitudinal rolls in near neutral atmosphere. *Mon. Wea. Rev.* **112** (1), 200–208.

WAMDI Group, 1988. The WAM model – A third generation ocean wave prediction model. *J. Phys. Oceanogr.* **18**, 1775–1810.

Wamser, C. and V. N. Lykossov, 1995. On the friction velocity during blowing snow. *Contr. Atmos. Phys.* **68**, 85–94.

Waseda, T., Y. Toba, and M. P. Tulin, 2001. On the adjustment processes of wind waves to sudden changes of the wind speed. *J. Oceanogr.* (in press)

Watson, K. M. and J. B. McBride, 1993. Excitation of capillary waves by longer waves. *J. Fluid Mech.* **250**, 103–119.

Weber, J. E., 1983. Steady-wind- and wave-induced currents in the open ocean. *J. Phys. Oceanogr.* **13**, 524–530.

Weber, J. E., 1990. Eulerian versus Lagrangian approach to wave-drift in a rotating ocean. Kung. Vetenskaps-og Vitterhets Samhallet, Gotebort, Acta: *Geophysica* **3**, 155–170.

Weber, J. E. and A. Melsom, 1993. Transient ocean currents induced by wind and growing waves. *J. Phys. Oceanogr.* **23**, 193–206.

Weber, S., 1994. Statistics of the air-sea fluxes of momentum and mechanical energy in a coupled wave-atmosphere model. *J. Phys. Oceanogr.* **24**, 1388–1398.

Webster, P. J. and R. Lukas, 1992. TOGA COARE: The Coupled Ocean Atmosphere Response Experiment. *Bull. Amer. Meteorol. Soc.* **73**, 1377–1416.

Wei, Y. and J. Wu, 1992. In situ measurements of surface tension, wave damping and wind properties modified by natural films. *J. Geophys. Res.* **97**, 5307–5813.

Weiler, H. S. and R. W. Burling, 1967. Direct measurements of stress and spectrum of turbulence in the boundary layer over the sea. *J. Atmos. Sci.* **24**, 653–664.

Weissman, D. E., K. L. Davidson, R. A. Brown, C. A. Friehe and F. Li, 1994. The relationship between the microwave radar cross section and both windspeed and stress: model function studies using Frontal Air-Sea Interaction Experiment data. *J. Geophys. Res.* **99**, 10087–10108.

Weissman, D. E., W. J. Plant and S. Stolte, 1996. Response of microwave cross section of the sea to wind fluctuations. *J. Geophys. Res.* **101**, C12149–12161.

Weller, R. A., M. A. Donelan, M. G. Briscoe, and N. E. Huang, 1991. Riding the crest: A tale of two wave experiments. *Bull. Amer. Met. Soc.* **72**, 163–183.

Wheless, G. H. and G. T. Csanady, 1993. Instability waves on the air-sea surface. *J. Fluid Mech.* **248**, 363–401.

Whitham, G. B., 1974. *Linear and nonlinear waves.* John Wiley, New York. pp. 636

Wilhelmsen, K. 1985. Climatological study of gale-producing polar lows near Norway. *Tellus* **37A**, 451–459.

Wills, J. A. B., 1984. *HEXOS – Model tests on the Noordwijk Tower.* R184, NMI Ltd., Teddington, UK, 8 pp.

Wilson, B. W., 1965. Numerical prediction of ocean waves in the North Atlantic for December, 1959. *Dtsch. Hydrogr. Z.* **18**, 114–130.

Wilson, J. D. and T. K. Flesch, 1993. Flow boundaries in random-flight models: enforcing the well-mixed condition. *J. Appl. Meteor.* **32**, 1695–1707.

Wippermann, P., 1972. A note on the parameterization of the large-scale wind stress at the sea surface. *Beitr. Phys. Atmos.* **45**, 260–266.

Woodcock, A. H., 1955. Bursting bubbles and air pollution. *Sewage Industr. Wastes* **27**, 1189–1192.

Wu, J., 1969. Wind stress and surface roughness at air–sea interface. *J. Geophys. Res.* **74**, 444–455.

Wu, J., 1979. Spray in the atmospheric surface layer: review and analysis of laboratory and oceanic results. *J. Geophys. Res.* **84**, No. C4, 1693–1704.

Wu, J., 1980. Wind-stress coefficients over sea surface near neutral conditions – A revisit. *J. Phys. Oceanogr.* **10**, 727–740.

Wu, J., 1983. Sea-surface drift currents induced by wind and waves. *J. Phys. Oceanogr.* **13**, 1441–1451.

Wu, J., 1989. Suppression of ocean ripples by surfactant: Spectral effects deduced from sunglitter, wave-staff, and microwave measurement. *J. Geophys. Oceanogr.* **19**, 238–245.

Wu, J., 1990. On parameterization of sea spray. *J. Geophys. Res.* **95**, 18269–18279.

Wucknitz, J., 1977. Disturbance of wind profile measurements by a slim mast. *Boundary-Layer Meteorol.* **11**, 155–169.

Wucknitz, J., 1979. The influence of anisotropy on stress estimation by the indirect dissipation method. *Boundary-Layer Meteorol.* **17**, 119–131.

Wucknitz, J., 1980. Flow distortion by supporting structures. In: *Air-sea interaction, instruments and methods*, Dobson, Hasse, Davis, Eds. Plenum Press, New York, 605–626.

Wuest, W., 1949. Beitrag zur Enstehung von Wasserwellen durch Wind. *Z. Angew. Math. Mech.* **29**, 239–252.

Wyngaard, J. C., 1973. On surface layer turbulence. In: *Workshop on micrometeorology.* D. A. Hangen, Ed. Am. Met. Soc., Boston, USA. 102–149.

Wyngaard, J. C., 1987. A physical mechanism for the asymmetry in top-down and bottom-up diffusion. *J. Atmos. Sci.* **44**, 1083–1087.

Wyngaard, J. C. and R. A. Brost, 1984. Top-down and bottom-up diffusion of a scalar in the convective boundary layer. *J. Atmos. Sci.* **41**, 102–112.

Wyngaard, J. C. and O. R. Cote, 1971. The budgets of turbulent kinetic energy and temperature variances in the atmospheric surface layer. *J. Atmos. Sci.* **28**, 190–201.

Wyngaard, J. C. and J. C. Weil, 1991. Transport asymmetry in skewed turbulence. *Phys. Fluids* **A3**, 155–162.

Yegorov, K. 1996. Analytical model of the upper-ocean layer. In: *The Air–sea interface*, M. A. Donelan, W. H. Hui, and W. J. Plant, Eds. 575–578.

Yelland, M. J. and P. K. Taylor, 1996. Wind stress measurements from the open ocean. *J. Phys. Oceanogr.* **26**, 541–558.

Yelland, M. J., P. K. Taylor, I. E. Consterdine and M. H. Smith, 1994. The use of the inertial dissipation technique for shipboard wind stress determination. *J. Atmos. Oceanic Technol.* **11**, 1093–1108.

Yelland, M. J., B. I. Moat, P. K. Taylor, R. W. Pascal, J. Hutchings and V. C. Cornell, 1998. Wind stress measurements of the open ocean drag coefficient corrected for air flow disturbance by the ship. *J. Phys. Oceanogr.* **28**, 1511–1526.

Yermakov, S. A., A. R. Panchenko and T. G. Palipova, 1985. Damping of high-frequency wind waves by artificial surfactant films. *Izvestiya Atmos. Oceanic Phys.* **21**, 54–58.

Yoshikawa, I., H. Kawamura, K. Okuda, and Y. Toba, 1988. Turbulent structure in water under laboratory wind waves. *J. Oceanogr. Soc. Japan* **44**, 143–156.

Yoshizawa, A., 1984. Statistical analysis of the deviation of the Reynolds stress from its eddy-viscosity representation. *Phys. Fluids*, 1377–1387.

Young, I. R. and R. J. Sobey, 1985. Measurements of the wind-wave energy flux in an opposing wind. *J. Fluid Mech.* **151**, 427–442.

Young, I. R. and L. A. Verhagen 1996. The growth of fetch limited waves in water of finite depth. Part 1. Total energy and peak frequency. *Coastal Eng.* **29**, Issue 1–2.

Young, I. R., S. Hasselmann and K. Hasselmann, 1987. Computations of response of a wave spectrum to a sudden change in wind direction. *J. Phys. Oceanogr.* **17**, 1317–1338.

Young, I. R., L. A. Verhagen and M. L. Banner, 1995. A note on the bi-modal directional spreading of fetch-limited wind waves. *J. Geophys. Res.* **100**, 773–778.

Zakharov, V. E. and N. N. Filonenko, 1966. Energy spectrum for stochastic oscillation of the surface of a liquid. *Doklady Akademii Nauk SSSR* **170**, 1291–1295.

Zemba, J. and C. A. Friehe, 1987. The marine atmospheric boundary layer jet in the coastal ocean dynamics experiment. *J. Geophys. Res.* **92**, 1489–1496.

Zhang, X., 1995. Capillary-gravity and capillary waves generated in a wind wave tank: observations and theories. *J. Fluid Mech.* **289**, 51–82.

Zilitinkevich, S. S., 1970. *Dynamics of the atmospheric boundary layer.* Gidrometeoizdat, Leningrad, 291 pp.

Zutic, V., B. Cosovik, E. Marcenko, N. Bihari and F. Krisinic, 1981. Surfactant production by marine phytoplankton. *Mar. Chem.* **10**, 505–520.

Index

adjustment time scales, 200
aerodynamic roughness, 4, 28, 207, 215
   Charnock's formula, 47
   length, 66
agestrophic angle, 228
agestrophic components, 65
agestrophic method, 37
agestrophic wind, 81
apparent frequency, 110
atmospheric boundary layer, 55, 218, 233
atmospheric rolls, 20, 21, 219
atmospheric stability, 67

basin width ratio, 271
blocking, 263
bottom friction, 271
Boussinesq hypothesis, 60
bulk aerodynamic method, 1
bulk Richardson number, 70
bulk transfer coefficients, 68
bursting of turbulence, 58

Charnock constant, 24, 52
cold fronts, 266
constant-flux layer, 55
convective boundary layer, 62
Coriolis force, 2, 63, 233
current fronts, 266
cutoff wavenumbers, 103

damping of capillary waves, 246
developing weather systems, 187
downdrafts, 71
drag, 175

   wave induced, 127, 143, 144, 149, 175
drag coefficients, 5, 35, 257
   geostropic, 28
   monthly mean, 29
   neutral, 257
   presence of spray, 8
   reduction, 79
   shallow water, 42, 271
drag measurement, 36, 155
   inertial dissipation method, 44, 155, 160
   profile method, 155, 165
   Reynolds stress method, 155
   scatterometer, 179
drift current(s), 127, 142, 273

eddy viscosity, 3
Ekman flow, 5
Ekman layer, 63, 76
Ekman spiral wind profile, 65
Ekman transport, 230
ergodic condition, 60
evaporation coefficient, 37

fetch, 17, 126
flow separation, 12, 132
flux coefficients
   heat, 38, 68
   momentum, 6, 35, 37, 38, 41, 43, 44, 68,
     166, 174, 197, 258
   water vapour, 38, 68
form drag, 12, 13,
frequency spectrum, 85, 91
friction velocity, 66, 116, 147, 195, 252
frictional decoupling, 58

305